Evaluation of Ambient Air Quality by Personnel Monitoring

Author:

A. L. Linch

Consultant
Environmental Health
Everett, Pennsylvania

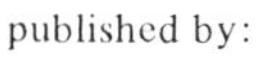

published by:

CRC Press, Inc.
18901 Cranwood Parkway, Cleveland, Ohio 44128

Third Printing, August 1976

International Standard Book Number 0-8493-5012-3

ormer International Standard Book Number 0-87819-020-1

Library of Congress Card Number 73-88623

Printed in the United States

THE AUTHOR

Adrian L. Linch has recently retired from E.I. du Pont de Nemours and Company as laboratory supervisor for the industrial hygiene and clinical laboratories of the medical division at Chambers Works in Deepwater, New Jersey.

Mr. Linch holds a bachelor's degree in chemical engineering (1933) and a master's degree in biochemistry (1934) from the University of Denver.

During his 39-year career with du Pont, Mr. Linch conducted research and development in such areas as the manufacture of dyes, dye intermediates, rubber chemicals, organic mercurials, tetraethyl lead, fluorocarbons, stabilization of aromatic amines, and the design of laboratory glassware. In addition to the medical laboratories, his career has included supervision of a Manhattan project laboratory for a plant producing fluorocarbons and uranium derivatives.

In 1952 he entered full-time practice in industrial hygiene. Biological monitoring procedures for the control of exposure to the cyanogenic aromatic nitro and amino compounds were developed under his supervision; personnel monitoring programs for alkyl lead and mercury derivatives, asbestos, radiation, silica-bearing dusts, and carbon monoxide also were established. Additional specialties involved development of air sampling and analysis techniques, design of personnel monitoring equipment, and protective clothing.

Mr. Linch is a member of the American Chemical Society, the American Industrial Hygiene Society, and the American Academy of Industrial Hygiene, and is a Fellow of the American Association for the Advancement of Science and the Franklin Institute.

Mr. Linch's bibliography includes over 70 publications. In addition to *Evaluation of Ambient Air Quality by Personnel Monitoring,* he has also written *Biological Monitoring for Industrial Chemical Exposure Control* for CRC Press.

PREFACE

Definition of Personnel Monitoring

Personnel monitoring is a term designating the determination of the inhaled dose of an airborne toxic material or of an air-mediated hazardous physical force by the continuous collection of samples in the breathing or auditory zone, or other appropriate exposed body area, over a finite period of exposure time. A personnel monitor is a self-powered device worn by the monitored individual to collect a representative sample for laboratory analysis, or to provide accumulated dose or instantaneous warning of immediately hazardous conditions by visible or auditory means while being worn.

Why Personnel Monitoring?

The case for control of exposure to toxic chemicals or injurious physical forces by personnel monitoring rests on the answers to six fundamental questions: (1) why personnel monitoring? (2) for whom? (3) when? (4) where? (5) for what? and (6) how? On the basis that with few exceptions even the most toxic substances have a no-effect level below which exposure can be tolerated by most workers for a lifetime without incurring any physiological impairment, and further that most hazardous agents are mediated by the air that surrounds us, then analysis of that surrounding air should provide sufficient warning to avoid dangerous concentrations of hazardous substances. Since the reliability of any analytical result is no better than the degree to which the sample analyzed truly represents the environment encountered by the worker, sampling technique is the controlling factor in exposure control. Direct continuous removal of a portion of the inhaled air as it enters the respiratory tract would satisfy the criteria for representative sampling. However, in most cases, physical hindrances to the performance of useful work render this approach impractical.

The concentration of most industrial air pollutants is a highly variable factor with respect to both time and space. Therefore, the workman in the course of his work assignments frequently encounters concentrations that may fluctuate in magnitude by orders of 100 or more over short time intervals. The dilemma faced by the industrial hygienist in attempts to assess exposure experience based on fixed station monitoring, or grab sampling, was illustrated by the results collected from a survey of the potential carbon monoxide (CO) hazard in a large warehouse in which a number of gasoline-powered forklift trucks operated continuously.

Even a cursory examination of the data clearly illustrated the futility of attempting to evaluate this typical industrial environment on the basis of one or two or even ten grab samples (3-min maximum sampling period) taken with the gas detector tubes now in common use.

In other cases the contribution of skin absorption to the total dose absorbed must be established. In a study of the physiological effects on man, exposure to methylene-bis-ortho-chloroaniline was clearly established by the urinary excretion of the amine and its metabolites. However, airborne amine concentrations disclosed by fallout plate, mobile fixed-station, and personnel monitor sampling were below levels that could account for the observed urinary concentrations. Therefore, assignment of a TLV (threshold limit value) for this compound in air would not be appropriate for health control. In this case biological monitoring was the program of choice.

Personnel monitoring for those substances that exert their effects on a cumulative basis (CT value) should be relegated to a supporting role for biological monitoring after the dose-response relationship is established. Among applications that may be cited are lead exposure control by blood and/or urine analysis, mercury by urine analysis, trichloroethylene by breath analysis, and many others well established in industry. On the other hand, those agents that have assigned "ceiling" TLV's, such as benzene, formaldehyde, the isocyanates, and nitrogen dioxide, by their very nature require indicator response time within a range of a few seconds, as in the NDIR or colorimetric detector for CO.

For Whom Is Personnel Monitoring Designed?

First and foremost, consideration must be given to conservation of the employee's health by establishing tolerable, respirable air levels that correlate directly with absorbed dose and medical history. For some hazardous agents that produce injury that becomes manifest only after a prolonged induction period in man, or that produce chronic effects in test animals, personnel monitoring must be continued for as long as the chemical is present in the work place or until its benign character has been established. The importance of the determination of total absorbed dose cannot be overemphasized.

Personnel monitoring is not necessarily limited to the industrial environment. Monitoring for CO exposure has been extended to traffic police in vehicular tunnels, taxicab drivers, and garage attendants. At least one governmental agency issued detector tabs to their employees to carry in their automobilies and homes to detect exhaust and furnace leaks.

Protection for industrial management also enters the overall exposure control program. To meet the increasing pressure from governmental enforcement agencies, management must demonstrate conformance with TLV's promulgated by the Occupational Safety and Health Administration (OSHA) and the Environmental Protection Agency (EPA). Definitive information relative to the 8-hr-day, 40-hr-week, time-weighted average TLV's can be obtained only by personnel monitoring. As an added incentive, data collected consistently over a period of months can be used with adequate medical histories to initiate the revision of unrealistic TLV's. The revision of the TLV's for tetraethyl and tetramethyl lead provides an example (Reference 96, Chapter I).

The production and engineering cost penalties to comply with a TLV set at an unreasonably low value can be sufficient to eliminate some products from the marketplace; to comply with the newer concepts of hazardous agents control, the development of TLV's, both airborne and biological, for new compounds is mandatory. Only with inhaled dose information can the contribution of skin absorption to the body burden be estimated. Adequate monitoring records also establish a favorable legal liability status on which successful defenses against unreasonable claims for alleged injury can be mounted.

When Should Personnel Monitoring Be Used?

General guidelines are suggested:

1. For preliminary surveys to define the problem, to delineate the limits of an airborne exposure problem, and to determine compliance with established TLV's. Short-term sampling, i.e., a minimum of 30 min, preferably 4 hr, or better, an 8-hr shift, would suffice.
2. For follow-up surveys to confirm the effectiveness of equipment or process revisions designed to correct deficient exposure control disclosed by the preliminary survey.
3. To confirm the relevancy of established TLV's by correlation with biological monitoring and medical records.
4. To establish new TLV'S.
5. For permanent routine monitoring programs such as the semiannual survey for asbestos fiber exposure control.

Where Should Personnel Monitoring Be Used?

The answer is simple and straightforward: wherever the potential for exposure to toxic chemicals might be encountered. Workmen assigned to manufacturing areas should be monitored; transportation facilities must not be forgotten; and the consumer is obligated to monitor his employees until the offending material has been converted to a benign form or relegated to an adequate disposal system. On delivery of drum quantities of toxic chemicals in boxcars or tractor trailers, prudence dictates analyzing the atmosphere within the enclosure for the presence of the material, both upon opening the vehicle and during removal of the cargo, to detect leakage before exposure of the cargo handlers has occurred. In addition, facilities should be immediately available for monitoring during decontamination and cleanup after spills and releases of toxic agents. Mechanics who must break into pipelines, enter process vessels, or weld on equipment used in toxic chemical service should be monitored unless they are totally isolated from their environment by protective clothing or unless respiratory protection suffices for exposure control.

What Should Be Monitored?

Again, the answer is clear and unequivocal: all substances known to be hazardous for man, those producing toxic effects in test animals, and those of unknown injury potential. However, from a practical consideration, attention would be focused on the most dangerous first. A cue can be taken from the National Institute of Safety and Occupational Health priorities published from time to time in the *Federal Register*. Not only the products themselves, but their combustion and pyrolysis derivatives generated by fire or overheating may generate serious exposure problems (e.g., phosgene from the pyrolysis of chlorine containing aliphatic hydrocarbons) and must be considered in a health surveillance program.

How Is Personnel Monitoring Carried Out?

Both the hardware and the technique employed will be determined by the purpose of the survey, the physical state of the offending material, the physiological effects of exposure, available equipment, and the analytical procedure to be applied after sample collection. The gravimetric determination of the total atmospheric dust loading by filtration provides the most direct air analysis.

For asbestos fibers a special membrane filter that "clears" (dissolves) when a mixture of dimethyl phthalate and dimethyl oxalate is applied to permit counting by phase contrast microscopy has been adopted by OSHA. A rechargeable battery-operated pump samples continuously for 8 hr in the 1 to 2 lpm range.

Not all dust entering the nose reaches the lungs, due to aerodynamic size limitations imposed by the nasal passages. The "respirable" fraction is defined as a percentage according to the size of the particles:

Aerodynamic diameter (μm)	% Passing selector
2	90
2.5	75
3.5	50
5.0	25
10	0

Therefore, if the monitor is intended to be used as a respirable fraction sampler, then all particles larger than 10 μ and 75% of those larger than 5 μ must be rejected before the sample reaches the filter.

A readily portable miniature cyclone separator that approximates the above criteria adopted by the Department of Labor for coal dust, silica containing dusts, and inert or nuisance dusts can be obtained commercially. Miniature cascade impactors also are commercially available for further classification of the particulates into discrete ranges (5 stages). The tangential flow filter provides uniform distribution of the dust collected over the entire filter surface, and the filter membrane need not be removed from the capsule for weighing.

Gases and vapors can be collected from the industrial atmosphere either by adsorption on a solid medium such as silica gel or activated charcoal for laboratory analysis (usually by gas chromatographic procedures) or in a liquid solvent or reagent system for either laboratory or field analysis. The personnel monitoring system developed to derive the data used successfully in a petition to revise upward the TLV's for tetraethyl and tetramethyl leads can be cited as a model (Reference 96, Chapter I). In this case collection of particulate inorganic lead separately from organic lead vapors was necessary.

Sampling was started when the employee left the change house and continued until he returned for lunch break. After lunch, the monitor was replaced and sampling continued until shift end. Six and seven monitors were operated simultaneously on each day shift in five work areas selected to include the widest possible range of activities. Each operator selected was monitored for 5 days on his day shift rotation. This rotation presented a maximum of individual differences in lead metabolism.

Linear and multiple regression analysis of daily individual results disclosed no correlation between fixed-station and personnel monitors, or between individual paired urine and monitor results on a daily basis. With few exceptions, personnel monitor results were significantly higher than corresponding, fixed-station results. Displacement of the urine results by 24- or 48-hr intervals to account for the lag in urinary excretion after exposure likewise produced no correlation.

An approximate linear relationship was found by converting the weekly averaged personnel monitor tetramethyl lead (TML) and inorganic lead results to TLV coefficients (Reference 96, Chapter I). The sum of the coefficients (TLVC's) then was related to the averaged urine analysis for the operator who wore the monitor.

A different approach was required for the solution of the potential carbon monoxide problem within a consolidated warehouse where a number of gasoline-powered forklift trucks are employed for drum storage within bays and for loading boxcars and truck trailers. Fixed point alarms, grab sampling, and infrared monitoring did not provide the data necessary to determine compliance with the TLV for CO. None of the available colorimetric or instrumental methods could be adapted to personnel monitoring. Therefore, the length of stain CO detector tube was recalibrated for 4-hr continuous sampling (Reference 53, Chapter I). Although the best accuracy that could be attained lay in the ± 25% range of the true value, the method sufficed to determine compliance with the TLV (0.5, 1, and 2 times 50 ppm).

For the first time the capability for at least semiquantitative correlation of CO body burden with industrial exposure in the TLV range has become available. In such surveys CO analysis of breath samples collected at the end of the 4-hr sampling period would provide correlation with the time-weighted average dose inhaled by the employee during the time spent in his work place. The most significant feature of this type of monitoring is the capability for instantaneous readout. At any time the accumulated dose can be quickly estimated from the 1-, 2-, 3-, and 4-hr calibration charts. Contact with a pocket of high CO concentration immediately turns the entire tube from amber to black to give the wearer ample warning to leave the area.

Surveys carried out during a 4-week period under restricted ventilation conditions imposed by cold weather confirmed CO levels that conformed with the 50-ppm CO TLV in this warehouse. Another survey completed in late 1972 again confirmed conformance with governmental regulations and the advantages derived from this type of monitoring with a minimum expenditure of time and cost. The recalibration of colorimetric length of stain detector tubes for other gases and vapors, such as SO_2, NO_2, H_2S, Cl_2, the halogenated hydrocarbons, aromatic hydrocarbons, solvents, etc., for 4-hr sampling could, to at least a minor degree, revolutionize the concept of personnel monitoring by providing both accumulated dose and ceiling value results on the spot in the field.

TABLE OF CONTENTS

Chapter III

Chapter I

I. GASES AND VAPORS

A. SELF-INDICATING DEVICES – SEMIQUANTITATIVE

1. Impregnated Paper

The use of a paper support for reagents that change color on contact with traces of reactive ions or compounds in air or water was undoubtedly among the first methods to detect and monitor hazardous air pollutants. Minimal cost, ready portability, color development without application of liquid reagents, rapidity of response, and simplicity continue to provide commercial incentive for exploitation not only for "go-no go" application in locations where acute exposure to highly hazardous gases or vapors is a threat to health, but also for ambient air monitoring outside of the industrial environment (Figure I.1). However, without some provision for measuring the air sample volume and calibration with known concentration standards, only semiquantitative results can be obtained. An uncalibrated detector furnishes little more information than the human nose, which is a notoriously ineffective quantitative analytical instrument unless the contaminant concentration falls below the odor threshold. For those highly toxic gases such as phosgene, for which the threshold limit value (TLV)[1] is well below the odor threshold, a qualitative response which discloses the presence of the hazard can serve as a warning device even though no positive information relative to concentration is available.

a. Convection Activated Detectors

A reasonably acceptable estimate for concentration can be obtained by application of the CT concept: the product of concentration multiplied by time, also referred to as "dosage" or "time-weighted average." Since the first appearance of color, rate of color development, and color intensity (also referred to as depth of color, optical density, or color saturation) are related to concentration, it follows that an approximate calibration can be attained by noting the time required to reach an arbitrarily selected color hue or intensity when the test paper is freely suspended in an atmosphere generated from a known concentration of the contaminant. A good example of this approach is illustrated by the commercially available hydrogen sulfide detectors. The test paper may be exposed either for a preselected time such as 3 to 5 min, and the spot matched with a segment of the color chart supplied with the paper (Figures I.1.1 and I.1.3),[3] or the stain may be compared to a graded saturation series that is related to a conversion table based on the CT or dosage value.[3] In this latter case, if the time is known, the concentration in ppm can be calculated. For example, a lead acetate impregnated circular pad mounted in the center of a surrounding color density chart imprinted on stiff paper is exposed for a finite time interval (Figure I.1.1). The color intensity (grade) read from the chart is then referred to a conversion table (Table I.1) to obtain the dosage (CT value), i.e.:

Color grade	=	2.5
Exposure Time	=	24 hr
Dosage (from Table I.1)	=	0.56 ppm x hr
0.56 ppm x hr/24 hr	=	0.023 ppm H_2S
or for an 8-hr day	=	0.07

The lower limit is 0.03 ppm, and complete saturation is reached at 4 ppm for a 1-hr exposure. On this basis, the TLV of 10 ppm would be reached in 24 min. Although intended primarily for stationary air pollution analysis, the detector could be used directly for personnel monitoring by attaching the tab to the workman's shirt under his chin. A transparent plastic mounting would be necessary to protect the rather fragile impregnated pad and surrounding color chart.

This principle also has been applied to the more durable ceramic (bisque) tiles as a substrate for the lead reagent. A chart of dosage (D) plotted vs. tile grade (color intensity) on semilogarithmic graph paper indicated a 90% probability that the true D will fall between 0.67 D and 1.5 D (Figure I.2). For example, D estimated for grade 2 is equivalent to 0.33 ppm x hr with the probability that the true dosage lies between 0.22 and 0.50 ppm x hr. In this case the precision has been further improved by quantitating the relative color intensity by reflectance measurement at 425 nm.[3] If the accuracy can be controlled consistently within these limits of ± 40% relative deviation

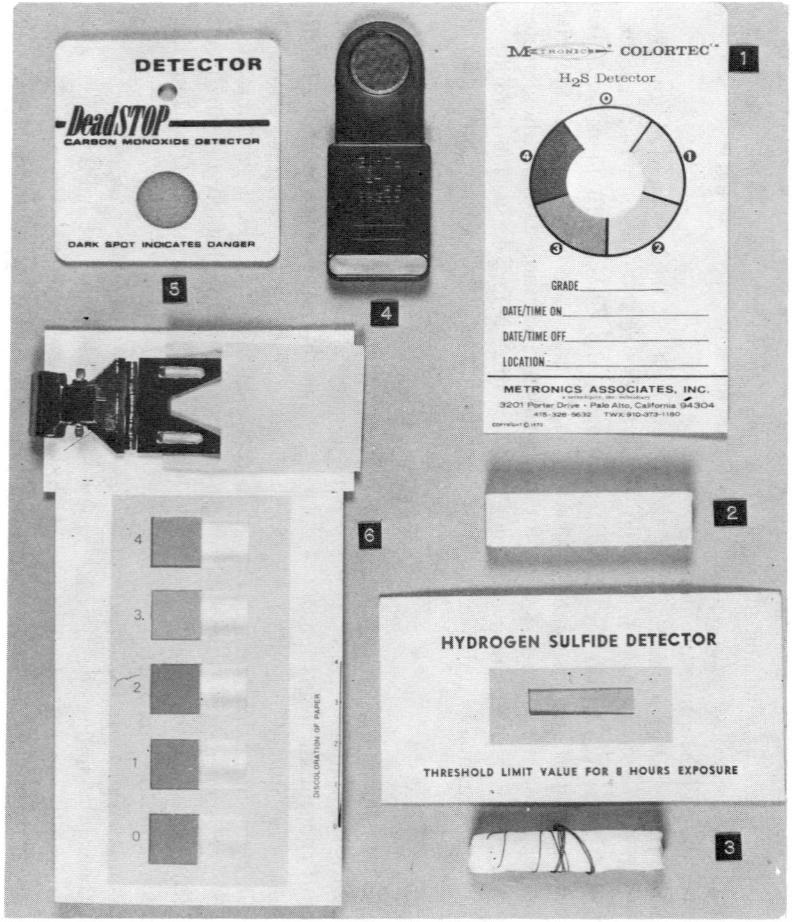

FIGURE I.1. Paper-supported, self-indicating hazardous gas detectors for personnel monitoring application. 1. Metronics Colortec® hydrogen sulfide detector. (Courtesy of Metronics Associates, Inc., Palo Alto, Cal.) 2. Metronics Ceramtec® tile hydrogen sulfide detector. (Courtesy of Metronics Associates, Inc., Palo Alto, Cal.) 3. Davis hydrogen sulfide detector.[8] 4. Portable carbon monoxide alarm, Unico® Model 555. (Courtesy of National Environmental Instruments, Inc., Pilgrim Station, Warwick, R.I.) 5. Deadstop carbon monoxide detector.[51] 6. Kitigawa® sulfur dioxide test paper.[2]

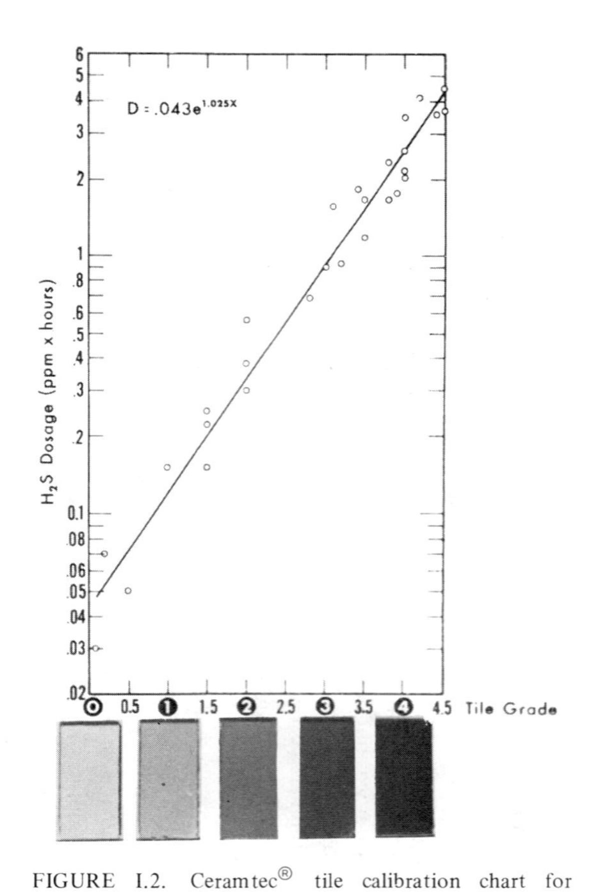

FIGURE I.2. Ceramtec® tile calibration chart for hydrogen sulfide. (Courtesy of Metronics Associates, Inc., Palo Alto, Cal.)

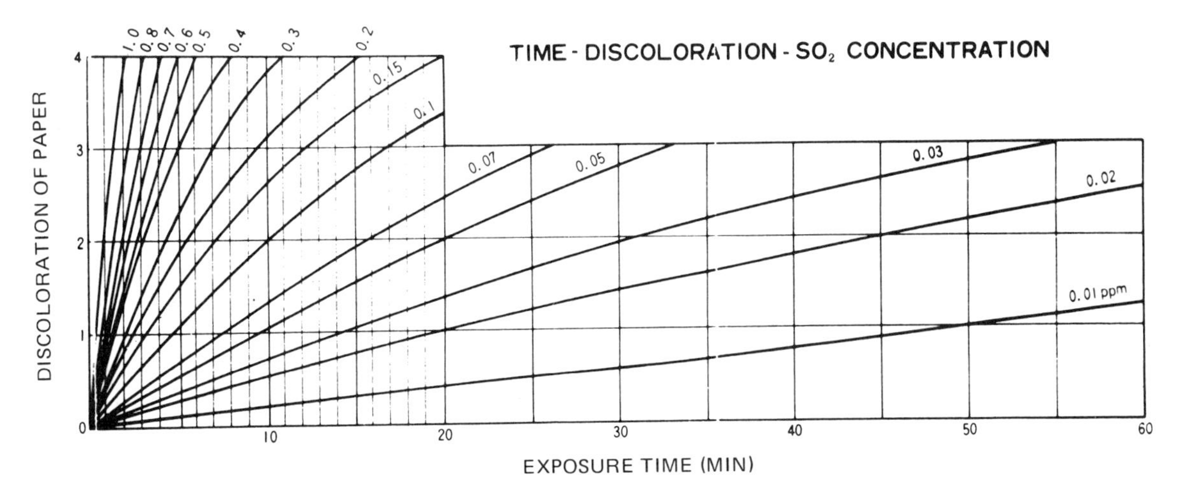

FIGURE I.3. Sulfur dioxide detector paper calibration chart showing the relationship between time, discoloration, and SO_2 concentration.[6]

TABLE I.1

Conversion Table – Relationship of Color Density to SO_2 Exposure Dosage (Concentration x Time)

Color grade	0	0.5	1.0	1.5	2.0	2.5	3.0	3.5	4.0	4.5
Dosage (ppm x hr)	0*	.07	.12	.20	.33	.56	.93	1.56	2.59	4.33

*Zero grade indicates dosage less than .03 ppm x hr.

Courtesy of Metronics Associates, Inc., Palo Alto, Cal.

(0.36 ± 0.14 ppm x hr), then the reliability of this system would be about equivalent to the performance of most length of stain detector tubes (interim performance level ± 50%, except for those which must meet American Conference of Governmental Industrial Hygienists-American Industrial Hygienists Association (ACGIH-AIHA) criteria of ± 25% for certification).[4,5]

A somewhat different approach has been developed for the estimation of sulfur dioxide (SO_2) in air based on the fading of a blue-red, pH-sensitive dye to a yellow-orange on filter paper in four graded steps (Figure I.1.6).[2] In this case, a gas phase titration can be employed, i.e., the time required to attain a definite end point color (grade 3 or 4) is related to the concentration in ppm of SO_2 (Tables I.2 and I.3). A humectant must be present to retain water for the acid reaction. A chart that relates other color grades and exposure times to SO_2 concentration is furnished also (Figure I.3). The colors can be reproduced by referring to Munsel's code (Table I.4).

TABLE I.2

Relationship Between Exposure Time and Concentration of SO_2 Required to Produce Color Grade 3

Exposure time (min)	SO_2 Concentration (ppm)
3	0.6
5	0.4
7	0.3
9	0.2
12	0.15
17	0.1
25	0.07
35	0.05
40	0.04
55	0.03

The potential utility of this approach to personnel monitoring was suggested by the report of a study conducted in Yokohama, Japan, during the summer of 1968. Although the mode of sampling was fixed station, the survey could have

TABLE I.3

Relationship Between Exposure Time and Concentration of SO_2 Required to Produce Color Grade 4

Exposure time (min)	SO_2 Concentration (ppm)
2.0	1.0
3.0	0.8
4.0	0.7
5.0	0.6
6.0	0.5
8.0	0.4
11.0	0.3
15.0	0.2
20.0	0.15
30.0	0.1

TABLE I.4

Color Change of Acidic Gas Detector Paper

Degree of color change	Munsel color code*	
0 (Original color)	7.5 RP	5/11
1	2.5 R	5.3/9
2	7.5 R	5.8/8
3	2.5 YR	6.2/8
4	7.5 YR	6.8/8.5

*R = red; P = purple; Y = yellow.

TABLE I.5

Comparison of the Results by West-Gaeke Method, Electroconductivity Method, and Test Paper Method

Date (June)	West-Gaeke (ppm)	Electroconductivity (ppm)	Test paper (ppm)
11	0.139	0.25	0.23
	0.148	0.35	0.30
14	0.058	0.15	0.15
16	0.020	0.01	0.02
	0.010	0.01	0.015
17	0.140	0.30	0.25
	0.095	0.28	0.23
	0.042	0.10	0.10
18	0.090	0.187	0.23
	0.072	0.160	0.15
	0.100	0.120	0.18
20	–	0.08	0.10
	0.150	0.13	0.15
	0.040	0.08	0.15
23	0.032	0.08	0.08
	0.030	0.04	0.05

been carried out just as well by selected individuals wearing detectors while engaged in their normal daily activities.[6]

Using pieces of the SO_2 paper developed by Prof. Dr. Tetsuzo Kitagawa and his collaborators in the Department of Engineering at Yokohama National University, the concentration of atmospheric SO_2 was measured eight times at each of 56 schoolgrounds in Yokohama City between 9 and 10 a.m., and between 12 and 1, 3 and 4, and at 5 p.m. on July 25 and August 3, 1968. On both days a clear sky prevailed, and the wind blew from the southeast, the south, or southwest at speeds up to 7.3 m/sec.

The usually employed methods of analysis for sulfur oxides in the atmosphere included the West-Gaeke method, the barium molybdate method, and electroconductivity measurement. These methods required heavy equipment and involved extensive manipulation, which was considered to be impractical for survey analysis. Therefore, the lightweight, self-indicating indicator paper was chosen to obtain rapidly results relative to: (1) geographical distribution of SO_2 pollution (Figure I.4) by simultaneous measurements at the various locations, (2) variations of pollution concentration with time, and (3) the efficiency of abatement equipment by comparison of before and after analyses. Comparisons between replicate determinations obtained from the classical wet chemical procedures confirmed the validity of the detector paper results (Figures I.5 and I.6 and Table I.5).

Three measurement techniques were used:

1. The paper strips were exposed continuously, observed at the end of consecutive, discrete time intervals (10 to 15 min), and the color graded 0, 1, 2, 3, or 4 by comparison with the reference color chart supplied with the kit. The color grades then were plotted vs. time on the calibration chart (Figure I.3) and the concentration located on one of the calibration curves or extrapolated between adjacent curves. If the plotted results disclosed scattered points, the last reading was used as a time-weighted average to compensate for the changes in SO_2 concentration that occured during the observation period. This technique gave the best accuracy.

2. The time required to produce a given stain density (grade 3 or 4, Tables I.2 and I.3) by gas titration was calculated. This approach was

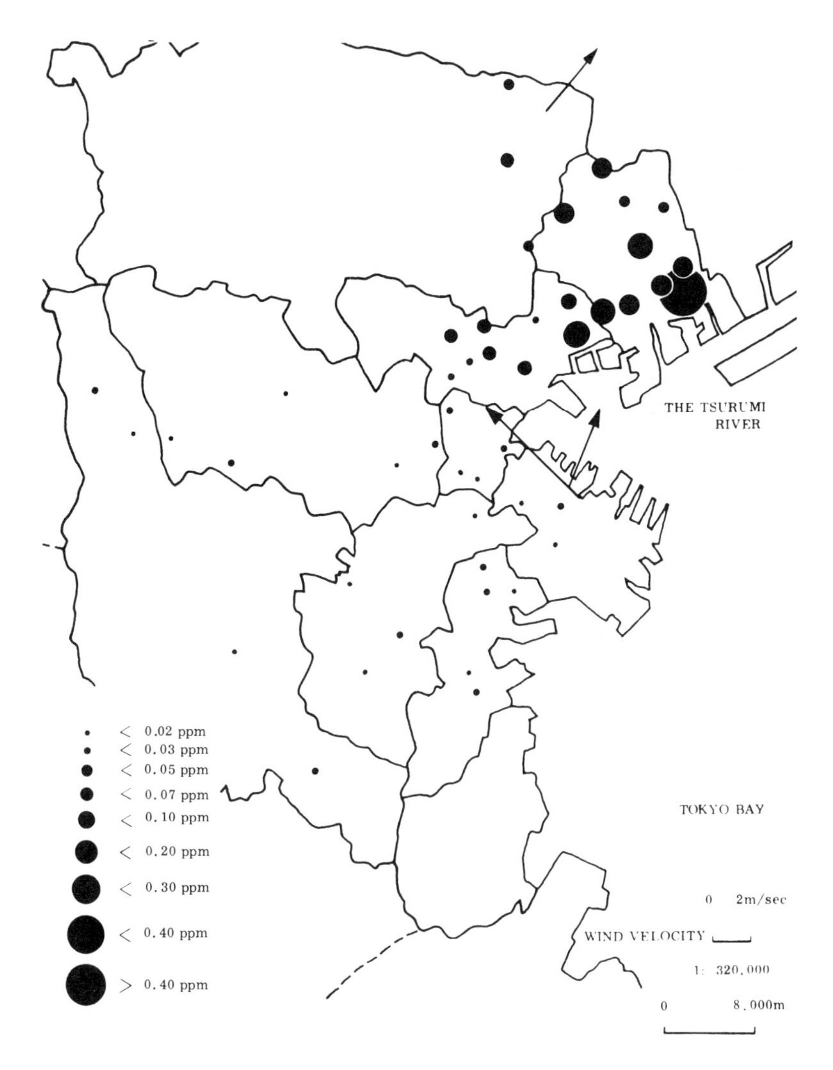

FIGURE I.4. SO_2 concentration distribution map from Yokohama, Japan, on July 25, 1968.[6]

considered suitable for concentrations above 0.2 ppm.

3. The degree of color change after a fixed exposure period was determined. The range of measurement varied with time. This was considered the best technique for SO_2 concentrations less than 0.03 ppm.

The variables that influenced the results included wind speed, temperature, and humidity. If the wind speed was less than 1.5 m/sec, low results were noted, but when the velocity exceeded 2 m/sec, the effect disappeared. Temperature was without influence in the range 0 to 40°C. Although a humectant was added to the paper, relative humidity below 30% RH contributed to low results. The magnitude of this error increased as the temperature rose.

Since the SO_2 test paper is not affected by the ionizable salt content of the ambient atmosphere, the results checked more closely with the West-Gaeke method than with electroconductivity (Figures I.5 and I.6, and Table I.5). When SO_2 or SO_2-containing aerosols were present, the color change on the test paper was not uniform, but rather presented a yellow speckled appearance. This SO_2-sensitive paper is available from a domestic source.[2]

b. Tape Samplers

A miniature portable tape sampler (20 mm x 58 mm x 125 mm) that provides over 1,000 spots 8

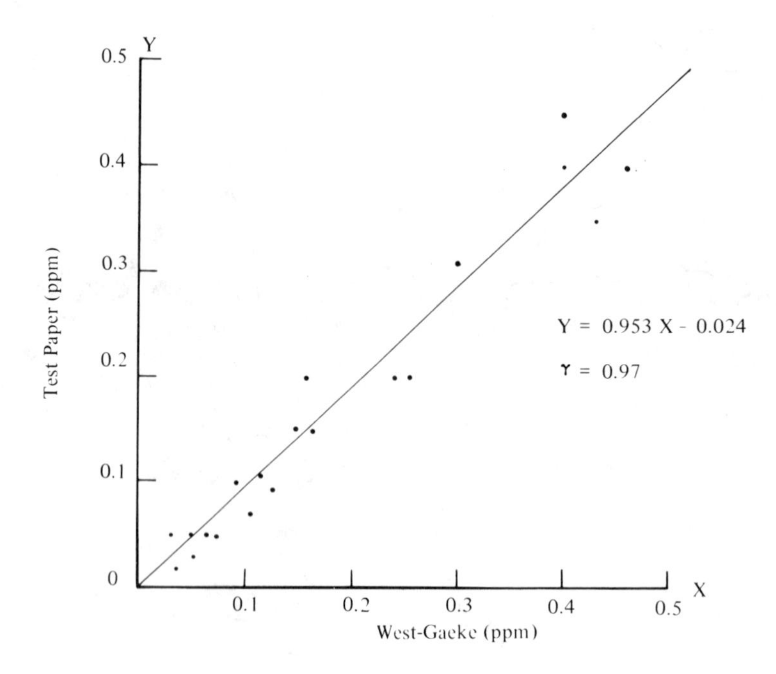

FIGURE I.5. Correlation of test paper discoloration with West-Gaeke analysis results at the Yokohama Marine Tower.[6]

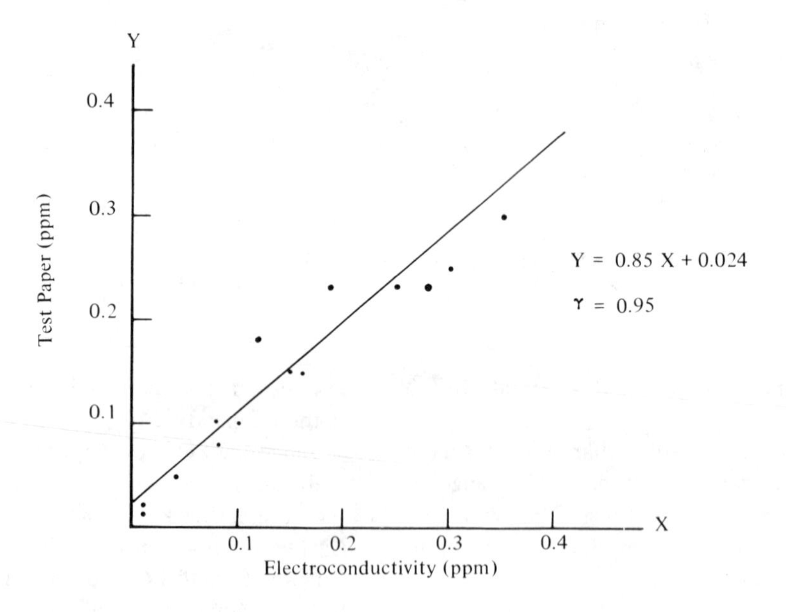

FIGURE I.6. Correlation of test paper discoloration with electroconductivity at the Yokohama Marine Tower.[6]

FIGURE I.7. Hydrogen sulfide-sensitive continuous paper tape detector. (Courtesy of Del Mar Products, Houston, Texas.)

mm in diameter on a single roll of filter paper has been developed and exploited commercially (Figure I.7).[7] The tape, instructions, and color charts are housed together in a transparent plastic case, and the sampler weighs 110 g complete. Only lead acetate impregnated tape for H_2S detection has been supplied. The unit is suitable for personnel monitoring if a mounting bracket is attached and the tape recalibrated in terms of dosage (CT values).

c. Reagent Ampule Activated Samplers

Although not strictly a paper supported reagent system, mention should be made of the wet wick method. An ampule that can easily be crushed is contained in a cotton covering (Figure I.1.3). Exactly 1 min after saturation of the cotton by the lead acetate solution, the color is compared with the standard by placing the ampule directly behind a color-bordered slot in a card. A lighter shade of brown indicates less than 10 ppm H_2S, a darker color, more than 10 ppm. The system could be adapted to personnel monitoring as an alarm device where intermittent or casual exposure would be expected, since the reagent-soaked wick remains active for approximately 6 days after the ampule is crushed.[8]

d. Applications to Continuous Monitoring

Although the performance of the foregoing gaseous pollutant detectors is at best, as stated in the introductory paragraph, only semiquantitative (deviation range greater than ± 25%), application to personnel monitor surveys designed to determine order of magnitude concentrations (0.5, 1, 2, and 4 times the TLV) will in many cases supply sufficient information to define the problem. With minimum cost and technical skill, large areas can be surveyed for long periods of time to locate pollution focal points and to indicate trends. The possibility that a major source of pollution could be corrected without establishing a costly, comprehensive air surveillance program adds further incentive to this approach.

If the results fall consistently below one half of the TLV, then a more sophisticated analytical system is not necessary. In the 0.5 to 1.5 x TLV range, an extensive, refined analytical survey must be undertaken to evaluate the compliance factor fully. When results collected from any given location in the area under investigation consistently exceed twice the TLV, then corrective action should be initiated and completed before additional funds are expended on improving the reliability of the air quality monitoring system.

Although, after a short briefing, the individual wearing the badge could read off the concentration values from the color comparator furnished with the detector, more consistent results would be expected if the color matching were limited to a few trained technicians to minimize variations in visual acuity. Of course, if the stains are not sufficiently stable to permit forwarding to a central collection point, then reliance must be placed on the veracity of the individual observer.

This type of survey could be applied to ambient air monitoring by the populace at large to delineate pollution areas, as well as for in-plant exposure control. Also, for localized plant conditions the badges could provide instantaneous warning (within reagent response time limits) of TLV ceilings, or excursions into immediately hazardous concentration ranges, in time to permit evacuation of the area before irreparable physiological damage can occur.

The approach of an accumulated dose limit (carbon monoxide, for example) can be anticipated by periodic observation of the stain intensity as it develops during the wearer's work assignment in time to take appropriate corrective action before excessive exposure can occur (by eliminating the pollutant at the source, or transferring the workman to an area of less exposure for the remainder of the shift).

e. Grab Sampling – Piston-type Pumps

A closer approach to quantification is attainable if the sample volume is known precisely. Earlier literature described manually operated piston type exhausting pumps that draw a known volume of air through test paper clamped in a circular metal holder. The most sophisticated design was developed by the Department of Scientific and Industrial Research in England during the period 1938 to 1943 for the semiquantitative detection of hydrogen sulfide,[9] hydrogen cyanide,[10] sulfur dioxide,[11] phosgene,[12] and arsine.[13] The pump had a 32-mm bore and a 126-ml volume, and could be fitted with a stroke counter. A 50-mm square of the test paper was clamped in the threaded aluminum holder under a rubber sealing gasket. The circular retaining ring encompassed a 38-mm diameter field for collection of the contaminant as the air sample was drawn into the pump through the detector paper. Equidiameter color standards printed on heavy paper were supplied with each copy of the method. The concentration range was extended by multiple strokes of the pump, i.e., in the case of SO_2, as many as 10 strokes to detect less than 4 ppm.[11] The sampling rate approximated to 2.5 l/min (3 sec/126-ml stroke).[11]

1. Phosgene

Phosgene derived considerable attention as a poison gas during World War I, and its detection in the ambient atmosphere has been rather intensively studied. A comprehensive literature review disclosed at least 48 references to procedures developed for its detection in the trace ranges.[14] To obtain sensitivity in the 1 ppm range for phosgene with the British hand pump, 85 piston strokes were required. On the basis of 4 sec/stroke (3 to retract, 1 to return) the sampling time would have been approximately 6 min and the volume 10.7. In 1966, the TLV was lowered from 1 to 0.1 ppm; therefore, this manual system became highly impractical. However, if the color generating reagent were sufficiently stable, this system could serve for personnel monitoring with a miniature battery driven pump to draw the air sample throughout the detector paper (see Chapter III). Then the reduced TLV would be within practical sampling limits, that is, 1 hr and a 107-l sample; or, if the sampling rate were reduced to 320 ml/min, 8-hr shift monitoring could be established. However, Harrison's reagent (diphenylamine + 4-dimethylaminobenzaldehyde) is not stable for more than 1 hr, and calibration of alternative reagents must be considered.[14]

Greater sensitivity was derived by substituting either *N,N*-dimethylaniline[15] or *N*-ethyl-*N*-2-hydroxyethylaniline (blue at 0.25 ppm)[16] in place of diphenylamine as the color-forming component with the 4-dimethylaminobenzaldehyde, but again, these detector papers were not sufficiently stable for extended sampling, and were sensitive to acidic gases and chlorine.[14] Probably the most promising reagent with respect to sensitivity and specificity for phosgene was described by Lamouroux in 1956.[17] The system, 4-(4′-nitrobenzyl)pyridine (NBP) + *N*-benzylaniline, produced a sharp, distinct, and easily detected color change from yellow to pink to brilliant red on contact with phosgene concentrations as low as 5 ppb.[14,18]

Whatman® No. 1 filter paper was saturated with 5% aqueous sodium carbonate, dried, then saturated with a solution of 4% *N*-benzylaniline and 2% NBP in benzene, and dried. The sodium carbonate was added to provide greater tolerance for volatile acids which would otherwise interfere at acid concentrations greater than 500 ppm. This paper, prepared, dried, and stored in an atmosphere free of acid vapors and phosgene, had a shelf life in excess of 5 months.[14] An airflow of 190 ml/min for each cm^2 of exposed filter paper area for 5 min ± 10 sec provided color response proportional to the amount of phosgene in the range 0.002 to 2 ppm. A somewhat higher range, 0.25 to 10 ppm, was obtained at a higher rate for a shorter period, 230 ml/cm^2/min for 40 sec. Permanent glass color standards for this procedure were available from abroad.[18] These results indicate that application to personnel monitoring would require only adjustment of rates and volumes and recalibration. The color formation related to sampling rate as well as to concentration; therefore, if the rate were reduced to 19 to 20 ml/cm^2/min, the time required for detection of the 0.1-ppm TLV within the range 0.05 to 0.2 ppm probably would fall in the range of 7 to 8 hr.

The NBP test paper may be applied also to gas titration by determining the volume of air required to produce the first perceptible pink coloration. This coloration, when monitored by a miniaturized photoelectric sensing circuit, provides an integrated alarm system in the personnel monitor (see Unico® Portable Personnel Carbon Monoxide Alarm[52]). For example, 5 ppb can be detected by passing a 10-l sample through an indicator disk. By equating the volume of air required to duplicate this color, estimates of unknown concentrations of phosgene can be made.[14]

2. Hydrogen Cyanide

Other detectors developed for use in the hand pump include hydrogen cyanide sensitive paper impregnated with either benzidine and copper sulfate (blue), or Congo red and silver nitrate (blue).[10,19,20] These papers must be moist when used, are not stable in storage, and are subject to interference from H_2S, SO_2, and acidic and alkaline reacting gases, however. The Prussian blue test (ferrous sulfate + sodium hydroxide) is specific for HCN, but requires immersion in 30% H_2SO_4 or dilute (1 to 4) HCl for color development (blue).[10,19,20,26] Although not sufficiently sensitive for grab sampling, on an 8-hr basis this detector could provide satisfactory cumulative results for personnel monitoring, since a 360-ml air sample is sufficient to detect 2.5 ppm HCN, and 10 ppm yields a bright blue color.[19] The major weakness of this method is the requirement for a wet reagent treatment to develop color.

3. Arsine

Arsine may be detected on filter paper impregnated with $HgCl_2$ (yellow to orange).[13,20] A hand pump-filter assembly which was designed to draw a reproducible volume of air through an arsine sensitive detector filter has been developed and offered for sale on the domestic market.[21]

4. Sulfur Dioxide

Sulfur dioxide on contact with a mixture of starch-potassium, iodate-potassium, and iodide-glycerine supported on filter paper will produce a brown stain in the 4 to 10 ppm range. However, close attention to details of preparation are necessary to ensure papers which are sensitive, yet stable in storage. Inasmuch as this reagent appears promising for personnel monitoring, but reference to its preparation is not readily available, details are included here.

To 60 ml distilled water, add 12 ml of approximately *N*/10 aqueous barium hydroxide. Heat to boiling and add 1 g potassium iodate and 2 g potassium iodide, both of AR grade.[22] When completely dissolved, add a thin paste of 1 g of soluble starch in 10 ml distilled water and continue boiling for a few minutes. Cool to room temperature, add 30 ml AR glycerine, dilute to 100 ml, and mix thoroughly. The pH should be approximately 9.5; if it is not, add more barium hydroxide solution dropwise. Cut extra-thick white filter paper (18 x 24 to 60 lb) into strips 50 mm wide, immerse 30 sec in the reagent, drain excess liquid, pre-dry by pressing between sheets of dry filter paper two or three times, and dry 10 min in a well ventilated oven at 50°C. The papers should not be limp, but should have dried sufficiently to regain some of their original stiffness; otherwise the stain will be spotty and not evenly distributed on the exposed paper. Discard 1 in. length from top and bottom of the strip and cut into 75-mm lengths. The test papers should be used preferably within a few hours, but may be stored in a well-stoppered bottle in the dark for 1 to 2 days. The reagent is stable for about 7 days.[11]

A more recent procedure[23] recommended by the British Ministry of Labor[24] employs ammoniacal zinc prusside reagent, which turns red on contact with SO_2. Between 1 and 20 ppm SO_2 can be detected in a 360-ml air sample (three pump strokes) at a rate of 90 to 180 ml/min. This method is highly specific for SO_2, and only very slight interference from H_2S is encountered. Color standards are available,[24,25] preparation is simple, and the papers may be stored for at least a month. After mixing 50 ml of 6% aqueous zinc sulfate heptahydrate solution with 50 ml freshly prepared 10% aqueous sodium nitroprusside solution, ammonium acetate is added under agitation until the zinc nitroprusside is dissolved. Then 20 ml glycerine is added and mixed thoroughly, and the filter strips are dipped immediately – before zinc prusside again precipitates. Excess reagent is drained off, and the strips are air dried at room temperature.[19]

If obtainable in sufficient size (area) from Kitigawa, the paper impregnated with the bluish-red dye, which bleaches to a brownish-yellow color (see Section I.A.1.a, "Convection Activated Detectors"), would provide an ideal personnel monitor with a battery-powered pump to draw through a known volume of air.[6] However, storage stability may be a problem that remains to be solved.

Alkali-treated filter paper has been used to collect SO_2 for quantitative analysis in the laboratory by the standard West-Gaeke procedure (reaction of SO_2 + pararosaniline + formaldehyde in the presence of sodium tetrachloromercurate).[19,26] The impregnating solution contains 20% potassium hydroxide and 10% triethanolamine or glycerine. The uptake, after removing excess reagent by dabbing with dry filter paper and drying at 110°C, is approximately 13 mg of solution per cm^2 of filter area. The air sample is drawn through at a rate not exceeding 1 $l/cm^2/min$. The air sample should contain not less than 30% relative humidity and not more than 5 ppm SO_2 (not more than 10% of the available alkali should be consumed) at this flow rate, which should be reduced proportionately for higher concentrations. After washing into dilute aqueous sodium tetrachloromercurate, the colorimetric analysis procedure is applied.[19,26] Recoveries that exceed 95% have been reported, and the method can be used for short-term or routine daily and weekly analyses.

5. *Hydrogen Sulfide*

Lead acetate impregnated papers also have been used for hydrogen sulfide (H_2S) estimation with the hand pump, and could be applied directly to personnel monitoring. In this application glycerine must be added as a humectant.

6. *Mercury*

Test paper for the semiquantitative detection of mercury vapor has been developed commercially for stationary monitoring (General Electric Co.).[20] The active ingredient is selenium sulfide powder applied by rubbing into the pores of the filter paper. The stability and sensitivity are excellent (0.12 ppm in 8 min). The color change from yellow to black is considered satisfactory; however application is limited to an optimum temperature of 70°C.[20]

Possible application of the dithizone colorimetric technique to the estimation of mercury was suggested by the development of quantitative (90+%) collection on filter paper impregnated with potassium triiodide (from aqueous $KI + I_2$). The excess iodine is reduced with sodium sulfite. In the reference procedure the mercury was quantitated by UV spectroscopy after reduction with $SnCl_2$ and collection on a copper coil. However, spraying with the dithizone reagent after destruction of the excess iodine would develop the red mercury dithizonate which could be compared visually with known standard stains.[27]

7. *Lead*

Although wet chemical color development methods are not entirely desirable for field application where laboratory facilities are not provided, their utility when available in readily portable kit form should not be ignored. In some cases, no other alternative is practical. For example, a colorimetric semiquantitative field method for the determination of inorganic lead fumes and dust in air was derived from the pink to red complex that forms when tetrahydroxyquinone comes in contact with lead.[28] The insoluble compounds such as lead sulfide, arsenate, carbonate, chromate, and oxides, as well as the water soluble salts, react quantitatively. The intensity of the red stain on the filter paper is proportional to the amount of lead collected and is estimated by visual comparison with color standards.

The lead-containing dust is collected on No. 42 Whatman filter paper by drawing the air sample through a 2.0 -cm diameter exposed area at a 4 to 5 l/min rate for 10 min (1.25 to 1.60 $l/cm^2/min$. To the center of the filter field then are added two drops of freshly prepared orange colored (0.04 ml) tetrahydroxyquinone reagent (0.3 g of the disodium salt dissolved in 0.5 ml distilled water + 1 ml acetone). After approximately 20 sec, two drops of pH 2.8 buffer, prepared by dissolving 15 g tartaric acid and 18.9 g sodium bitartrate in 100 ml distilled water, are added. Since the air sample volume is known, the concentration can be estimated by comparing the color density of the developed stain with color standards prepared by analyzing a series of exposed filters by the dithizone method. This calibration technique conpensates for the typical speckled effect produced by insoluble lead dust, but not by soluble lead compounds, when applied by the usual methods for standard preparation.

The original reference[28] supplies the following description:

> A series of atmospheric samples was collected close to a source of lead fumes at different intervals of time. Stains were developed on each filter as described above, and grouped according to the color intensity. Six filters showing the same stain intensity were grouped together. Five of these were analyzed for lead content by the dithizone procedure and the sixth was set aside for a visual color standard. The lead in air concentration was calculated from the volume of air sampled, and from these groups four standards corresponding to 0.100, 0.200, 0.300 and 0.400 mg Pb/m^3 were selected. For

convenience in estimating lead concentrations, the four color standards were arranged in a radial pattern on a circular disk which contained a circular center aperture. Unknown stains were positioned in this aperture for color comparison. These standards were found to be stable for at least six months. However, in the commercially available field kits, the color standards are made up from stable printing inks.[29]

Experience in the field indicated that the error of visual estimation of lead concentrations was not greater than 0.05 (± 0.025) mg/m^3. Beyond 0.4 mg/m^3 the stain intensity becomes too dark for visual comparisons; therefore, the lead content must be determined by laboratory procedures or a smaller air sample taken. The Whatman No. 42 filter dissolved easily by heating in a mixture of 0.2 ml H_2SO_4 and 1 ml HNO_3 in a silica evaporating dish for analysis by either atomic absorption,[30] polarographic analysis,[31,32] dithizone colorimetry,[33] or the non-destructive x-ray emission spectrography[34] methods. In this fashion the colorimetric estimation could be confirmed or precisely quantitated at a later date if desired.

Blanks should be carried out along with the samples to confirm satisfactory level of contamination control in handling the filter discs and for background lead in the paper itself. Concentrations of lead intermediate between the 0.10 color standard intervals were estimated by interpolation in 0.025 increments. No interference was detected from Zn, Sb, As, Cr, Ca, Bi, Cu or Fe in concentrations equal to the lead content. When concentrations of Cd were twice the lead content, a slight interference was observed.*

The need for adjustment of the sample volume for personnel monitoring is obvious, as the color intensity otherwise would be saturated in 10 min (0.40 mg/m^3 limit). Therefore, the sampling rate must be reduced to 52 $ml/min/cm^2$ (equivalent to 164 ml/min for a 2.0-cm diameter filter) for a 4-hr sample or 25 $ml/min/cm^2$ (82 ml/min for a 2.0-cm diameter filter) for an entire 8-hr day in order to bracket the current TLV for inorganic lead dust in air (0.15 mg/m^3).[1]

The red color of lead dithizonate also has been adapted to colorimetric paper analysis. Lead-containing dust is collected on No. 54 Whatman filter paper, solubilized by treatment with a mixture of *n*-butanol saturated with 3 *N* aqueous hydrochloric acid, and then sprayed with 0.1% dithizone in chloroform reagent and 0.1 *N* aqueous sodium carbonate solution in turn. The change in background color from green to pale orange provides a built-in indicator for full development of the red lead dithizonate complex. The stains are stable for more than 6 months after development. Other heavy metal ions that also produce colored complexes include Bi, Cd, Co, Mn, Hg, Ag, and Sn. If lead or any one of these other metals occur singly in the air sample, this procedure can be applied directly. For mixtures, a paper chromatographic separation directly on the paper with the butanol-HCl developer has been described.[28] Furthermore, when the sample is concentrated onto a narrow paper strip, the stain length becomes proportional to the amount of lead (or any other color-forming metal) present. Then concentration can be estimated by reference to calibration charts in the same fashion as employed for length of stain detector tubes described in Section I.B.l, "Impinger Systems – Liquid Reagents,"[35] and confirmed by laboratory analysis as described previously for the tetrahydroxyquinone procedure.[28]

8. Fluorides

In at least one case, a double impregnated paper has been recommended for collecting separately the particulate and gaseous pollutant phases. The upper filter is treated with sodium citrate to retain particulate fluorides and the lower (downstream) paper with sodium hydroxide to absorb gaseous fluorides (HF).[36,37] Although no specific reference to a direct color development system for either layer was described, references to support direct reading reagents based on the fading of the violet zirconium alizarin lake to the yellow color of the dissociated dye could be applied directly.[38,39] Another color development system is based on the bleaching by fluoride ion of the intense red color of alizarin sulfonate absorbed on thorium hydroxide, which can be equated to concentration.[40]

A squeeze bulb type hand pump fitted with a filter holder has been offered for detection of fluorides in grab samples. The air sample is drawn across a replaceable impregnated paper that changes color on contact with HF. Concentration is determined by comparison with the reference chart delivered with the kit. However, the reagent was not identified, nor was a reference to extended sampling mentioned.[41]

9. Dimethylhydrazine, Unsymmetrical

This squeeze bulb filter assembly has been adapted also to the detection of unsymmetrical

*From Quino, E. A., *Am. Ind. Hyg. Assoc. J.*, 20, 134, 1959. With permission.

dimethylhydrazine, which was developed for use as a fuel in rocket engines.[42] Again, application to personnel monitoring was not suggested. The color-forming reagent (purple on exposure) was not identified, but probably was bindone – ($\Delta^{1,2}$-biindon)-1,3,3′-trione – which was described in a later reference by authors affiliated with the manufacturer of these kits.[43] This reagent was applied in the development of a dosimeter designed to be worn as a detector badge.

10. Boron Hydrides

An adaptation of the constant-volume piston hand pump developed in England during the 1930's has been offered on the domestic market for chromic acid mist and boron hydride detection. In this model, the filter paper retaining assembly became an integral part of the pump cylinder. The boron hydride kit was developed to meet the needs for monitoring in locations where this highly toxic rocket fuel creates a serious health problem for industries manufacturing and handling volatile boranes.[44] The procedure would be classified as a gas titration since 600-ml aliquots of air were drawn through the freshly impregnated paper (wet reagent system) until the color change matched a reference standard and the concentration determined by the number of piston strokes. A sensitivity of 1 ppm (pentaborane) to 3 ppm (diborane) was claimed. The reagent composition was not identified, but at least three have been described.[45] Adaptation to personnel monitoring in this case probably would require only calibration for 4- or 8-hr sampling.

11. Chromic Acid

This same equipment has been offered in a kit form for quick check for chromic acid mist over the range 0.1 to 0.2 mg/m^3 in electrolytic chrome-plating rooms. The colorimetric reagent has been identified as the usual phthalic anhydride + *S*-diphenylcarbazide couple, which turns blue to violet (lilac color) on contact with chromates. The shelf life of the impregnated paper appeared to be approximately 6 months from date of manufacture.[46]

12. Aniline

Aniline vapor in concentrations of about 5 ppm (by volume) in air is known to be toxic. Current applications of aniline make it imperative that a fast and simple field method be developed. The concentration of aniline vapor in the air can be determined easily and rapidly by use of paper strips impregnated with the liquid or vapor of a furfural-acetic acid mixture (4%, by volume, in glacial acetic acid). The test paper turns from white to pink or red in the presence of aniline vapor, having a range of determination from 5 to approximately 150 ppm. The speed and intensity of color development increases with the concentration of aniline vapors. The test is specific for aniline in the presence of many other organic vapors. The reagent mixture has a life of at least 1 month at room temperature.[171]

f. Quantitative Photometric Instruments

Photometric instruments that have been designed to determine the color density of stains, dyes, and light-absorbing dusts, such as smoke, on paper and other filter media can be applied to personnel monitoring paper impregnated with color-generating reagents. Nearly quantitative results are obtained when the paper is sufficiently uniform in texture and thickness to deliver consistently precise blank values required to correct the total absorbance for the absorbance of the unexposed paper. Usually absorbance measurements are made through either the maximum or the average density area of the color zone and the unexposed background directly adjacent to the boundary of the color zone. The blank is then subtracted either mechanically, by adjusting the photometer (densitometer) to 0% absorbance, or arithmetically in the final calculation. In some cases optical density ratios are more convenient to apply.

Several easily portable precision densitometers that provide a high degree of sensitivity have been adapted to the scanning of paper chromatograms.[47–49] Tape samplers that automatically draw successive air samples through 25- or 50-mm wide paper tape at predetermined intervals through a sampling head incorporate an integral photometer and provide a zero reference for each sample.[36,37] For applications where transmittance measurements are impractical, reflectance measurement is available.[37] Since this is a modular instrument, the detector unit could be removed or purchased separately for reading personnel monitor spots in the field.

2. Inorganic Support Systems

a. Portable Alarms for Hazardous Gases

Paper as a support for chromogenic reagents has very definite limitations. Strong acids, alkalines, or oxidizing agents are not compatible. Parasitic side reactions and rapid deterioration of the reagents often render the system impractical. Packaging in sealed containers to prevent premature contact with the ambient atmosphere presents further problems often difficult to resolve. Fortunately, several inert, highly adsorptive solid support materials are commercially available in granular forms that are readily adaptable to a wide variety of color-generating techniques. Silica and alumina gels probably are the most widely used, on the basis of high adsorption capacity, mechanical strength, and an essentially white background which provides maximum contrast for visual evaluation.

However, with the exception of the impregnated bisque tiles developed for monitoring H_2S, as described in Section I.A.1, "Impregnated Paper," application to diffusion (spontaneous) monitoring has been restricted to reagents available for the detection of carbon monoxide (CO), H_2S, and phosgene. The potassium pallado sulfite on silica gel reagent developed by Silverman for CO detection[50] has been adapted to a wafer prepared by cementing a layer of this impregnated gel on a 15-mm diameter recess in a 55-mm square by a 2-mm thick plastic tab (Figure I.1.5).[51] The tab may be worn either on the clothing within the person's line of vision or may be suspended in fixed locations such as on the instrument panel of a truck or crane cab. As supplied, the detector is a "dead stop" warning device that indicates dangerous concentrations of CO when the orange-brown "spot" turns black.

These carbon monoxide detector tabs were used quite successfully by Argonne National Laboratory personnel to monitor nonoccupational exposures. Of the 454 persons who received these tabs and reported results, 124 (27%) reported positive indications and 21 (nearly 5%) detected sources that were emitting sufficient quantities of carbon monoxide to have some physiological effect.[51a] Sixteen of this group followed through with corrective actions which included:

Automobile exhaust system maintenance
Automobile manifold heater repairs
Automobile replacement
Home furnace replacement
Cleaning, maintenance, or replacement of furnace heat exchangers
Improved exhaust ventilation of home gas heaters

When exposed over a period of "a few hours," the nominal ranges of response in known atmospheric concentrations of CO were found to be

Slight darkening	– 30 to 70 ppm
Gray	– 80 to 120 ppm
Black	– 130+ ppm

Unexposed tabs were furnished as a "reagent blank" reference for visual detection of the first stage, slight darkening. Although a 50-ppm TLV has been assigned to CO for an 8-hr time-weighted average, in a 40-hr work week the nonoccupational exposure for the remaining 128 hr should not exceed 30 ppm if adverse health effects, as manifested by impaired time interval discrimination, certain other psychomotor functions, and physiological distress for patients with heart disease, are to be avoided.[51b] Therefore, even the first stage (30 to 70 ppm) must be considered hazardous for individuals who also encounter occupational CO exposure. "Normal traffic" in many cases produced stage one conditions for passengers in automobiles. In those cases for which a follow-up confirmatory analysis was conducted with an indicator tube, the tab results were found to indicate somewhat higher CO concentrations, i.e., slight darkening indicated 30 to 70 ppm, whereas the detector tube showed 15 to 50 ppm. No instance in which a tab showed less than the true concentration, or in which the tab failed to respond to an environment where the presence of detectable CO was known or highly suspected was reported. On this basis the detector tab would be classified as a fail-safe device. Although considerable variance between samples taken at separate times in different ways, especially if one sample was integrated over a period of time and the other was a grab sample of short duration, was expected, the data collected from the two methods showed considerable agreement.

Automation has been applied to the detection of the color change. By retaining the impregnated gel between two closely spaced fine mesh metal screens, a beam of light can be passed through the CO sensitive "spot" to a phototube which monitors the beam intensity (Figure I.1.4).[52] When the light intensity is reduced to a preset value by darkening of the translucent silica gel reagent, a relay circuit is tripped, and a visual as well as an audible alarm is activated (Figure I.8). The alarm point may be adjusted by means of a potentiometer in the detector circuit and is usually set at the TLV (50 ppm for CO). Under low humidity conditions, the reaction rate of CO with the palladium complex decreases to a level which prevents effective functioning of the alarm. Therefore, a humidifying accessory (water reservoir and wick) is furnished as an integral part of the phototube assembly.

Some idea of the CO concentration can be obtained by calibrating the monitor as a gas titration end point detector. That is, the time required to reach the alarm point is directly related to the CO concentration. Therefore, calibration can be carried out in atmospheres of known concentration (glove box) by recording the time to reach the alarm point. Then, in field use, as soon as the alarm is activated, a fresh tab is immediately inserted, time until it alarms again is noted, and the approximate CO concentration read off from a calibration chart. Field experience has indicated that a good level of reliability can be attained in this manner (see Section I.A.2.d, "Direct Reading Length of Stain Gas Detecting Tubes").[53]

This monitor is small enough (16.2 cm x 9.5 cm x 5.1 cm) to be carried as a personnel monitor and is powered by a 10-V rechargeable nickel-cadmium battery (Figure I.8). The detector tabs regenerate themselves (the black color "bleaches" back to the original orange-brown color) on standing in a CO-free atmosphere (usually overnight) unless overexposed to high CO concentrations. However, after repeated cycles, the regenerative capacity deteriorates to a point of no return when balancing of the optical-electronic system is no longer possible.

Tabs are available also for H_2S detection in the TLV range (down to 5 ppm with Model 40).[52] Tabs could be made up from any one of the granular reagents used in filling the length of stain gas and vapor detector tubes currently available on the open market (see Section I.A.2.d, "Direct Reading Length of Stain Gas Detecting Tubes"). Each of these granular reagents would require individual calibration within the TLV range for the particular contaminant for which the detector tube was designed. The nonspecificity of several of the detector systems imposes limitations on their use. Chromic acid, for example, would respond to most reducing agents, and therefore, would be highly nonspecific. This approach offers an attractive solution involving a minimum expenditure of time and money for development, to the alarm problem associated with the nearly instantaneous detection of those hazardous gases and vapors for which ceiling TLV's have been assigned.

A novel design for a colorimetric personnel dosimeter for hydrazine fuel handlers incorporated a filter to protect a light sensitive reagent adsorbed on a thin layer chromatogram sheet.[43] A No. 8 Wratten® filter was mounted as a window flush with the upturned edges (6 mm) of the badge at a distance sufficiently above the hydrazine sensitive plate to permit free circulation of air through the annular space (approximately 4 mm). The filter absorbed the destructive actinic light while allowing the color development to be observed. The window could be raised for direct comparison of the developed color with the three color standards that surrounded the reagent area. The sensitive plate (14 x 41 mm) was cut from an Eastman® chromatogram sheet type K301R2 (a 100-μ layer of silica gel bonded to a film base with polyvinyl alcohol binder) that had been impregnated with bindone and dried at 65°C. The strip of sensitive material was held in place with springs so that an 8 x 14-mm rectangle was exposed, and an equivalent area was held tightly against a clear plastic (polyacrylate) retainer plate to serve as an unexposed blank section for visual detection of very faint stains. The color change from off-white to a purple provided a very distinct and sensitive response to hydrazine and methyl hydrazine exposure in very low concentration ranges. However, the detector will respond to most alkaline-reacting, volatile bases, including cigarette smoke.

Corrosion and darkening of metallic silver by H_2S have been suggested for ambient air quality measurements.[54] Air contaminated with H_2S or mercaptans on contact forms silver sulfide on the surface of the silver in the presence of oxygen and moisture. The decrease in reflectance of the silver surface as determined in a reflectance meter (Photovolt, Model 610)[47] is proportional to the H_2S-mercaptan dose. When air is drawn through the filter, a cellulose membrane filter is inserted upstream to prevent darkening and light scattering from dust, which would otherwise collect on the

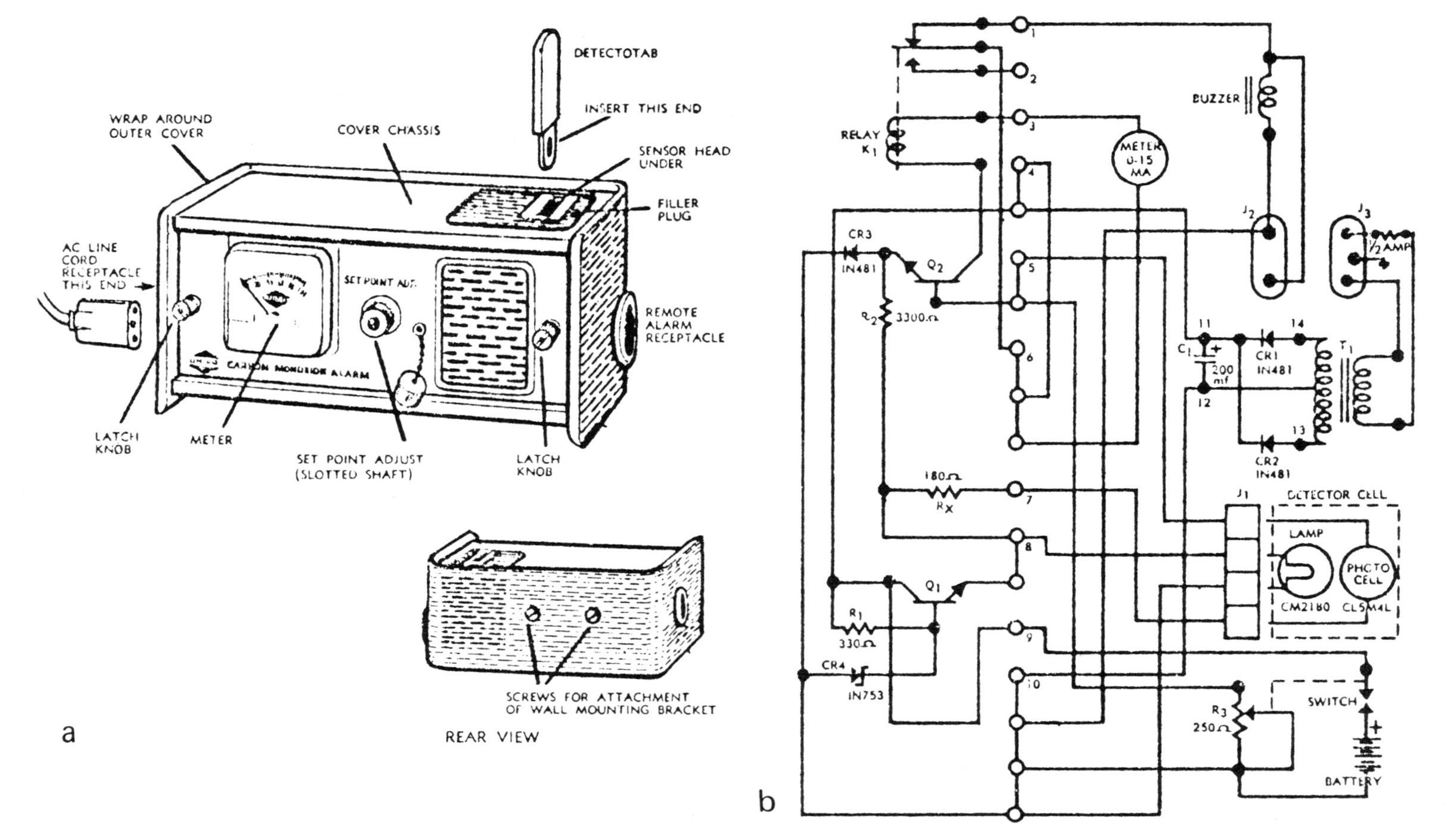

FIGURE I.8. Use of a carbon monoxide detector tab in an alarm system. a. Assembly. b. Schematic. c. The detector head. The Unico® Detectotab® (U.S. Patent 3,245,917) is the activating sensor in all Unico carbon monoxide detector alarms. Molded of high impact polystyrene, it provides a chamber and reservoir (A) for the sensitized silica gel reagent. This chamber (C) is defined by two parallel screens of corrosion resistant mesh which allows light to pass through. Constructed to assure proper seating, an indentation in the Detectotab (B) together with a mechanical detent (F) in the module sensor cell retains the Detectotab securely in the proper position relative to the optical path between lamp (D) and photocell (E). When exposed to carbon monoxide, the chemical reagent changes from its initial translucent straw color to a brownish black stain. The color change in the detecting silica gel is measured photoelectrically in a sensor head containing a zinc cadmium photocell and an exciter lamp (ordinary tungsten filament light bulb). The rugged, molded Zytel® sensor head is a plug-in module and can be remotely located up to 1,000 ft from the instrument. The molded chamber in the sensor head which holds the Detectotab is so constructed that heat from the exciter lamp creates a "chimney" effect drawing air over the reactive chamber surface of the Detectotab. An airflow of 200 cc/min is obtained assuring proper sampling with no maintenance or possibility of pump failure or wear. (Courtesy of National Environmental Instruments, Inc. Pilgrim Station, Warwick, R.I.)

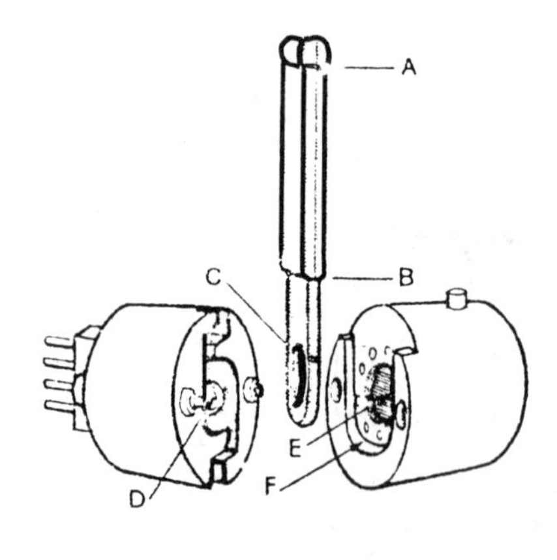

Figure 8c.

filter. This method is specific, is not sensitive to the usual interferences (O_3, SO_2, NO_X, etc.), and avoids the inaccuracy due to the fading of PbS stains on exposure to light in the lead acetate procedures.[55]

b. Chalk and Crayons

Detector chalk and crayons have been formulated from active chromogenic agents, inert binders, and fillers. When these chalks or crayons are rubbed on a suitable surface, a mark that will change color in the presence of the air contaminant for which the mixture was sensitized will be deposited. Detector paper can be produced by rubbing with chalk, or a powder can be prepared by crushing the chalk. Since the crayons present a relatively small surface area, the stability is superior to that of impregnated paper, they are easily carried, and they can be used to prepare on-the-spot personnel monitoring tabs.

Development work in this field has been confined to detector crayons for the war gases – phosgene, hydrogen cyanide, cyanogen chloride, Lewisite, and the "G gases."[56,57] The following description of the crayon preparation is a summary of the work reported by Witten and Prostak.[57]

1. Cyanogen Chloride

Ingredients

4-Benzylpyridine – 10%
Barbituric acid – 4%
Blanc fixe (precipitated barium sulfate) – neutral, amorphous, and dry – 86%

Procedure

After thorough mixing, 15-g aliquots of this dry mixture were compacted to 422 kg/cm^2 compression in a 16-mm diameter cylinder held in a Carver laboratory hydraulic press. The piston and cylinder were lubricated with a thin film of silicone grease. White marks from this crayon turned first red and then blue on contact with traces of cyanogen chloride or bromide. In 1 min a concentration of 0.4 ppm was detectable. Storage stability for 3 years was indicated.

2. Hydrogen Cyanide

Hydrogen cyanide was first converted to cyanogen chloride with a chlorinating agent such as Chloramine-T, and then detected by cross-marking with the cyanogen chloride crayon.

Ingredients

Chloramine-T (*N*-Chloropara-toluene-sulfonamide sodium salt) – 14%
Blanc fixe – neutral, amorphous, and dry – 86%

Procedure

Preparation followed the cyanogen chloride crayon technique. The mixed mark formed by superimposing this crayon's mark on the cyanogen chloride mark turned red-blue in the presence of HCN. A 1-min exposure detected 5 ppm. The Chloramine-T crayon was stable for at least 3 years at room temperature, but decomposed at 65°C. By substituting dibromantin (1,3-dibromo-5,5-dimethylhydantoin), stability at 65°C was attained.

3. Phosgene

The highly sensitive reagent system, 4-(para-nitro-benzyl)pyridine + *N*-phenylbenzylamine, previously described for detector papers in Section I.A.1, "Impregnated Paper," was employed for this system.

Ingredients

4-(para-nitro-benzyl)pyridine – 2%
N-phenylbenzylamine – 5%
Sodium carbonate – 5%
Blanc fixe – neutral, amorphous, and dry – 88%

Procedure

A benzene solution of the first two ingredients was mixed with the blanc fixe. Sufficient benzene to completely wet the inert substrate was suggested. After drying overnight, an aqueous solution of the sodium carbonate was thoroughly mixed in. Again, sufficient solvent (water) to completely wet the blanc fixe mixture was specified, and cooling was applied to avoid melting the *N*-phenylbenzylamine. The dried mixture was reground and pressed into crayons as described above.

The light yellow marks from this crayon turned bright red in the presence of phosgene. High concentrations of acid-producing gases (HCl) interfered by consuming the alkaline reserve (Na_2CO_3) required for color formation. At room temperature, storage stability for at least 1 year was indicated. A 1-min exposure detected 5 ppb of phosgene, but the color tended to fade at high concentrations. However, the color returned in fresh air.

4. Lewisite [Dichloro(2-chlorovinyl)arsine]

Thio-Michler's ketone reacts with Lewisite or high concentrations of phosgene to produce a green color.

Ingredients

4,4-Bis(dimethylamino)thiobenzophenone – 5%
Blanc fixe – neutral, amorphous, and dry – 95%

Procedure

A benzene-chloroform solution of the organic component was mixed thoroughly with blanc fixe, and most of the solvent was allowed to evaporate spontaneously overnight. Drying was completed under vacuum for 12 hr. The dried powder was reground and pressed into crayons as previously described. Light tan marks from this crayon exhibited color changes for several irritant gases as follows:

Lewisite: intense green-blue
Ethyldichloroarsine: green-blue
Phosgene: purple (high concentrations)
Chlorine: gray
Cyanogen bromide: gray

5. G Gases

In U.S. Patent 2,929,791, a crayon for detecting the G agents, chemical warfare gases – GB or sarin (isopropyl-methyl-phosphorofluoridate), GA or tabun (ethylmethylphosphoramide cyanidate), GD (pinacolylmethylphosphonofluoridate), and GF (cyclohexylmethylphosphonofluoridate) – was described.[56] The chromogenic detector system included a mixture of disodium diisonitroacetone and tolidine that turned to a magenta or purple color on contact with traces of one or the other of the G agents. The recommended composition:

Disodium diisonitrosoacetone	20%
o-Tolidine	20%
Urea	15%
Lithium stearate	27%
Lithium chloride	3%
Calcium oxide	15%

was compressed into crayons by the usual procedure. The CT value was given as CT = 6 μg min/l.

The concept of CT values (concentration x time) has been applied to crayon marks as well as to paper detectors and ceramic tiles, as discussed in Section I.A.1, "Impregnated Paper." If the crayon marks are exposed to equivalent air turbulence, temperature, humidity, and incident light, the time lapse to the first detectable color change was inversely proportional to the concentration of the contaminant within approximate limits. In other words, the CT value equals a constant that is a measure of sensitivity. In most cases the chromogenic reaction rate is both temperature and humidity dependent, and corrections are required when working outside of normal room temperature and RH limits. The CT values developed by Witten and Prostak for the chemical warfare gases are summarized in Table I.6. The crayons are commercially available.[58]

C. Oxygen Deficiency Alarms and Electrical Conductivity Meters

Although not strictly classified as hazardous gas or vapor detectors, some mention should be made of the pocket-size oxygen deficiency meters. The threat to life for workmen who must enter locations such as sewers, mine tunnels and shafts, silos, wells, fermentation vessels, storage tanks, chemical process equipment areas, oil refineries, etc., where oxygen may be reduced below respirable levels by bacterial action or spontaneous oxidation (coal mines), can be more serious than contact with toxic gases. Without the assistance of

TABLE I.6

Detector Crayon Sensitivity at 32°C*

Gas detected	Gas concentration (mg/m^3)	CT	Gas concentration (mg/m^3)	CT	% RH	Air movement
Cyanogen chloride	0.369		1.045			
		1.2		1.3	20	Turbulent
		1.0		0.7	69	Turbulent
		2.2		2.1	20	Static
Hydrogen cyanide	2.5		120			
		6		8	20	Turbulent
		3		4	70	Turbulent
		8		15	20	Static
Phosgene	0.064		0.611			
		0.02		0.02	20	Turbulent
		–		0.03	70	Turbulent
		–		0.04	20	Static

*From data presented by Witten and Prostak.[57]

a reliable instrument, man's senses cannot provide a warning in time to avoid the fatal consequences incurred by entering an oxygen-deficient atmosphere.

The model developed by Bio-Marine Industries[59] incorporates most of the desirable features that a personnel monitor should furnish: displays a continuous % O_2 readout; provides audible as well as visual alarm when O_2 concentration falls below 20%; is of small size (5.1 cm x 6.4 cm x 12.7 cm) and light weight (294 g); has rapid response (90% in less than 20 sec), high accuracy (± 1% of full scale), and negligible temperature compensation error (maximum ± 5% in the range O to 40°C); is not affected by humidity or CO_2; is rugged, remote sensing as well as self-contained; requires no warm-up; requires no battery or other electrical supply (two small batteries necessary for audible alarm model); and provides safe operation in explosion hazard areas (intrinsically safe per ISA-S.P.12.2 – Factory Mutual approval). Although the sensing element is expendable, replacement after 6,000 hr at 20% O_2 exposure (warranty) poses no problems (Figure I.9).

The principle of operation is derived from the galvanic action of oxygen in a cell composed of a gold anode and a lead cathode in an alkaline electrolyte. The entire cell is encapsulated in an inert plastic with a fluorocarbon polymer face that permits O_2 to diffuse into the electrolyte. The redox reaction at the lead cathode generates a minute current proportional to the O_2 partial pressure. A temperature-compensated circuit converts the current to a proportional voltage that is displayed directly on a meter as oxygen concentration. In another model, the signal is amplified through a battery-powered circuit.[60] From the standpoint of compactness, light weight, ruggedness, continuous readout, calibration simplicity, and reliable operation with a minimum of attention, this instrument would be classified as an ideal personnel monitor.

An entirely different principle for oxygen monitoring has been described recently.[61] Thin films of zinc oxide are used to measure oxygen pressure as a function of conductance. The assembly, which is simple, requires low power (approximately 7 W) and covers a wide dynamic range (1 mm to over 760 mm Hg); consists of a heater, electrical readout, and temperature controller; weighs about 30 g; and occupies about 30 ml of enclosure space. No failures were observed over several months of laboratory tests. Although this instrument has not gone into commercial production, the capabilities indicate that its exploitation as a personnel monitor would seem to offer advantages at least as favorable as the galvanic cell type.

The principle of thin metal film dosimeters has been extended to the detection of fluorine, nitrogen dioxide, and unsymmetrical dimethylhydrazine (UDMH).

The detection system is based upon the change of electrical resistivity of thin metal films when exposed to these gases. Silver metal films coated with appropriate salts proved to be applicable to

Dimensions:	2 x 2½ x 5 in. (5.1 x 6.4 x 12.7 cm)
Net Weight:	18 oz
Scale Reading:	0–40%
Response:	90% in less than 10 sec
Accuracy:	±1% of full scale
Sensor Warranty:	180,000% hr or 1 year

FIGURE I.9. Galvanic cell continuous oxygen monitor for personnel monitoring. (Courtesy of Bio-Marine Industries, Inc., Devon, Pa., and Mine Safety Appliances Co., Pittsburgh.)

the detection of all three gases; however, the following sensitized metal films were found to be optimum: for N_2O_4, silver; for F_2, copper; and for UDMH, gold. Using the best film and salt combinations found to date, N_2O_4 can be monitored over the range of 0.1 to 50 ppm, F_2 over the range 1.0 to 50 ppm, and UDMH over the range 10 to 100 ppm, with a standard deviation of about 20%. The effects of temperature over the range 50 to 90°C, and of humidity from 10 to 90%, on the response characteristics of the thin film sensors were found to be significant but within the tolerance limits. Means for reducing these effects were suggested which, if successful, would, in effect, make this detection system practically independent of changes in the environment. A portable breadboard readout instrument was designed and fabricated for use with the sensors to form an integrated detection system for personnel protection.[169]

A gas-detecting device was invented that uses a metal oxide semiconductor element, the resistance of which varies when exposed to a reducing gas such as hydrogen, carbon monoxide, alcohol vapor, gasoline vapor, or smoke. It also contains an alarm circuit for producing an alarm in response to a predetermined reduction of resistivity of the semiconductor element. Another component is an impedance circuit coupled to a pair of electrodes through a power supply. A thermal switch operates in response to a predetermined temperature rise for the semiconductor. The switch consists of a bimetal-operated, single-pole, double-throw switch with two fixed contacts: the first resistor and the second resistor. The first resistor has a positive temperature coefficient of resistivity and is connected to the impedance circuit. The second resistor has a negative temperature coefficient of resistivity and is connected to the alarm circuit.[170]

d. Direct Reading Length of Stain Gas Detecting Tubes

1. Historical

Glass tubes filled with a solid, granular material, such as silica or alumina gel, which has been inpregnated with a color-forming reagent, are by far the most universally used portable toxic gas or vapor detectors.

Hoover and Lamb designed the first detector tube at Harvard University during 1917 for the rapid analysis of CO with Hoolamite mixture (IO_5 + SO_3). This detector tube probably was the first of the color-indicating devices, and it survived the test of time for over 25 years. During the interval from World War I to 1935, few detectors were developed for field use. In 1935 the U.S. Bureau of Mines designed the first length of stain type tube to appear on the open market for the determination of H_2S in coal mines. The reliability and sensitivity were improved in 1946 by reducing the diameter of the glass tube. At the National Bureau of Standards, Martin Shepherd and his staff produced the first high sensitivity detector tube for determining CO concentrations in military vehicles and on aircraft carrier decks during World War II. This CO detector is still used, but it is being superseded rather rapidly by the length of stain reagent system, developed by the Royal Air Force Research Establishment during World War II and later improved by Leslie Silverman in 1965.[50]

The first detector tube for organic solvents (benzene, toluene, xylene, and other aromatic solvents) derived from research conducted at Harvard by Silverman's group.[62a] Since then, other tubes have appeared in Japan, Sweden, Germany, and the United States.

2. General Operating Procedure

The use of detector tubes is quite simple. After the two sealed ends are snapped off, the tube is inserted in the holder, which is fitted with a calibrated squeeze bulb, bellows, or piston pump. The specified volume of the air sample then is drawn through the tube by the operator; and after adequate time is allowed for each stroke to draw completely its full volume of air, the contaminant concentration is determined by examining the exposed tube. The concentration can be determined by one of three procedures:

1. By comparing the shade of color observed with a set of color standards. The accuracy of this visual judgment is limited by the color acuity of the observer and by the lighting conditions.
2. By measuring the length of stain produced and relating the length to a calibration graph or chart. The scale may be printed directly on the tube or on a chart which is provided.
3. By measuring the volume of air required to produce a visible color change and relating the sample volume to a calibration chart (gas phase titration).

The length of stain technique (2) is used most commonly and has with a few exceptions displaced the other two methods. However, under the right conditions the gas phase titration (3) is fundamentally capable of producing the best precision and accuracy. All three types have at one time or another over the past 25 years been commercially available.

3. Availability and Limitations

Examples of commercially available kits are shown in Figures I.10 to I.17. These direct-reading indicator tubes provide a very rapid, convenient, and inexpensive analytical tool with which a semiskilled operator can perform air analysis. However, this very simplicity creates problems when inexperienced, untrained operators misuse the technique. The design of commercially available, field-type equipment has been simplified to a degree that almost everyone considers himself qualified to carry out analyses and interpret the data. The results may be dangerously misleading unless the sampling procedure is supervised and the findings interpreted by an adequately trained industrial hygienist.[62]

4. Applications

A review of the common applications of grab sampling with detector tubes will convey some idea of their utility and broad problem-solving potential.[64] Even greater application potential can

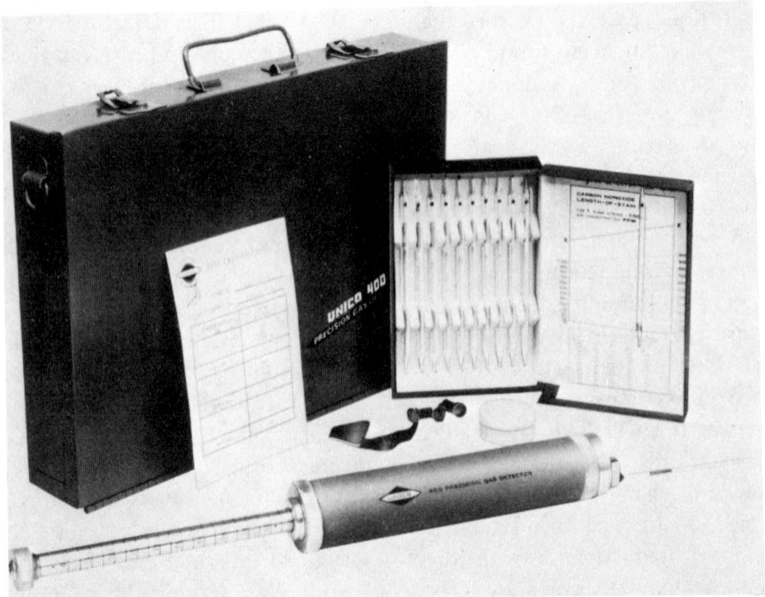

FIGURE I.10. The Kitagawa® gas detector tube kit. (Courtesy of National Environmental Instruments, Inc., Pilgrim Station, Warwick, R.I.)

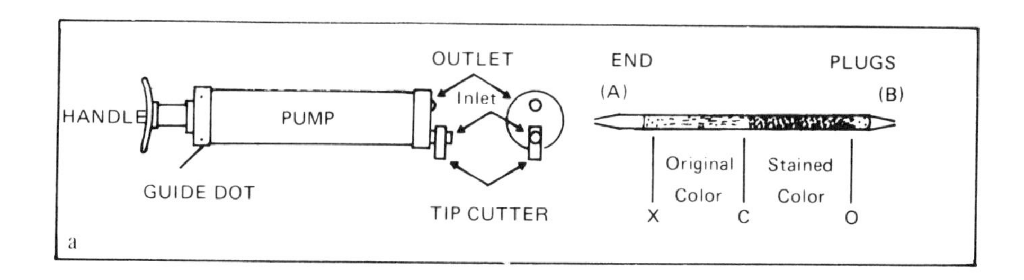

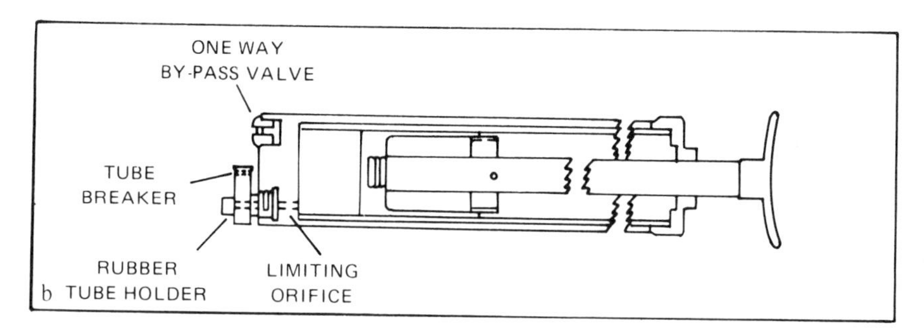

1. Cut off the tips (A) and (B) of a fresh tube by turning each tube end in the tip cutter.

2. Insert tip (A) marked with a red dot securely into the pump inlet. If the handle is not in all the way, push it in.

3. Line up the pump handle with the red dots on the end of the pump. Pull the handle all the way out, locking it with a half turn.

4. Wait 3 min. During this time, the air being sampled flows through the detector tube and constant-flow orifice into the evacuated pump. At the end of the sampling period, release the handle with a half turn. The handle should not move back more than 5cc (about ¼ in.). Occasionally a small particle becomes lodged in the constant-flow orifice, restricting the flow. If this happens, the handle will be drawn back more than 5cc when released, and the measurement should be discarded and repeated with a fresh tube.

5. As soon as the sample is drawn, remove the tube from the pump and place it vertically on the concentration chart. Position the tube with the stained end down so that the boundary between the end plug and the resin at the inlet (point O) is over line O-O and point X is over line X-X. Read the air concentration corresponding to the length of the stain O-C.

FIGURE I.11. Unico® Model 400 precision gas detector pump – cutaway section. (Courtesy of National Environmental Instruments, Inc., Pilgrim Station, Warwick, R.I.)

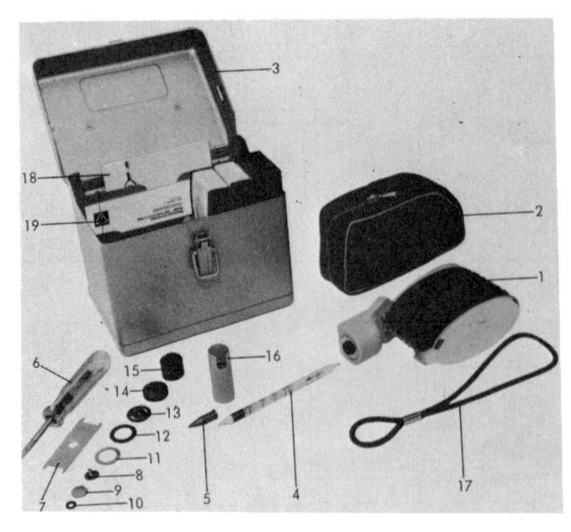

1 Gas detector pump, model 31
2 Protective bag
3 Carrying case
4 DRÄGER tube
5 Rubber cap
6 Special screwdriver
7 Special spanner
8 Valve disc
9 Sieve
10 Sealing ring
11 Gasket
12 Rubber ring
13 Washer
14 Valve seat
15 Rubber bung
16 Break-off husk
17 Carrying strap
18 Instructions for Use
19 Table

FIGURE I.12. The Dräger gas detector tube kit. (Courtesy of Drägerwerk AG, Lübeck, West Germany.)

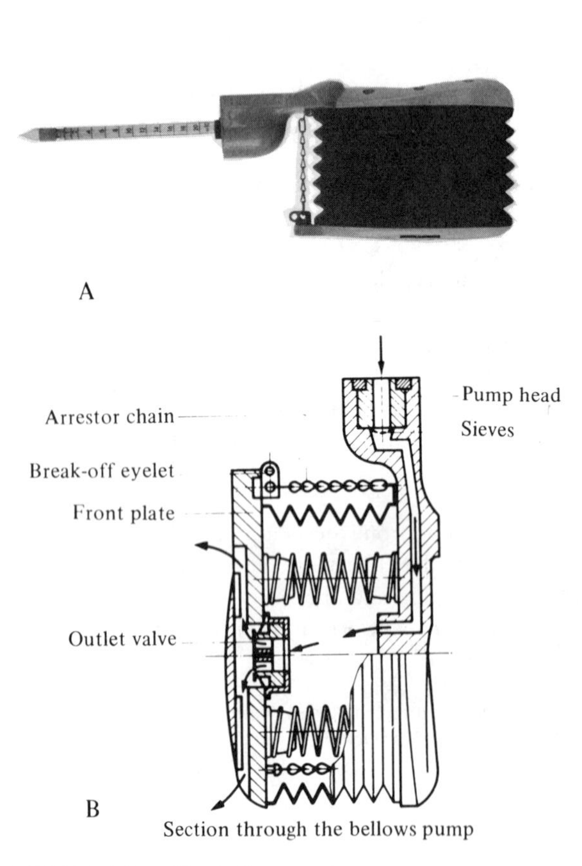

FIGURE I.13. The Dräger gas detector pump. A. The Dräger gas detector consisting of the gas detector pump and a Dräger tube. B. Cross-section through the gas detector pump. (Courtesy of Drägerwerk AG, Lübeck, West Germany.)

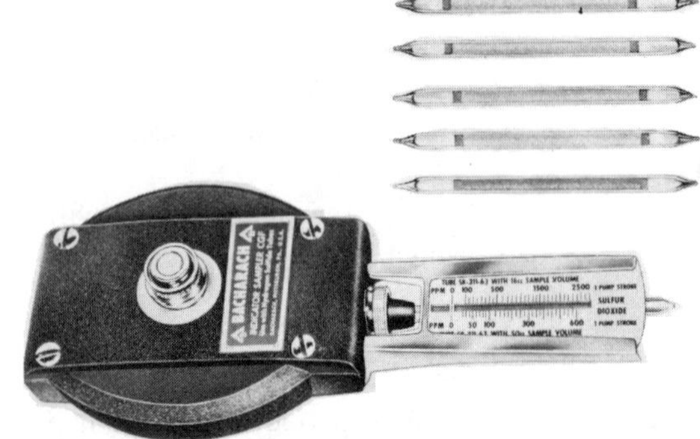

FIGURE I.14. The Bacharach gas hazard indicator. (Courtesy of Bacharach Instrument Co., Pittsburgh.)

FIGURE I.15. The M-S-A® universal detector tube kit. (Courtesy of Mine Safety Appliances Co., Pittsburgh.)

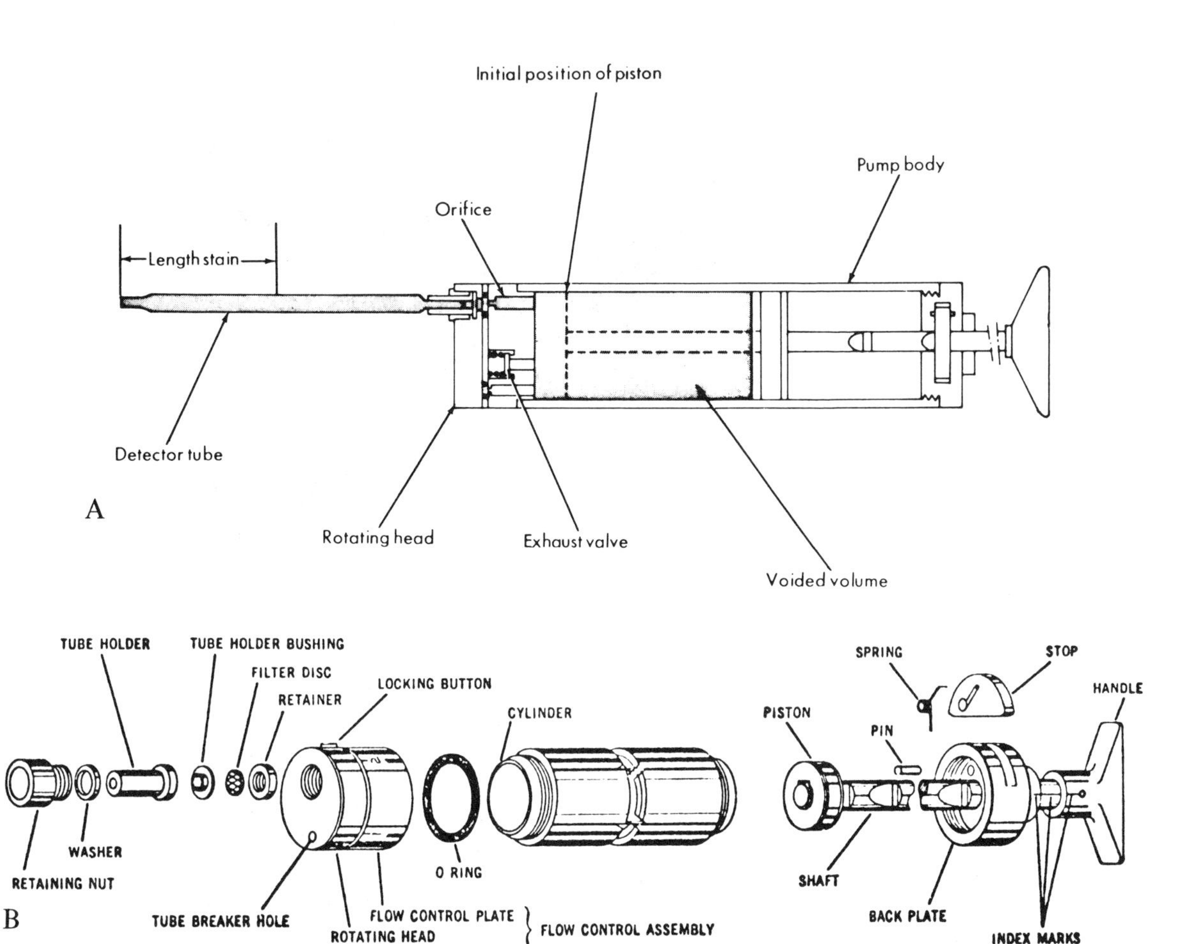

FIGURE I.16. The M-S-A® sampling pump. A. Cutaway view. Sampling begins with adjustment of orifice index marks to assure proper flow rate. Tips then are broken off glass detector tube, and it is fitted into rubber tube holder. Handle is drawn back until proper volume number appears on pump shaft. At this point spring-loaded stop will drop into place, preventing handle from being drawn forward into evacuated pump body. After minimum time has elapsed, handle is rotated 90° and is pushed forward to exhaust sample air. Concentration of contaminant is indicated by length (or, in some cases, color) of stain. B. Exploded view. Sampling pump components are of stainless steel. Pump body is plastic covered to eliminate possible damage from knocks encountered in ordinary service. Interesting design feature is use of Teflon® cap over standard O-ring in piston seal. O-ring provides required resiliency while Teflon provides minimum friction and good wear characteristics. Piston stop is of nylon. Bottom hole in rotating head is for breaking tips of sealed glass detector tubes. (Courtesy of Mine Safety Appliance Co., Pittsburgh.)

be expected when recalibration for continuous personnel monitoring is established.

a. Testing Atmospheres Within Vessels and Manholes

Clearance for entering enclosures that may contain toxic gases, explosive vapors, or may be deficient in oxygen, before a workman enters is required by law in some states (New Jersey, for one). Handheld, clothing or harness supported, self-indicating field-type instruments that provide quick, on-the-spot results offer the most practical means for testing atmospheres inside vessels, pipes, manholes, or other confined spaces into which entrance is necessary.

b. Work Area Monitoring

Problems that arise in this traditional area for industrial hygiene surveys and exposure control programs can best be solved by application of readily portable, field test equipment, which is described in other sections.

c. Emergency Air Sampling

Almost without exception, the rapidly reacting,

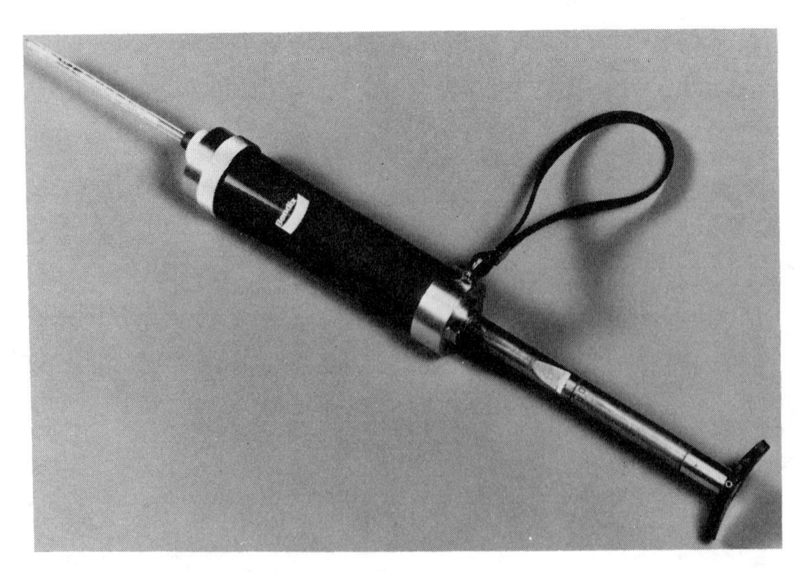

FIGURE I.17. The Gastec® gas and vapor detector – pump and tube. (Courtesy of Bendix Environmental Science Division, Baltimore.)

self-indicator type field instrument must be available to delineate the high-risk hazard area created by a chemical spill, accidental release of toxic gas or vapor, or abnormal operating conditions. Serious psychological problems can be encountered if the sensitivity range of the instrument chosen does not include the odor threshold, or if the instrument sensitivity is greater than the odor threshold. Hydrogen sulfide furnishes an example of the first condition: the TLV is 10 ppm[1] and the odor threshold less than 1 ppm. As a result, many unjustified complaints must be investigated only to confirm the nuisance value of H_2S in concentrations below 10 ppm. Phosgene may be cited as an example of the alternative condition. In this case the TLV is 0.1 ppm,[1] whereas the odor threshold is above 0.5 ppm. Persistent odor of phosgene, then, indicates a hazardous condition that requires correction.

d. Personnel Monitoring

Although the current application of detector tubes to short interval or grab sampling (3 min or less per sample) by drawing a discrete volume of air through the tube by means of a manually operated piston or bellows pump cannot be classified strictly as personnel monitoring, the technique falls in a gray zone between fixed station and mobile industrial atmosphere sampling. The equipment is readily portable, in some cases pocket size; results are obtained quickly on the spot; and these tubes can be used for practically instantaneous warning of hazardous gas concentrations before ceiling limits are reached. If a sufficient number of samples are taken in the breathing zone during residence in an exposure hazard area, the procedure does in fact approach continuous personnel monitoring.

However, grab sampling will reduce efficiency because the operator or mechanic must interrupt his assigned duties to draw the sample through the sampling device and then interpret the results. However, for those gases and vapors that exhibit an odor threshold below the TLV, this factor alone does not detract from the advantages that can be derived from quick assessment of the unusual condition. One might consider this a "tool box" instrument to be carried on the job for confirmation of safe or hazardous conditions when abnormal odors are encountered. However, olfactory fatigue becomes a limiting factor whenever the sense of smell is relied upon for protection.

The following procedure is used to determine compliance with the 8-hr, time-weighted average standard based on a small number of instantaneous (grab) samples collected at random intervals during the workday.[64b]

Given: the results of n samples with a mean m and a range (difference between least and greatest) r.

If, for from three to ten samples, m is greater than the total of:

A. The standard

B. The percentage of systematic instrument error multiplied by the standard

C. $\frac{t \times r}{n}$

then the true average concentration exceeds the standard ($p < 0.05$).

The value of t is taken from the following table:

n (number of samples)	t (student's "t" test value)
3	2.35
4	2.13
5	2.01
6	1.94
7	1.89
8	1.86
9	1.83
10	1.81

A more detailed example of this procedure is given by the following reprint of "OSHA SAMPLING DATA SHEET #1."[64b]

Substance:

Carbon monoxide.

Standard: 29 CFR 1910. 93(b) Table G-1, of August 13, 1971

8-hr time-weighted average: 50 ppm (55 mg/m^3).

Sampling Equipment:

1. CO meter (Portable, Hopcalite – reagent type) – Follow directions. (Primary Method.)

2. Certified or NIOSH approved detector tubes and a hand operated pump (not squeeze bulb) may be used for compliance purposes. (Secondary Method.)

Analytical Procedure:

1. Direct reading instrument based on measurement of heat or reaction produced by oxidation of CO to CO_2. Follow instructions.

2. Detector tubes are read for length of stain, strokes of pump are noted and appropriate corrections made for pressure (or altitude) above 5000 feet (1500 m). The tubes are not to be used when below 0°F(-18°C) or above 125°F(52°C).

Sampling Period:

1. Hopcalite: maximum feasible number of readings should be averaged to determine average exposure. Time required dependent on variability and cyclical nature of exposures.

2. Detector tubes: direct reading, length-of-stain detector tubes (certified or NIOSH approved) may be used for compliance purposes, if at least one sample per hour is taken per exposed worker or close group of workers; in an 8-hr day more frequent samplings per hour are preferred, especially for exposures over 50 ppm. Eight different samplings (one per hour is considered tne minimum acceptable number).

Recent studies have indicated that at least some of these detector tubes can be recalibrated and adapted directly to personnel monitoring for periods of at least 4 hr.[53] Tubes for NH_3, CO_2, CO, Cl_2, HCl, H_2S, Hg, NO_2, and SO_2 have been offered for this purpose, but information relative to calibration details and operational limits has not been made available either in technical publications or in answer to requests for assistance.[63] Therefore, the user must remain skeptical with respect to the reliability of these tubes until he has carried out his own recalibration.

e. Product Control Testing[64]

In some cases a contaminant in an otherwise inert gas, e.g., phosphine in acetylene, is the controlling hazard. Self-indicating devices offer a satisfactory solution to this problem, provided the "carrier" gas or vapor does not interfere with the detector's reagent system.

f. Troubleshooting and Technical Development[64]

Detector tubes are quite suitable for leak testing. However, if the tube selected does not include the TLV range, potential problems for health protection can be created by their misuse. A nickel carbonyl tube, for example, has been available for the concentration range 20 to 700 ppm, which is adequate for source emission (leak) surveying, but must not under any circumstances be used for health conservation[65] (TLV = 0.001 ppm).[1]

g. Fire Fighting[64]

Detector tubes have demonstrated usefulness for establishing the presence or absence of certain toxic gases derived from the thermal decomposition or burning of combustible materials, especially plastics (HCN, $COCl_2$, Cl_2, CO, SO_2, etc.). Fire fighters especially could benefit from routine use of these tubes to detect near lethal concentrations of those hazardous combustion products that offer little or no odor warning, or that exhibit a strong olfactory fatigue potential.

h. Detection of Explosive Hazards

Detector tubes for detection of flammable gases and vapors in the explosive mixture range have been designed.[62] This particular application should be exploited further as it shows considerable promise since no electrical circuits which might provide an ignition source are involved.

Such tubes also have been used to detect gasoline in soil and to clarify cases of suspected incendiarism.[62]

i. Breath Analysis – Personnel Monitoring and Medical Diagnosis

Carbon monoxide poisoning may be confirmed by determining CO in the exhaled breath.[66-68] Also, indicator tubes have been employed for law enforcement purposes to determine the amount of alcohol in the breath, the presence of which is then related to blood concentration by means of well-established charts and time decay curves.[67,69-71]

j. Blood Analysis

Volatile organic compounds such as CO or alcohol, after release by acidification, oxidation, or other appropriate chemical treatment, have been analyzed in the blood by means of detector tubes.[62,65]

k. Ambient Atmosphere Analysis for Air Quality Criteria Enforcement

To date, no tubes have been developed that are sufficiently sensitive to determine the pollutants for which criteria have been established (SO_2, CO, photochemical oxidants, hydrocarbons, lead, and nitrogen oxides).[72]

l. Remote Sampling

When sampling an inaccessible location, the detector may be placed directly in the location to be examined, and the pump operated at some conveniently remote spot. A connecting flexible (rubber) tube of the same internal diameter as the inner diameter of the glass tube connects the pump to the exhaust end of the detector tube. Under no circumstances should the sample be drawn through the flexible hose, as serious errors can ensue from losses of the contaminant on the walls of the sampling line. Lengths as great as 60 ft can be used without appreciable error, provided more time is allowed between strokes of the pump to compensate for the reservoir effect and obtain the full sample volume. This method has a major disadvantage; the detector tube cannot be observed during the sampling.

m. Source Emission Sampling

Cooling the sample is essential when analyzing hot gases from stacks or engine exhaust. Otherwise the calibration will be invalid and the sample volume uncertain. A glass or metal probe may be attached to the inlet end of the detector tube with a short length of flexible tubing (silicone rubber or polyperfluoroethylene). If this probe is cold initially, a length as short as 4 in. outside of the hot duct will cool the gas sample from 250°C to about 30°C.[62] This equipment assembly must be used with caution since under some conditions serious absorption losses occur on the tube walls or in the condensed moisture. The dead volume of the probe should be negligible with respect to the sample volume. Condensible vapors should not be sampled by this method, as serious condensation interferences and losses will be encountered when the dew point (saturated vapor temperature) is above the temperature of the sample as it is drawn through the detector tube.

n. Oxygen Deficiency Determination

Although the detector tube for oxygen analysis is no longer available commercially, mention should be made of it in passing as the principle was unique and showed great promise as a "vest pocket" instrument that required no electrical circuiting for operation. In this case, the sample (50 ml) was pushed rather than pulled through the detector to avoid complications introduced by a significant decrease in sample volume, i.e., 21% in the case of normal ambient atmosphere sampling (see Reference 2 for commercial source).

The precise identity of the reagent system was never completely divulged, but the author's understanding was that the oxygen-consuming reagent comprised pyrogallol and potassium hydroxide adsorbed on alumina gel, which turned from gray to dark brown on oxidation. The following brief description is reprinted by permission of the American Industrial Hygiene Association.

> The technique for determination of trace atmospheric contaminants by passing a small sample through a column of a granular solid supporting color-forming reagents and measuring the stain length to estimate concentration had not been applied to oxygen analysis before 1962. A detector tube which produces near-quantitative measurements from a 50-ml sample passed through at a uniform rate of 0.5 ± 0.05 ml/sec (90 to 100 seconds) was released in May, 1964, after a two-year shelf-life study. Laboratory delineation of flow rate, ambient temperature, and accuracy limits confirmed the reliability of the method.
>
> At room temperature, normal atmospheric concentration, and flow rates of 0.3 to 0.6 ml/sec, results deviated

less than 5% from the accepted value (21% ± 1% O_2). Subnormal temperatures in the range 25° to 50°F tended to yield low values, with the majority falling below normal (21% + 0.5% to −1.5% O_2). Results were unaffected by reaction time in the range 64 to 144 seconds. Correction charts for the 32° to 104°F range were added to the kits later. At temperatures below 20°F, some over-all staining occurred which reduced oxygen values 10% to 20% below 21% oxygen before quantitative absorption started. Preheating the indicator zone as recommended corrected this deficiency. Four individuals unskilled in the use of laboratory equipment obtained sixteen acceptable analyses. No preliminary heating was used, air temperature ranged from 35° to 75°F, and chart readings fell within 21% to 22% limits. Analysis for oxygen in the range 8% to 15% agreed within 5% of the calculated dilutions. Again, reaction time variations within 77 to 130 seconds did not significantly affect the results.

Stain from diffusion of oxygen through the open ends of the tubes introduced significant errors unless the tubes were used promptly after the tips were broken. The open ends must be capped immediately after use if a preservation of the stain length is required for purposes of record. The diffusion stain front reached the 21% level in 10 hours from one open end or in 2.6 hours with both ends open.*

This novel application suggests that the principle perhaps could be applied to major constituents in other gas mixtures, and that a volumetric readout could be substituted for measurement of length of stain. Absorbing columns to eliminate interferences have been coupled with certain detector tubes,[62] especially for removal of water to protect moisture sensitive reagents (see "Specificity and Interferences," Section I.A.2.d.6.h.).

5. Certification Criteria

These tubes are in a continuing state of development, and highly variable results have been obtained. Close attention to seemingly trivial details is required to obtain even semiquantitative results. The failure to understand the limitations of the technique, to meticulously follow instructions, to be aware of interferences, and to apply corrections when necessary have accounted for most of the unsatisfactory performances turned in by semi- or unskilled users. To attain the ultimate precision and accuracy, frequent recalibration must be carried out by the user.

Accuracy ultimately depends on the attainment of a high degree of skill on the part of the manufacturer who, even more than the user, must be aware of the controlling variables and limitations, as no one else is in a better position to eliminate the sources of variability and error that limit the application of currently available tubes to semiquantitative range finding. The best accuracy that can be expected lies in the range of ± 20%,[62] and many tubes do not deliver a performance better than ± 50%. The reliability of these tubes has so often been questioned that the U.S. Public Health Service, Bureau of Occupational Safety and Health, was prompted to undertake a performance study of the four most commonly used tubes: CO, benzene, SO_2 and CCl_4.[73-77] Of these, only the SO_2 tubes consistently delivered results that conformed to the criteria established by the ACGIH-AIHA Joint Committee on August 9, 1965: "Each calibration point listed by the manufacturer shall be correct to within ± 25% of the stated concentration over the working range of the tube".[76,78] The calibrations were checked at 0.5, 1.0, and 5 times the TLV for the gas under investigation. Some of the tubes could not meet an alternate performance level of ± 50% at the 95% confidence limit (CCl_4),[69] and others (CO and benzene) did conform[73-75] to the alternate ± 50% criteria, but not to the recommended ± 25% limits. Later testing indicated that the Unico®-Kitagawa® tubes (catalog no. 100) did indeed perform within the ± 25% criteria (unpublished reports). Similar reports relative to the inaccuracies encountered with styrene vapor,[79] ozone,[80] and other detector tubes[81] have appeared in the technical literature.

Further testing by the National Institute for Occupational Safety and Health (NIOSH) confirmed the accuracy of the Unico-Kitagawa and Dräger tubes for perchloroethylene to be within the ± 25% criteria limits.[78,81a,83]

As a further consequence of these findings, the mounting dissatisfaction, uncertainty voiced by professional industrial hygienists,[64,82] and recommendations presented by the Joint ACGIH-AIHA Committee,[78] the creation of a Federal Certification Agency was initiated during June, 1971.[83] The preliminary draft of the Detector Tube Certification Schedule included:

1. Purpose
2. Gas detector unit
3. Construction and performance

*Reprinted from Linch, A. L., *Am. Ind. Hyg. Assoc. J.*, 26, 645, 1965. With permission from the American Industrial Hygiene Association.

4. Conduct of tests: demonstrations
5. Applications
6. Certification of approva
7. Approval tables
8. Material required for record
9. Changes after certification
10. Withdrawal of certification
11. Quality Control

6. Accuracy-controlling Factors

a. Uniformity of Granular Support

In general one of three particle-size ranges (Tyler mesh 20-40, 40-60, or 60-80) of silica or alumina gel is selected for the reagent support. Pumice, ground glass, and organic resins have been restricted to specific applications in which some special property is required. The stain length increases as the particle size increases, but the demarcation of the stain front becomes obscure. Conversely, demarcation of the stain front becomes more definite as the mesh size decreases, but the sensitivity also decreases due to shortening of the reaction zone (stain length) and the pressure drop (ΔP) increases to a point beyond which airflow practically ceases. The distribution of particle sizes in the chosen range should be sufficiently uniform to provide on a unit-length basis uniform ΔP, homogeneous reagent distribution, and stability in transport. When the size analysis includes an appreciable range, the fines may segregate to one side of the tube with the result that the cross-sectional airflow resistance will not be uniform and airflow will vary from side to side within the granular column (channeling). This nonuniform sample velocity frequently produces oval, spiked, or trailing (sometimes referred to as "tailing") stain fronts that are not perpendicular to the tube bore and are particularly difficult to delineate visually. If the gel support is friable, attrition may produce troublesome fines.

For those detectors which depend on ΔP rather than on a limiting orifice to control the flow rate, the uniformity of particle size distribution is especially critical. The diameter of the column has been increased as partial compensation for this factor, but this alternative approach involves a greater risk of channeling. By variations in the "charging preparation" used, a range of 3 to 40 sec in sampling time can be obtained with a constant tension spring bellows hand pump[6,7] (Figure I.18).

The ideal solid support system should contribute no color of its own, that is, it should be a "pure white" background. Refinements in gel production have for all practical purposes permitted attainment of this goal. However, very wide and unpredictable variations in the physical and chemical properties of different batches of

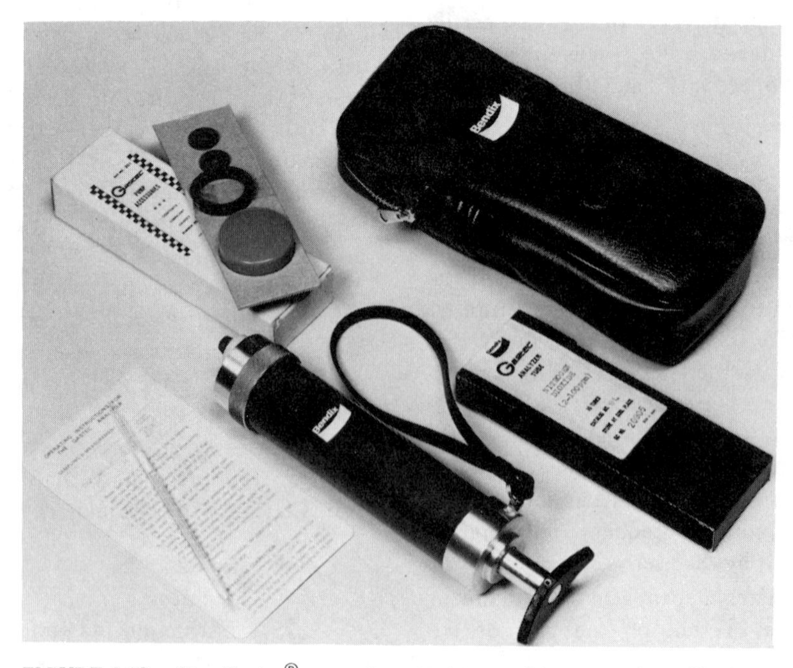

FIGURE I.18. The Gastec® gas and vapor detector kit – complete. (Courtesy of Bendix Environmental Science Division, Baltimore.)

indicating gels is the rule rather than the exception. Since the major portion of the chemical reaction occurs on the surface, the number of active centers that are highly sensitive to traces of impurities affects the reaction rate. These problems are well recognized in the manufacture of catalysts.

Here the manufacturer must exercise very close quality control on the purity and activity of materials going into the processes, the method of preparation, the cleanliness of the atmosphere in the production and assembly areas, and the particle size distribution of the finished product. Each batch of indicating gel must be individually calibrated, and adequate quality control procedures must be applied (see Chapter IV). The preparation of "active" gels that satisfy specific requirements is largely empirical, and details of their manufacture are closely guarded trade secrets. In some cases the properties of the gel itself contribute as much as the color-forming reagent to the success of the detector.

Moisture may have a significant effect on both the formation of the colored derivative and the physical properties of the gel itself. As an example, consider the lengthening of the colored zone produced by arsine in contact with mercuric chloride adsorbed on silica gel as the water content of the gel increases[82a] (Figure 19). When the moisture content decreases, not only does the stain length diminish, but the color reaction changes from dark brown to yellow. Furthermore, above about 120% water content (by solid absorbant weight), the gel tends to shrink in storage, thereby loosening the column and creating voids. A compromise between maximum stain response and gel shrinkage was reached by maintaining water content close to 120%.[82a]

b. Tube Construction

In general, the narrower the column, the better the definition of the stain front and the smaller the risk of channeling. However, here again, as in decreasing the particle size to sharpen the leading edge, this advantage is attained at the expense of increasing ΔP. Therefore, a compromise between grain size, column diameter, channeling, and ΔP must be reached. One manufacturer has successfully reduced the diameter to 2 to 4 mm (inside diameter of the glass tube),[64a,65] while others have continued to use the larger diameter (8 to 10 mm).[63,67]

In the manufacture of the narrow indicator tubes, variations of internal diameter produce

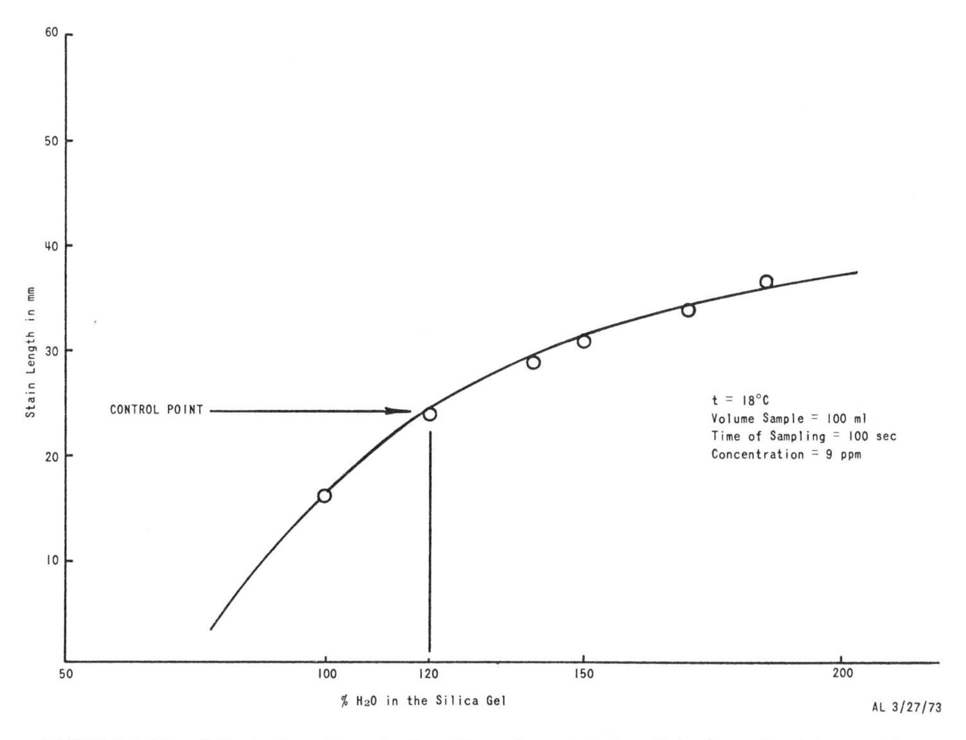

FIGURE I.19. Effect of moisture in the silica gel on stain length in the arsine detector tube. (Data taken from Reference 82a.)

significant variations in cross-sectioned (transverse) area that result in calibration errors as great as 50%. This error is possible because the volume of sample per unit cross-section differs from that established under standard test conditions. In addition, constant flow rate through the variable cross-sectioned areas cannot be maintained within critical limits. If an exactly equal quantity – by weight or by volume – of indicating gel is added to each tube, then variations in the transverse area will be disclosed by corresponding deviations in total column lengths. One manufacturer furnishes correction charts, which reduce this error to within ± 5%, on which the tube is positioned according to column length before reading the stain length from the compensated scale[65] (see Figure I.20). Although the theoretical corrections are rather complex, practically linear approximations can provide corrections that reduce the error to within 10%. In the larger diameter tubes, errors from tube diameter variations are ignored. A generalized correction factor can be derived from the geometry of the system:

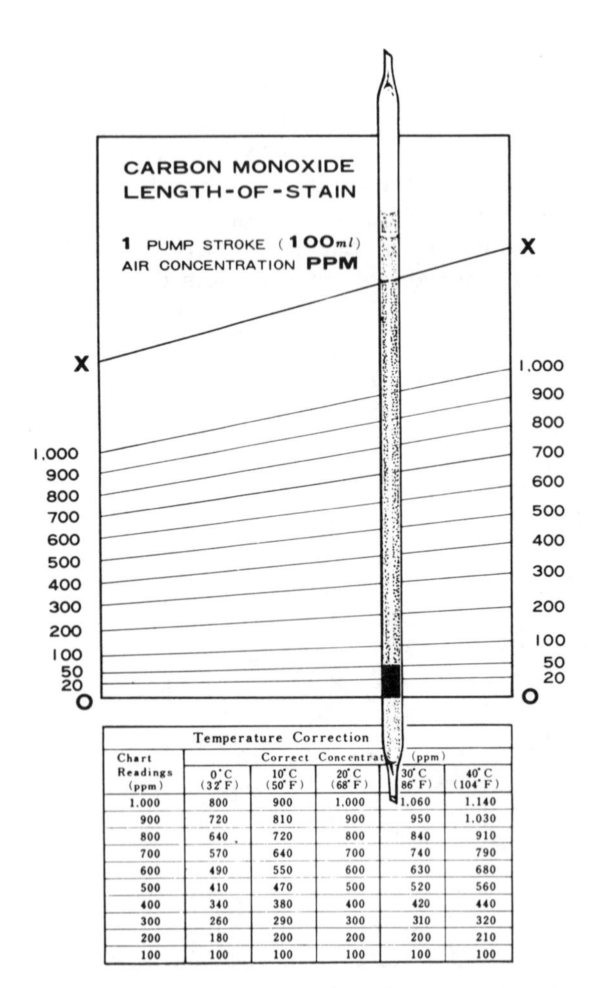

Temperature Correction					
Chart Readings (ppm)	Correct Concentrat (ppm)				
	0°C (32°F)	10°C (50°F)	20°C (68°F)	30°C 86°F)	40°C (104°F)
1,000	800	900	1,000	1,060	1,140
900	720	810	900	950	1,030
800	640	720	800	840	910
700	570	640	700	740	790
600	490	550	600	630	680
500	410	470	500	520	560
400	340	380	400	420	440
300	260	290	300	310	320
200	180	200	200	200	210
100	100	100	100	100	100

FIGURE I.20. Carbon monoxide length of stain detector tube calibration chart. (Courtesy of National Environmental Instruments, Inc., Pilgrim Station, Warwick, R.I.)

$$l_0 = l\,(L + a)/(L + a)$$

in which

- l = stain length in mm
- l_0 = corrected stain length in mm
- L = total length of detecting reagent column in mm
- L_0 = standard total reagent column length in mm
- a^0 = a constant determined experimentally for each detecting reagent and each kind of gas to be measured[65]

and

$$L = w/\pi r^2 d$$

in which

- w = total weight of detecting reagent in g
- r = radius of the tube in mm
- d = density of the detecting reagent in g/ml
- π = 3.1416

and

$$l = w/\pi r^2 d$$

in which

- w = weight of stained detecting reagent in mg

If the sample volume is fixed (100 ml) and the sampling rate held constant, then:

$$l/L = w/W = Kc/w$$

in which

- K = a constant
- c = gas concentration,

when w is directly proportional to c.

Correction charts can be constructed also from calibration data.[53]

Glass tubing with internal diameters controlled within tolerances that would eliminate the need for a correction factor (± 0.01 mm) is available,[84] but the cost is prohibitive for use in detector tubes.

c. Packing Variations – Pressure Drop Effects

Not only is the relationship of particle size distribution to pressure drop (Section 1.A.2.d.6.a., "Uniformity of Granular Support") a critical factor in controlling variations in flow rate, but the skill exercised in filling and packing the reagent granules in the tubes, and in inserting the reagent granule retainers (usually cotton or glass wool plugs), becomes a major factor when improvements in precision and accuracy are considered. On this basis, the use of reagent systems that, for reasons of stability of the chemical reagent system, must be mixed and loaded into the glass tubes by the user just before use in the field cannot be condoned for even "approximation analysis." Also, the addition of a liquid reagent to an already packed tube before sampling constitutes a highly questionable practice.

Although the major source of sampling rate variability is located in the tube itself, the overall characteristics of the system must be considered whenever grab sampling is carried out. In most of these devices, the flow of air created by the vacuum is controlled by a limiting orifice. The notable exception is the Dräger system, which relies on the tube packing controls to regulate the flow rate.[67] The dimensions of these fine orifices are critical both in the manufacturing precision, to ensure reproducible rates from one instrument to another, and in the reduction of effective cross-section by particulate matter, grease, or corrosion. Fine-pore porous polyethylene or fluorocarbon prefilters can prevent mechanical plugging. Only recalibration or microscopic examination will reveal minor changes due to corrosion, erosion, or mechanical damage. Extreme care must be exercised when cleaning these orifices to avoid causing enlarging by reaming with a fine wire or by etching in cleaning solutions.

Contrary to widely held opinion, these single-stroke manual pumps do not deliver a uniform, constant sample flow through the detector tubes. The sampling characteristics of a 100-ml metal syringe-type hand pump,[65] as determined by the soap bubble technique,[83a-83d] is illustrated in Figure I.21 (HP curves). With no coupled resistance, the rate remained linear (upper curve) for approximately 70 ml, but then became exponential as the total 100-ml volume was approached. With a benzene detector tube[65] in place, (lower curve) linearity occurred only over a limited span near the midpoint. A plot of sampling rate vs. time more graphically illustrates this variable flow

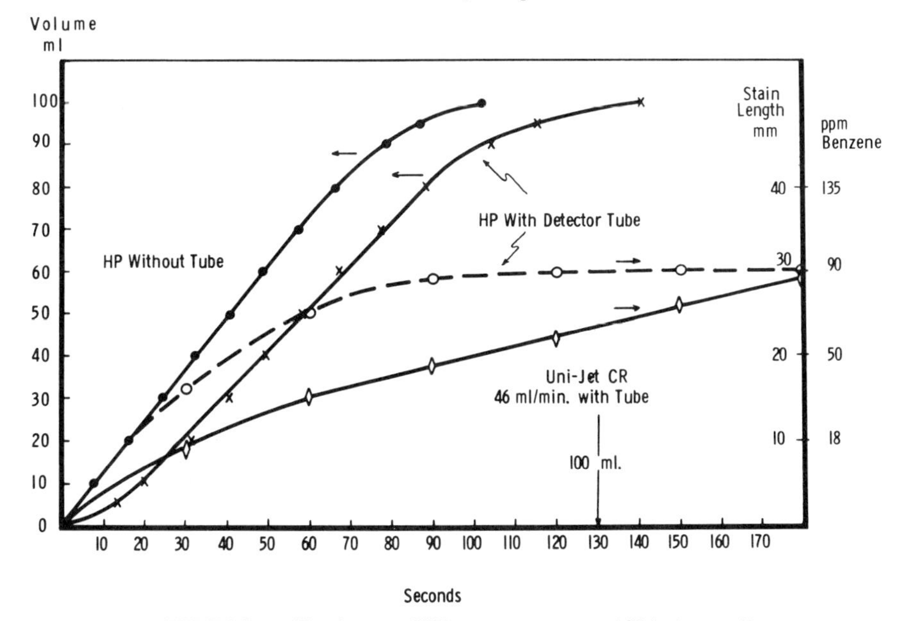

FIGURE I.21. Hand pump (HP) vs. constant rate (CR) air sampling.

pattern (Figure I.22, curve 1). The rate reached a maximum within 60 sec, then remained relatively constant only for 20 sec. At the point of maximum rate, only 40% of the sample had passed through the tube and the pressure drop had reached a maximum equivalent to 27 in. of water.

The resistance measured as pressure drop across the indicator tube msut be held constant if reproducible airflow precision is to be attained. However, this is not a realistic goal in routine manufacturing operations. As an example, the variations in pressure drop across benzene detector tubes selected at random from the same lot of one commercially available type were within ± 10% (95% C.L.) of the averaged results (Figure I.23). This range of flow resistance would be expected on the basis of known variations in tube diameter and packing density.[65] However, variables other than pressure drop must influence the response of this system, as recalibration by an independent agency indicated an accuracy of only ± 50% (95% C.L.).[74] The chemical nature of the reagent system (formaldehyde + concentrated H_2SO_4), rather than the geometry of the system, probably limits the performance.

d. Sampling Rate

Sampling rate is the most critical controlling variable encountered in the calibration and use of these detectors and should not exceed ± 5% relative deviation from tube to tube.[83]

Again referring to the Kitagawa benzene detector tube as an example, the critical effects of sampling rate can be demonstrated by reference to changes in stain front velocity (Figure I.21). As a result of the changing flow rate, the stain front progressed at a nonlinear rate to a maximum at midpoint, then decelerated rapidly as a linear function of time (Figure I.22). During this deceleration phase, movement of the stain front ceased when the sample flow dropped to 37 ml/min. The "effective" sample volume turned out to be 90 ml, and the "effective" time to draw this "effective" sample fell in the range 100 to 110 sec (Figure I.21, HP – dashed curve). A plot of stain front velocity vs. time (first derivative) accentuated this nonlinear response (Figure I.22, curve 1 – stain front).

The effect of air sample velocity was further amplified if the rate was held constant (Figure I.21, lower curve – Uni-jet® CR). A 138-ml sample

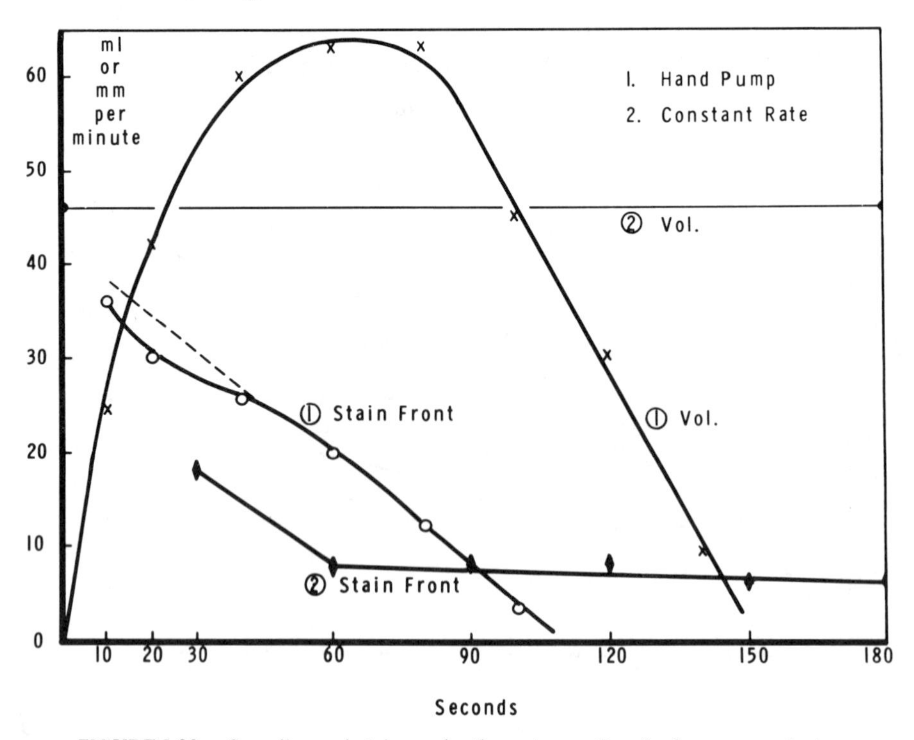

FIGURE I.22. Sampling and stain production rates vs. time for benzene analysis.

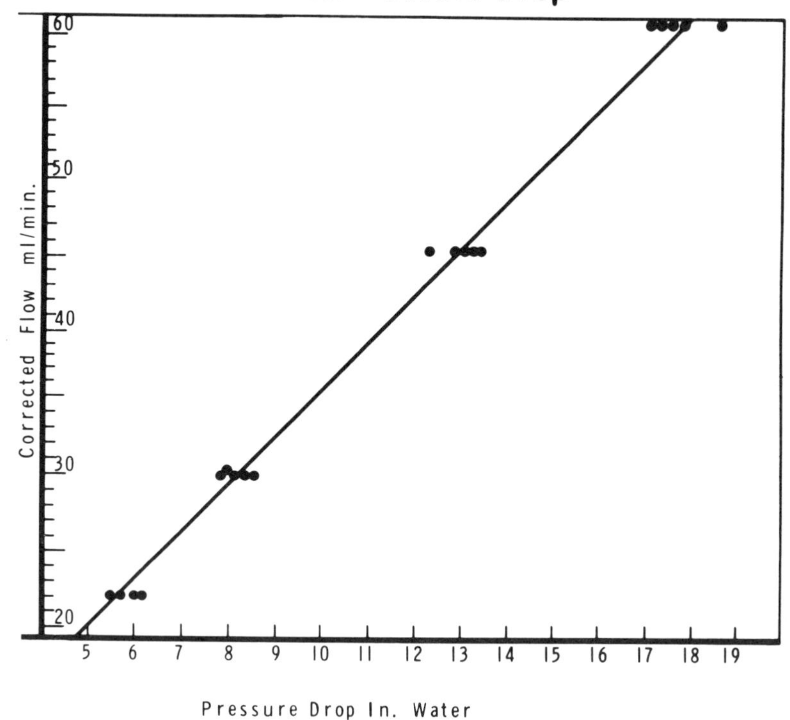

FIGURE I.23. Kitagawa® benzene detector tube – airflow vs. pressure drop.

was required to match the stain length of the 90-ml aliquot drawn by hand pump, and the time was extended to 180 sec. However, after a 60 sec "conditioning" period, the stain front advanced at a near linear rate (Figure I.22, curve 2 – stain front). The effect of a change in sampling rate expressed as a deviation from the supplied calibration provides the most significant incentive to operate within the calibration parameter limits (Figure I.24).

The same consideration must be paid to precision control of these variables when constant rate sampling is considered. When the data are presented in terms of the effect of sampling rate on stain length development, a family of curves can be derived either to determine the precision of flow rate control required to reduce the analytical deviation range within acceptable limits or to select conditions for monitoring for extended periods (Figure I.25 and Table I.7). For example, at a 20 ml/min rate, the sampling period can be extended to 9 min without deviating more than ± 10% from the calibration chart supplied with the hand pump.

Included in Figure I.25 are data published by Kinosian and Hubbard for a nitrogen dioxide detector tube.[85] These samples were drawn by squeeze bulb under controlled conditions. The critical effect of concentration is worthy of note. At 5 ppm NO_2, the stain front ceases to move after about 4 min, whereas the 25 ppm curve continues to rise steeply.

The variable sample flow rate produced by hand pump (Figures I.21 and I.22) has been claimed to be an advantage since the initial high rate produces a long stain and the final tapering off sharpens the stain front. However, even if this effect is valid, accuracy requires precise reproduction of the flow rate pattern to maintain valid calibrations.[62]

Serious pressure drop defects can be detected easily in the field by testing for residual vacuum in the hand pump, or incomplete filling of a squeeze bulb or bellows pump after the specified sampling time interval has elapsed. For example, after sampling a 49-ppm-CO-in-air calibration mixture, the piston of the hand pump, when rotated from the locked position at the end of the designated 3

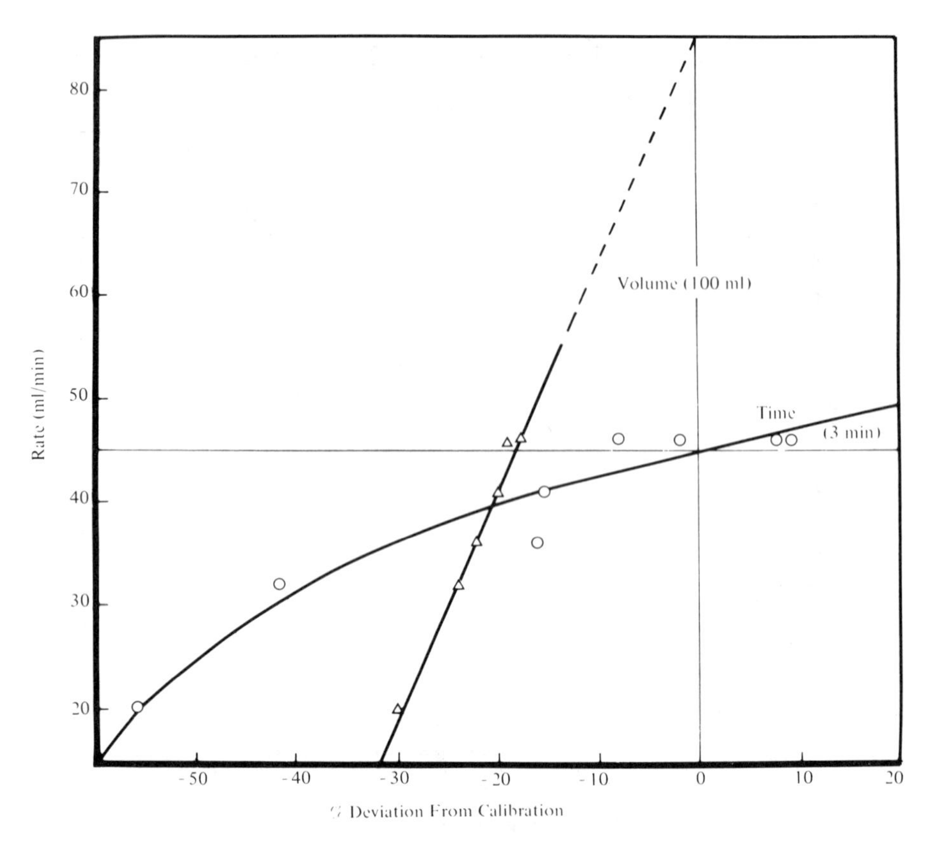

FIGURE I.24. Sampling rate vs. deviation from hand pump calibration (benzene).

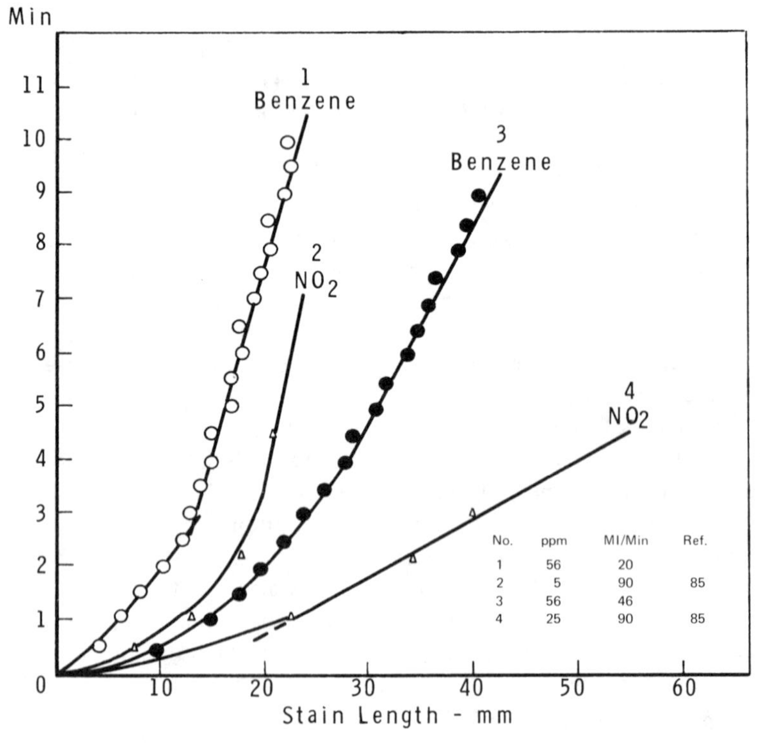

FIGURE I.25. Sampling rate vs. stain development rate (benzene and nitrogen dioxide).

TABLE I.7

Calculated Slopes for Log-log Curves for the Plots of Stain Length vs. ppm Benzene

Key:
(log Y) = $b_0 + b_1$ (log X)

Y = Benzene concentration
X = Stain length
b_1 = Slope
b_0 = Intercept

CR = Constant rate
CCR = Continuous CR
HP = Hand pump

	b_0	b_1	95% C.L. for b_1	
First sample (CR and HP)	−0.117	1.396	±0.171	
Second and third (CR)	−0.165	1.437	±0.270	Same curve
(Composite)	−0.196	1.453	±0.138	
Second sample (CR and HP)	0.435	1.165	±0.165	
Third sample (CR and HP)	0.664	1.008	±0.304	Same curve
(Composite of second and third)	0.665	0.982	±0.159	
Nitrogen dioxide (Reference 6)	−2.010	2.231	±0.730	

min was pulled back into the cylinder by the residual vacuum a distance equivalent to 10 ml. The 90-ml of sample which had been drawn through the tube at a nonstandard rate indicated a 30 to 35 ppm CO concentration instead of the 45 to 50 ppm range registered by the other tubes of the same lot.

e. Volume of Air Sampled

The volumetric error of repetitive aliquots delivered by the sampling pump should be less than ± 5%, and with less than ± 5% deviation from the volume used to calibrate the detector tubes.[83] Saltzman established the mathematical basis that equates volume and flow rate with the other variables required for calibration. The following excerpt is quoted from his article prepared for the ACGIH air sampling instruments handbook:

The variables which can affect the length of stain are: concentration of test gas, volume of air sample, sampling flow rate, temperature, and pressure, as well as a number of factors related to tube construction. There is a striking similarity in the fact that most of the calibration scales are logarithmic with respect to concentration in spite of the widely differing chemicals employed in different tube types. Although very little data are available for these relationships, a basic mathematical analysis was made by Saltzman.[86] These theoretical formulae discussed below will, of course, have to be modified as more data become available.

In the usual case, although the test gas is completely sorbed, equilibrium is not reached between the gas and the absorbing indicator gel, because the sampling period is relatively short and the flow rate relatively high. The length of stain is determined by the kinetic rate at which the gas either reacts with the indicating chemical or is adsorbed on the silica gel. The theoretical analysis shows that the stain length is proportional to the logarithm of the product of gas concentration and sample volume as shown in the following simplified equation:

$$L/H = \ln(CV) + \ln(K/H) \quad (1)$$

where

L is the stain length in cm.
C is the gas concentration in parts per million (ppm)
V is the air sample volume in ml.
K is a constant for a given type of indicator tube and test gas
H is a mass transfer proportionality factor having the dimension of centimeters, and known as the height of a mass transfer unit.

The factor H varies with the sampling flow rate raised to an exponent of between 0.5 and 1, depending on the nature of the process that limits the kinetic rate of sorption. This process may be diffusion of the test gas through a stagnant gas film surrounding the gel particles, the rate of surface chemical reaction, or diffusion in the solid gel particles. If the indicator tube follows this mathematical model, a plot of stain length, L, on a linear scale, versus the logarithm of product CV (for a fixed constant flow rate) will be a straight line of slope H. The equation indicates the importance of controlling flow rate, as it may affect stain lengths more than gas concentration.

If larger samples are taken at low concentrations and the value of L/H exceeds 4, the gel approaches equilibrium saturation at the inlet end, and calibration relationships are modified. The solution to the equations for this case has been presented by Saltzman[86] graphically in a generalized chart. However, there is little advantage to be gained in greatly increasing the sample size, since the stain front is greatly broadened and various errors are greatly increased.

For some types of tubes such as hydrogen sulfide and ammonia the reaction rate is fast enough so that equilibrium can be attained between the indicating gel and the test gas. Under these conditions there is a stoichiometric relationship between the volume of discolored indicating gel and the quantity of test gas absorbed. In the simplest case the stain length is proportional to the product of concentration and volume sampled:

$$L = K'CV \qquad (2)$$

If adsorption is important, the exponent of concentration may differ from unity:

$$L = K''C^{(1-n)}V \qquad (3)$$

The value of n is the same as that in the Freundlich isotherm equation for equilibrium adsorption, which states that the mass of gas adsorbed per unit mass of gel is proportional to the gas concentration raised to the power n. If the value of n is unity, which is not unusual, equation 3 indicates that stain length is proportional to sample volume, but is independent of concentration. The physical meaning of this is that all concentrations of gas are completely adsorbed by a fixed depth of gel. Such a tube is obviously of no practical value.

Equilibrium conditions may be assumed for a given type of indicator tube if stain lengths are directly proportional to the volume of air sampled (at a fixed concentration), and are not affected by air sampling flow rate. A log-log plot then may be made of stain length versus concentration for a fixed sample volume. A straight line with a slope of unity indicates equation 2 applies; if another value of slope is obtained, equation 3 applies.*

f. Temperature Effects

In addition to the volumetric correction required by the gas laws (PV = NRT) to compensate for temperatures significantly different from the calibration standard, 25°C and 760 mm Hg pressure, correction for changes in reaction kinetics must be applied in some cases. If the manufacturer furnishes temperature correction tables or charts, the recommended practice should be closely followed. For example, the reaction system employed for CO determination is temperature dependent, and the correction factors should be applied whenever samples are taken below 15°C or above 40°C (Figure I.20). However, at low concentrations the magnitude of error is small enough to be ignored. The effect on the response of mercuric chloride to arsine is similar.[82a] Failure on the part of the supplier to include temperature corrections does not necessarily mean that a significant temperature coefficient is not involved. The magnitude of the coefficient should be determined by the user before assuming that this variable can be ignored.[72a]

When the reaction velocity between the reagent and the contaminant sought is relatively low, colorimetric rather than stain length measurement must be applied. The first CO tube was in this category (color change from buff yellow to blue through a graded series of greens). A relatively large temperature coefficient was involved. The rate of color change was doubled for each 10°C rise in temperature. Therefore, even small deviations from the calibration temperature produced significant errors if correction was ignored. The colorimetric reagents for methyl and ethyl alcohol behave in a similar fashion.

g. Stability of the Chemical Reagent System – Storage Life

The significance of this factor can best be appreciated by the importance assigned to prominently displaying the expiration date on each box of tubes.[83] Very few of the reagents currently employed have an indefinite shelf life. The decomposition of the reagents during storage has had a profound effect on the production of gas detectors and has severely limited the variety of reagents that can be considered for air analysis. In general, the most sensitive colorimetric systems fail in this respect. Attempts to avoid this limitation have entailed either mixing the two components of an inpregnated gel system just before use, or adding a liquid reagent either before or after sampling. As previously stated, this expedient is not without its pitfalls and cannot be considered good practice.

Storage temperature has a profound effect on shelf life as would be expected. For example, the temperature coefficients for the Dräger carbon monoxide tubes are shown here:

*Reprinted from Saltzman, B. E., Direct Reading Colorimetric Indicators, in *Air Sampling Instruments for Evaluation of Atmospheric Contaminants,* 4th ed., ACGIH, Cincinnati, O., 1972. With permission.

Temperature (°C)	25	50	80	100	125	150
Shelf life	2 years	6 months	6 weeks	1 week	3 days	1 day

Note: The time indicates the period for which calibration accuracy remains within ±30%.

This data, which reveals a rate typical of a simple chemical reaction, describes an approximately straight line when the logarithm of time is plotted against a linear scale of the reciprocal of absolute temperature. Of course, more complex relationships will be encountered with other systems.[62] From this relationship, it is obvious that tube life can be greatly extended by storage under refrigeration. Indeed, the life expectancy of most of the otherwise unstable detector tubes can be at least doubled by storage at temperatures below 4°C. In the absence of stability data from the manufacturer, the user of these tubes should determine for himself the useful life by setting aside controls of each lot for observation and recalibration. Most of the limited shelf life reagents indicate decomposition by definite, characteristic color changes. "Off-color" tubes should be discarded. In any case, tubes should not be used after the expiration date without confirmation of the calibration.

A shelf life of at least 2 years is highly desirable from the standpoints of both the distributor and the customer. It is not reasonable to expect a distributor to absorb the financial losses from the storage of short life detectors. In most cases the supplier can estimate the service life of new reagent systems by accelerated aging at elevated temperatures and extrapolation to room temperature or refrigeration based on the relationship of time to temperature for simple chemical reactions (see Dräger CO tube data above). Small variations in gel composition, such as moisture content, may exert a significant effect on stability of a given reagent system from batch to batch. The responsibility for this quality control rests with the manufacturer, but the customer should "beware."

When multilayer tubes are involved, shorter shelf life can be expected, as diffusion of the "guard" reagent may accelerate deterioration. This condition usually is quite apparent from a "pseudo" stain front or liquid (tar) droplets that develop in the sealed tube. Again, the importance of calibration and recalibration cannot be overemphasized. The user of colorimetric analysis systems must satisfy himself that what is seen actually is the result produced by the material that is sought, and that the color nuances are not due to extraneous causes.

h. Specificity and Interferences

Probably no other factor in the application and interpretation of results from detector tube surveys taxes the expertise of the investigator more than the recognition and allowance for the contribution of coexistant components that are present in addition to the one under study in the contaminated environment. Few industrial pollution problems involve a single contaminant, and a majority of the detector tube types are nonspecific. For example, not less than 35% of the detector tubes offered by one manufacturer are based on the reduction of either a chromate or chromic acid to a colored chromium salt (Tables I.8A to I.8D). It is rather obvious that most reducing agents will produce this effect. Therefore, the investigator must determine what components are to be expected from an inventory of chemicals entering the area under investigation.

A preliminary survey carried out in a small paint and varnish manufacturing facility to detect the presence of health hazards and establish conformity with TLV's illustrates the difficulties that can be encountered when using a chromate reagent system.

Procedure – An inventory of the volatile solvents incorporated into the products disclosed more than twenty organic compounds that would reduce hexa-valent chromium (Table I.9). Detector tubes calibrated for twelve of the solvents were available, but except those for toluene and xylene, all were derived from the same chromic acid reagent. However, both toluene and xylene depend upon the reaction with formaldehyde in the presence of sulfuric acid to produce a brown stain, and therefore these tubes are not specific either.

Under the circumstances, a decision was reached to analyze only for the vapors of the major components, methylethylketone (MEK or 2-butanone) and toluene, in use on the day of the survey, and to ignore the contribution of the very minor amounts of the other solvents. Furthermore, since the concentrations of toluene vapor

TABLE I.8A

Classification of Detecting Reactions in Kitagawa® Gas Detector Tubes

1. Production of colored complex compounds.

105A	Ammonia – high range – cobaltous salt – complex of NH_3
142	Mercury vapor

2. Production of metallic compounds (salts).

103A	Sulfur dioxide – high range – bichromate – chromic compound
120A	Hydrogen sulfide – high range – lead salt
120B	Hydrogen sulfide – low range – lead salt
120HH	Hydrogen sulfide – extra-high range
121A	Phosphine – high range – $CuSO_4$ + mercuric compound
121B	Phosphine – low range
164	Methyl mercaptan
165	Ethyl mercaptan
200	Sulfide ion in water – lead salt
201	Chloride ion in water

3. Ionic color change by oxidation-reduction of metallic salts.

103B	Sulfur dioxide – middle range – KIO_3 – iodine
103BF	Sulfur dioxide in flue gas
129	Nickel carbonyl – auric compound
167	Hydrogen selenide
140	Arsine – auric compound

4. Production of molybdenum blue from ammonium molybdate and palladium sulfate.

101	Acetylene (palladium catalyzed)
106A	Carbon monoxide (palladium catalyzed)
106B	Carbon monoxide and ethylene coexisting (palladium catalyzed)
106C	Carbon monoxide, olefin hydrocarbon, and nitrogen oxides coexisting – chromic acid guard layer
108B	Ethylene (palladium catalyzed)
120C	Hydrogen sulfide and sulfur dioxide coexisting (palladium catalyzed)
168B	Butadiene – low range

5. Ionic color change with reduction of chromic acid.

102A	Acetone
104A	Ethyl alcohol
107	Ether
111	Ethyl acetate
113	*n*-Hexane
115	Cyclohexane
119	Methyl alcohol
122	Ethylene oxide – acidified bichromate
123	Dimethyl ether
128A	Acrylonitrile – high range
128B	Acrylonitrile – low range
132	Vinyl chloride

TABLE I.8A (continued)

Classification of Detecting Reactions in Kitagawa® Gas Detector Tubes

138	Butyl acetate
139B	Methyl ethyl ketone
148	Methyl acetate
149	Isopropyl acetate
150	Isopropyl alcohol
151	Propyl acetate
153	Isobutyl acetate
154	Dioxane
155	Methyl isobutyl ketone
161	Furan
162	Tetrahydrofuran
163	Propylene oxide
168A	Butadiene – high range

6. Color change of pH indicators.

105B	Ammonia – low range – thymol blue
112B	Hydrogen cyanide – low range
126A	Carbon dioxide – high range – NaOH + indicator
126B	Carbon dioxide – low range – NaOH + indicator
133	Acetaldehyde – hydroxylamine hydrochloride – thymol blue

7. Bleaching of dyestuff.

103C	Sulfur dioxide – low range
103D	Sulfur dioxide, D-type

8. Formation of dyestuffs,

109	Chlorine – *o*-tolidine
112A	Hydrogen cyanide – high range – alkaline picrate – isopurpurate
114	Bromine – *o*-tolidine
116	Chlorine dioxide – *o*-tolidine
117	Nitrogen dioxide – nitroso-*o*-tolidine
141	Carbon disulfide
145	1,2-Dichloroethylene
134	Trichloroethylene and perchloroethylene
146	Phosgene
157	Methyl bromide
160	Methyl chloroform
166	Ethylene dibromide

9. Formation of colored condensed polymer from formaldehyde + H_2SO_4 reagent.

118A	Benzene
118B	Benzene and other aromatic hydrocarbons coexisting
124	Toluene
143	Xylene
158	Styrene

10. Oxidation of basic pyrogallol.

159	Oxygen

TABLE I.8B

Classification of Detecting Reactions in Dräger Analyzer Tubes

Detector tube	Reaction principle	Measuring range	Number of strokes
1 Acetone 100/b	Dinitrophenylhydrazine	100–12,000 ppm	10
2 Acrylonitrile 5/a	Via hydrogen cyanide	5–30 ppm	5
3 Alcohol 100/a	Chromate and sulphuric acid	100–3,000 ppm	10
4 Ammonia 5/a	Bromphenol blue and acid	5–70 ppm	10
		50–700 ppm	1
5 Ammonia 25/a	Mercuric nitrate	25–700 ppm	10
6 Ammonia 0.5%/a	Bromphenol blue and acid	0.5–10 vol %	1
		0.05–1 vol %	10
7 Aniline 5/a	Furfural + acid	1–20 ppm	25–5
8 Arsine 0.05/a	Gold salt	0.05–3 ppm	20
		1–60 ppm	1
9 Benzene 5/a	Formaldehyde + sulphuric acid	5–40 ppm	15–2
10 Benzene 0.05	Formaldehyde + sulphuric acid	15–420 ppm	20–2
11 Carbon dioxide 0.01%/a	Hydrazine + crystal violet	0.01–0.3 vol %	10
		0.5–6 vol %	1
12 Carbon dioxide 0.1%/a	Hydrazine + crystal violet	0.1–1.2 vol %	5
13 Carbon dioxide 0.5%/a	Hydrazine + crystal violet	0.5–10 vol %	1
14 Carbon dioxide 1%	Hydrazine + crystal violet	1–20 vol %	1
15 Carbon dioxide 5%/A	Hydrazine + crystal violet	5–60 vol %	1
16 Carbon disulphide 0.04	Copper salt + amine	13–288 ppm (0.04–0.9 mg/l)	23–1
17 Carbon disulphide 30/a	Copper salt + amine	32–3.200 ppm (0.1–10 mg/l)	6
18 Carbon monoxide 5/c	Iodine pentoxide + fuming sulphuric acid	5–150 ppm	10
		100–700 ppm	2
19 Carbon monoxide 8/a (only for CO in H_2)	Iodine pentoxide + fuming sulphuric acid	8–150 ppm	10
20 Carbon monoxide 10/a	Iodine pentoxide + fuming sulphuric acid	10–300 ppm	10
		100–3,000 ppm	1
21 Carbon monoxide 10/b	Iodine pentoxide + fuming sulphuric acid	10–300 ppm	10
		100–3,000 ppm	1
22 Carbon monoxide 0.1%/a	Iodine pentoxide + fuming sulphuric acid	0.1–1.2 vol %	1
23 Carbon monoxide 0.3%/a	Iodine pentoxide + fuming sulphuric acid	0.3–4 vol %	1
24 Carbon monoxide 0.5%/a	Iodine pentoxide + fuming sulphuric acid	0.5–7 vol %	1
25 Carbon tetrachloride 10/b	Via phosgene	10–100 ppm	3
26 Chlorine 0.2/a	*o*-Tolidine	0.2–3 ppm	10
		2–30 ppm	1
27 Chlorine 50/a	*o*-Tolidine	50–500 ppm	1
28 Chloroprene 5/a	Permanganate	5–60 ppm	3
		7.5–90 ppm	2
29 Cyanogen chloride 0.25/a	Pyridine + barbituric acid	0.25–5 ppm	20–1
30 Cyclohexane 100/a	Chromic acid	100–1,500 ppm	10
31 Diborane 0.05/a	Gold complex	0.05–3 ppm	20
32 Diethyl ether 100/a	Chromate + sulphuric acid	100–4,000 ppm	10
33 Dimethylformamide 10/b	Via amine	10–40 ppm	10
34 Ethyl acetate 200/a	Chromate + sulphuric acid	200–3,000 ppm	20
35 Formaldehyde 0.002	Xylene + sulphuric acid	2–40 ppm (0.002–0.05 mg/l)	5

TABLE I.8B (continued)

Classification of Detecting Reactions in Dräger Analyzer Tubes

Detector tube	Reaction principle	Measuring range	Number of strokes
36 Hydrazine 0.25/a	Bromphenol blue + acid	0.25–3 ppm	10
37 Hydrocarbon 0.1%	Iodine pentoxide + fuming sulphuric acid	0.1–1 vol %	–
38 Hycrocarbon 2	Selenium dioxide + fuming sulphuric acid	2–25 mg/l	–
39 Hydrochloric acid 1/a	Bromphenol blue	1–10 ppm	10
		2–20 ppm	5
40 Hydrogen 0.5%/a	Via H_2O	0.5–3 vol %	5
41 Hydrogen cyanide 2/a	$HgCl_2$ + methyl red	2–30 ppm	5
		10–150 ppm	1
42 Hydrogen fluoride 1.5/b	Zirconium quinalizarinate	1.5–15 ppm	20
43 Hydrogen sulphide 1/c	Lead salt	1–20 ppm	10
		10–200 ppm	1
44 Hydrogen sulphide 5/b	Lead salt	5–60 ppm	10
		50–600 ppm	1
45 Hydrogen sulphide 100/a	Lead salt	100–2,000 ppm	1
46 Hydrogen sulphide 0.2%/A	Copper salt	0.2–7 vol %	1
47 Hydrogen sulphide + sulphur dioxide 0.2%/A	Iodine	0.2–7 vol %	1
		0.02–0.7 vol %	10
48 Mercaptan 2/a	Copper salt + sulphur	2–100 ppm	10
49 Mercury vapor 0.1/b	Copper(I)-iodide	0.1–2 mg/m³	20–1
50 Methyl bromide 5/b	Via bromine	5–50 ppm	5
51 Methylene chloride 100/a	Iodine pentoxide + selenium dioxide + fuming sulphuric acid	100–2,000 ppm	10
52 Monostyrene 10/a	Sulphuric acid	10–200 ppm	15–2
53 Monostyrene 50/a	Sulphuric acid	50–400 ppm	11–2
54 Natural gas test	Via CO	Qualitative	2
55 Nickel carbonyl 0.1/a	Iodine + dioxine	0.1–1 ppm	20
56 Nitrogen dioxide 0.5/c	Diphenylbenzidine	0.5–10 ppm	5
57 Nitrogen dioxide 2/c	Diphenylbenzidine	2–50 ppm	10
		5–100 ppm	5
58 Nitrous fumes 0.5/a	Diphenylbenzidine	0.5–10 ppm	5
59 Nitrous fumes 2/a	Dipheylbenzidine	2–50 ppm	10
		5–100 ppm	5
60 Nitrous fumes 20/a	*o*-Dianisidine	20–500 ppm	2
61 Nitrous fumes 100/c	*o*-Dianisidine	100–1,000 ppm	5
		500–5,000 ppm	1
62 Olefin 0.05%/a	Permanganate	1–50 mg/l	20–1
63 Ozone 0.05/a	Indigo	0.05–1,4 ppm	10
		0.5–14 ppm	1
64 Ozone 10/a	Indigo	10–300 ppm	1
65 Oxygen 5%/A	Via CO	5–21 vol %	1
66 *n*-Pentane 100/a	Chromic acid	100–1,500 ppm	5
67 Perchloroethylene 10/a	Via chlorine	10–400 ppm	3
68 Phenol 5/a	Dibromoquinonimide	5 ppm	10
69 Phosgene 0.05/a	Dimethylaminobenzaldehyde + diethylaniline	0.05–1.2 ppm	26–1
70 Phosgene 0.25/b	Dimethylaminobenzaldehyde + dimethylaniline	0.25–15 ppm	5
		1.25–75 ppm	1

TABLE I.8B (continued)

Classification of Detecting Reactions in Dräger Analyzer Tubes

Detector tube	Reaction principle	Measuring range	Number of strokes
71 Phosphine 0.1/a	Gold salt	0.1–4 ppm	10
		1–40 ppm	1
72 Phosphine 50/a	Gold salt	50–1,000 ppm	3
		15–300 ppm	10
73 Polytest tube	Iodine pentoxide + fuming sulphuric acid	Qualitative	5
74 Simultaneous tube	See carbon dioxide and carbon monoxide tubes	200–2,500 ppm CO) 2–12 vol % CO_2)	2
75 Sulphur dioxide 0.1/a	Disodium tetrachloromercurate	0.1–3 ppm	100
76 Sulphur dioxide 1/a	Iodine + starch	1–20 ppm	10
77 Sulphur dioxide 20/a	Iodine	20–200 ppm	10
		200–2,000 ppm	1
78 Systox 1/a	Chloramide + gold salt	0.5 mg/m^3	20
79 Toluene 5/a	Iodine pentoxide + sulphuric acid	5–400 ppm	5
80 Toluene 25/a	Iodine pentoxide + sulphuric acid	25–1,860 ppm (0.1–7 mg/l)	10
81 Toluylenediisocyanate 0.02/A	Pyridylpyridiniumchloride	0.02–0.2 ppm	25
82 Triethylamine 5/a	Bromphenol blue + acid	5–60 ppm	5
83 Trichloroethane	Via chlorine	50–350 ppm	10
		100–700 ppm	5
84 Trichloroethylene 10/a	Via chlorine	10–400 ppm	5
85 Vinyl chloride 100/a	Permanganate	100–3,000 ppm	18–1
86 Water vapor 0.1	Selenium dioxide + sulphuric acid	0.1–40 mg/l	10

Courtesy of Drägerwerk AG, Lübeck, West Germany.

TABLE I.8C

List of Gastec® Detector Tubes

Catalog tube no.	Detector tube description	Calibration scale	Measuring range
81	Acetic Acid	2–40 ppm	0.2–80 ppm
151	Acetone	0.01–0.8 vol %	0.01–2 vol %
171	Acetylene	0.2–2.0 vol %	0.1–4 vol %
191**	Acrylonitrile	10–500 ppm	10–500 ppm
3H	Ammonia – high range	1–16 vol %	0.2–32 vol %
3M	Ammonia – middle range	50–500 ppm	5–1,000 ppm
3L	Ammonia – low range	2–30 ppm	1–60 ppm
121*	Benzene	10–200 ppm	10–200 ppm
142	Butyl acetate	0.01–0.8 vol %	0.01–0.8 vol %
2H	Carbon dioxide – high range	1.0–10 vol %	0.1–20 vol %
2L	Carbon dioxide – low range	0.2–3.0 vol %	0.02–6 vol %
2LL	Carbon dioxide – extra low range	300–5,000 ppm	300–5,000 ppm
13**	Carbon disulfide	5–50 ppm	5–50 ppm
1H	Carbon monoxide – high range	0.2–5.0 vol %	0.1–10 vol %
1L	Carbon monoxide – low range	50–1,000 ppm	5–2,000 ppm
1La	Carbon monoxide – low range	50–600 ppm	5–1,200 ppm
1LL	Carbon monoxide – extra low range	5–50 ppm	5–50 ppm
8L	Chlorine – low range	1–30 ppm	0.5–60 ppm
103	Cyclohexane	0.03–0.6 vol %	0.015–1.2 vol %
141	Ethyl acetate	0.04–1.5 vol %	0.04–1.5 vol %
112	Ethyl alcohol (Ethanol)	0.1–2.5 vol %	0.05–5 vol %
125	Ethylbenzene	20–250 ppm	10–500 ppm
161	Ethyl ether (Ether)	0.04–1.0 vol %	0.04–1.5 vol %
163	Ethylene oxide	0.1–3.0 vol %	0.1–3.0 vol %
91**	Formaldehyde	2–20 ppm	2–20 ppm
101	Gasoline	0.03–0.6 vol %	0.015–1.2 vol %
102H	*n*-Hexane – high range	0.03–0.6 vol %	0.015–1.2 vol %
102L	*n*-Hexane – low range	50–1,200 ppm	50–1,200 ppm
14L***	Hydrogen chloride – low range	2–20 ppm	0.2–40 ppm
12H	Hydrogen cyanide – high range	0.05–2 vol %	0.05–2 vol %
12L*	Hydrogen cyanide – low range	5–15 ppm	2.5–30 ppm
4H	Hydrogen sulphide – high range	100–1,600 ppm	10–3,200 ppm
4L	Hydrogen sulphide – low range	10–120 ppm	1–240 ppm
113	Isopropyl alcohol (isopropanol)	0.04–2.5 vol %	0.02–5 vol %
111	Methyl alcohol (methanol)	0.02–1.5 vol %	0.01–3 vol %
136**	Methyl bromide	10–100 ppm	10–200 ppm
135**	Methyl chloroform	100–500 ppm	100–500 ppm
152	Methyl ethyl ketone	0.02–0.6 vol %	0.02–0.6 vol %
153	Methyl isobutyl ketone	0.01–0.6 vol %	0.01–0.6 vol %
15L***	Nitric acid – low range	2–20 ppm	0.2–40 ppm
9L*	Nitrogen dioxide – low range	2–100 ppm	2–100 ppm

* Tubes must be kept below 10°C (50°F) for stability of reagent contained.
** Consisting of an analyzer tube for indicating gas concentration and a primer tube for pretreating sample air. A pair of the tubes must be connected using a rubber tubing (contained in each box).
***Containing five analyzer tubes and five water vapor tubes (No. 6). Since the indication of the tubes is based on water content in the test air, measurement of water content is required before test for humidity correction.

TABLE I.8C (continued)

List of Gastec® Detector Tubes

Catalog tube no.	Detector tube description	Calibration scale	Measuring range
11L*	Nitrogen oxides – low range	2–100 ppm	2–100 ppm
133	Perchloroethylene	50–200 ppm	5–200 ppm
16	Phosgene	0.1–10 ppm	0.1–10 ppm
7*	Phosphine	50–500 ppm	5–1,000 ppm
124*	Styrene	20–500 ppm	10–1,000 ppm
5H	Sulphur dioxide – high range	0.5–4.0 vol %	0.05–8 vol %
5M	Sulphur dioxide – middle range	100–1,800 ppm	10–3,600 ppm
5L*	Sulphur dioxide – low range	5–100 ppm	0.5–200 ppm
122*	Toluene	20–250 ppm	10–500 ppm
132H	Trichloroethylene – high range	50–200 ppm	5–200 ppm
132L	Trichloroethylene – low range	2–25 ppm	1–50 ppm
131	Vinyl chloride	0.05–1.0 vol %	0.025–2 vol %
6	Water Vapor	2–18 mg/l	1–36 mg/l

* Tubes must be kept below 10°C (50°F) for stability of reagent contained.

TABLE I.8D

Classification of Detecting Reactions in Gastec® Detector Tubes

1. No. 1H Carbon Monoxide – High range

   CO reduces iodine pentoxide to liberate iodine, which produces a blackish green stain. Coexisting hydrocarbons are eliminated in the brown precleaning layer consisting of chromic acid and sulfuric acid.

   $CO + I_2O_5 \qquad I_2 + CO_2$

2. No. 1L Carbon Monoxide – Low range

   CO reduces potassium palladosulfite to liberate metallic palladium, which produces blackish brown stain.

   $CO + K_2Pd(SO_3)_2 \qquad K_2 + Pd + SO_2 + CO_2$

3. No. 1La Carbon Monoxide – Low range

   See point 2, "No. 1L Carbon Monoxide – Low range."

4. No. 1LL Carbon Monoxide – Extra-low range

   See point 2, "No. 1L Carbon Monoxide – Low range."

5. No. 2H Carbon Dioxide – High range

   CO_2 neutralizes potassium hydroxide and changes the color of pH indicator (allizarine yellow).

   $CO_2 + KOH \qquad K_2CO_3 + H_2O$

6. No. 2L Carbon Dioxide – Low range

   See point 5, "No. 2H Carbon Dioxide – High range."

TABLE I.8D (continued)

Classification of Detecting Reactions in Gastec® Detector Tubes

7. No. 2LL Carbon Dioxide – Extra-low range

   CO_2 reacts with hydrazine to form carbonic acid monohydrazide, which changes the color of redox indicator (crystal violet).

   $CO_2 + N_2H_4$ $NH_2 \cdot NH \cdot COOH$

8. No. 3H Ammonia – High range

   NH_3 reacts with cobalt chloride to form a complex or a coordinate compound.

   $NH_3 + CoCl_2$ $aCoCl_2 \cdot bNH_3$

9. No. 3M Ammonia – Middle range

   NH_3 neutralizes phosphoric acid to change the color of pH indicator (thymol blue).

   $NH_3 + H_2PO_3$ $(NH_3)_2PO_3$

10. No. 3L Ammonia – Low range

    NH_3 neutralizes sulfuric acid to change the color of pH indicator (thymol blue).

    $NH_3 + H_2SO_4$ $(NH_4)_2SO_4$

11. No. 4H Hydrogen Sulfide – High range

    H_2S reacts with lead acetate to form lead sulfide.

    $H_2S + Pb(CH_3COO)_2 \cdot 3H_2O$ $PbS + CH_3COOH$

12. No. 4L Hydrogen Sulfide – Low range

    See point 11, "No. 4H Hydrogen Sulfide – High range."

13. No. 4LL Hydrogen Sulfide – Extra-low range

    See point 11, "No. 4H Hydrogen Sulfide – High range."

14. No. 5H Sulfur Dioxide – High range

    SO_2 reduces chromic ion of orange yellow to trivalent chromous ion (green).

    $SO_2 + CrO_3$ $Cr_2(SO_4)_3 \cdot 18H_2O + K_2O$

15. No. 5M Sulfur Dioxide – Middle range

    SO_2 neutralizes sodium hydroxide to change the color of pH indicator (phenol red).

    $SO_2 + NaOH$ $Na_2SO_3 + H_2O$

16. No. 5L Sulfur Dioxide – Low range

    See point 15, "No. 5M Sulfur Dioxide – Middle range."

18. No. 6 Water Vapor

    H_2O is absorbed by magnesium perchlorate to produce an alkaline, which changes the color of Hammett indicator (crystal violet).

    $H_2O + Mg(ClO_4)_2$ $Mg(ClO_4)_2 \cdot aH_2O$

TABLE I.8D (continued)

Classification of Detecting Reactions in Gastec® Detector Tubes

19. No. 7 Phosphine

PH_3 reacts with mercuric chloride to produce mercuric phosphochloride of yellow color.

$PH_3 + HgCl_2 + H_2O$ → $Hg_3P_2 \cdot 3HgCl_2 \cdot 3H_2O$

20. No. 8L Chlorine – Low range

Cl_2 reduces *o*-tolidine to form nitroso-*o*-tolidine of orange color.

$Cl_2 + H_2O$ + *o*-tolidine → Nitroso-*o*-tolidine + HCl

21. No. 9L Nitrogen Dioxide – Low range

NO_2 reduces *o*-tolidine to form nitroso-*o*-tolidine of bright yellow color.

$NO_2 + H_2O$ + *o*-Tolidine → Nitroso-*o*-tolidine

22. No. 10 Nitrogen Oxides (NO and NO_2)

NO_2 is absorbed in the first tube (NO_2 tube) and reduces *o*-tolidine to form nitroso-*o*-tolidine of bright yellow color. NO passes through the first tube and is absorbed in the second tube (NO tube), then oxidized by orange prelayer consisting of chromic and sulfuric acids to NO_2, which reduces *o*-tolidine to form nitroso-*o*-tolidine of bright yellow color.

NO_2 tube–(first tube):

NO_2 + *o*-Tolidine + H_2O → Nitroso-*o*-tolidine

NO tube–(second tube):

$NO + CrO_3 + H_2SO_4$ → NO_2

NO_2 + *o*-Tolidine → Nitroso-*o*-tolidine

23. No. 11L Nitrogen Oxides ($NO + NO_2$) – Low range

Coexisting NO is oxidized by chromic and sulfuric acids in orange prelayer to form NO_2, which combining with NO_2 in sample, reduces *o*-tolidine in white indicating layer to form nitroso-*o*-tolidine of bright yellow color.

$NO + CrO_3$ → NO_2

$NO_2 + H_2O$ + *o*-Tolidine → Nitroso-*o*-tolidine

24. No. 12H Hydrogen Cyanide – High range

HCN reacts with potassium palladosulfite to form a white complex or coordinate compound, the chemical formula of which is still unknown.

$HCN + K_2Pd(SO_3)_2$ → White compound

25. No. 12L Hydrogen Cyanide – Low range

HCN reacts with mercuric chloride to liberate HCl, which changes the color of pH indicator (methyl orange) to pink. Other acidic gases are eliminated by benzeneazodiphenylamine in precleaning layer.

$HCN + HgCl_2$ → $HCl + Hg(CN)_2$

TABLE I.8D (continued)

Classification of Detecting Reactions in Gastec® Detector Tubes

26. No. 13 Carbon Disulfide

CS_2 is oxidized by nascent oxygen, which is yielded by the reaction of chromic and sulfuric acids in brown prelayer in the primer tube, to form SO_2. This SO_2 neutralizes sodium hydroxide to change the color of pH indicator (phenol red) to dull yellow.

Coexisting H_2S is removed by cupric sulfate in white cleaning layer in the primer tube.

Primer tube:

$CrO_3 + H_2SO_4$ (O)

$(O) + CS_2$ $SO_2 + CO_2$

Analyzer tube:

$SO_2 + NaOH$ $Na_2(SO_3) + H_2O$

27. No. 14L Hydrogen Chloride – Low range

HCl is absorbed in silica sand to change the color of Hammett indicator (4-phenylazodiphenylamine).

28. No. 15L Nitric Acid – Low range

HNO_3 is absorbed in silica sand to change the color of Hammett indicator (4-phenylazodiphenylamine).

29. No. 16 Phosgene

$COCl_2$ reacts with Nitrobenzylpyridine to form urea derivative, which reacts with benzylaniline to produce a red dye. Chemical formulas of the produced compounds are still unknown.

$COCl_2$ + Nitrobenzylpyridine Urea derivative

Urea derivative + Benzylaniline Red dye

30. No. 81 Acetic Acid

Acetic acid neutralizes sodium hydroxide to change the color of pH indicator (phenol red).

$CH_3COOH + NaOH$ $Na_2(CH_3COO)_2 + H_2O$

31. No. 91 Formaldehyde

HCHO reacts with *o*-xylene in the primer tube and when dehydrated by sulfuric acid forms a condensation polymer in the analyzer tube.

$(HCHO)_n + [2 \cdot C_6H_4(CH_3)_2]_n$ $[C_6H_4(CH_3)_2-CH_2-C_6H_4(CH_3)_2]_n + H_2O$

32. No. 100A LP Gas (propylene)

Propylene or butane reduces bichromic ion of potassium bichromate to trivalent chromous ion.

C_3H_6 or $C_4H_{10} + K_2Cr_2O_7 + H_2SO_4$ $Cr_2(SO_4)_3$

33. No. 101 Gasoline

Gasoline reduces chromic ion of chromic acid to trivalent chromous ion.

$C_nH_m + CrO_3 + H_2SO_4 + H_2O$ $Cr_2(SO_4)_3$

TABLE I.8D (continued)

Classification of Detecting Reactions in Gastec® Detector Tubes

34. No. 102H *n*-Hexane – High range

n-Hexane reduces chromic ion of chromic acid to trivalent chromous ion.

$C_6H_{14} + CrO_3 + H_2SO_4 + H_2O$ → $Cr_2(SO_4)_3$

35. No. 102L *n*-Hexane – Low range

n-Hexane reduces bichromic ion of bichromic acid to trivalent chromous ion.

$C_6H_{14} + K_2Cr_2O_7 + H_2SO_4$ → $Cr_2(SO_4)_2$

36. No. 103 Cyclohexane

Cyclohexane reduces chromic ion of chromic acid to trivalent chromous ion.

$C_6H_{12} + CrO_3 + H_2SO_4 + H_2O$ → $Cr_2(SO_4)_3$

37. No. 104 Butane

Butane reduces bichromic ion of bichromic acid to trivalent chromous ion.

$C_4H_{10} + K_2Cr_2O_7 + H_2SO_4$ → $Cr_2(SO_4)_3$

38. No. 111 Methyl Alcohol (methanol)

Methyl alcohol reduces bichromic ion of bichromic acid to trivalent chromous ion after water content in sample is removed by the white precleaning layer consisting of potassium carbonate. The indicating reagent is not reduced by other organic vapors except alcohol.

$CH_3OH + K_2Cr_2O_7$ → $Cr_2(SO_4)_3$

39. No. 112 Ethyl Alcohol (ethanol)

Ethyl alcohol reduces bichromic ion of bichromic acid to trivalent chromic ion. Water content in sample is eliminated by the white precleaning layer consisting of potassium carbonate. The indicating reagent is not reduced by other organic vapors except alcohol.

$C_2H_5OH + K_2Cr_2O_7$ → $Cr_2(SO_4)_3$

40. No. 113 Isopropyl Alcohol (isopropanol)

Isopropyl alcohol reduces bichromic ion of bichromic acid to trivalent chromous ion after water content in sample is eliminated in the white precleaning layer consisting of potassium carbonate. The indicating reagent is not reduced by other organic vapors except alcohol.

$(CH_3)_2CHOH + K_2Cr_2O_7$ → $Cr_2(SO_4)_3$

41. No. 121 Benzene

Benzene reduces iodine pentoxide to liberate iodine, which produces a green stain.

$C_6H_6 + I_2O_5 + H_2SO_4$ → $I_2 + CO_2$

42. No. 122 Toluene

Toluene reduces iodine pentoxide to liberate iodine, which produces a blackish brown stain. Water content in sample is eliminated in the rose precleaning layer of molecular sieves.

Classification of Detecting Reactions in Gastec® Detector Tubes

$C_6H_5CH_3 + I_2O_5 + H_2SO_4$ → $I_2 + CO_2$

43. No. 123 Xylene

Xylene reduces iodine pentoxide to liberate iodine, which produces a black stain. Water content in sample is removed in the rose precleaning layer of molecular sieves.

$C_6H_5(CH_3)_2 + I_2O_5 + H_2SO_4$ → $I_2 + CO_2$

44. No. 124 Styrene

Styrene is dehydrated by sulfuric acid to form a condensation polymer.

$C_6H_5CH{:}CH_2 + H_2SO_4$ → $C_6H_5CH{:}CH{-}CH{:}CHC_6H_5$
$C_6H_5CH{=}CH_2$

45. No. 125 Ethylbenzene

Ethylbenzene reduces iodine pentoxide to liberate iodine, which produces a blackish brown stain. Water content in sample is removed in the rose precleaning layer of molecular sieves.

$C_6H_5O_2H_5 + I_2O_5 + H_2SO_4$ → $I_2 + CO_2$

46. No. 131 Vinyl Chloride

Vinyl chloride reduces chromic ion of chromic acid to trivalent chromous ion.

$CH_2{=}CHCl + CrO_3 + H_2SO_4 + H_2O$ → $Cr_2(SO_4)_2$

47. No. 132H Trichloroethylene – High range

Trichloroethylene is oxidized by nascent oxygen, which is generated by the reaction of lead dioxide and sulfuric acid in the black oxidizing agent, to liberate HCl. This generated HCl changes the color of Hammett indicator (benzeneazodiphenylamine) to pink.

$PbO_2 + Pb(NO_3)_2 + H_2SO_4$ → (O)

$CHCl{:}CCl_2 + (O)$ → HCl

48. No. 132L Trichloroethylene – Low range

See point 48, "No. 132H Trichloroethylene – High range."

50. No. 133 Perchloroethylene

Perchloroethylene is oxidized by nascent oxygen, which is generated by the reaction of lead dioxide and sulfuric acid in the black oxidizing agent, to liberate HCl. This HCl changes the color of Hammett indicator (benzene-azodiphenylamine).

$PbO_2 + Pb(NO_3)_2 + H_2SO_4$ → (O)

$CCl_2{:}CCl_2 + (O)$ → HCl

51. No. 135 Methylchloroform (1,1,1-trichloroethane)

Methylchloroform is oxidized by nascent oxygen, which is liberated by the reaction of iodine pentoxide and sulfuric acid in the white oxidizing agent, to generate chlorine in the primer tube. This generated chlorine reduces *o*-tolidine to form nitroso-*o*-tolidine of yellow color.

Classification of Detecting Reactions in Gastec® Detector Tubes

Primer tube:

$I_2O_5 + H_2SO_4$ → (O)

$CH_3CCl_3 + (O)$ → $Cl_2 + H_2O + CO_2$

Analyzer tube:

Cl_2 + *o*-Tolidine → Nitroso-*o*-tolidine + HCl

52. No. 136 Methyl Bromide

Methyl bromide is oxidized by nascent oxygen, which is liberated by the reaction of iodine pentoxide and sulfuric acid in white oxidizing agent, to generate bromine in the primer tube. This bromine reduces *o*-tolidine to form nitroso-*o*-tolidine in the analyzer tube.

Primer tube:

$I_2O_5 + H_2SO_4$ → (O)

$CH_3Br + (O)$ → $Br_2 + CO_2 + H_2O$

Analyzer tube:

Br_2 + *o*-Tolidine → Nitroso-*o*-tolidine + HBr

53. No. 138 Chloropicrine

Chloropicrine is decomposed by nascent oxygen, which is generated by the reaction of iodine pentoxide and sulfuric acid in white pre-oxidizing agent, to liberate phosgene in the primer tube. This liberated phosgene reacts with nitrobenzylpyridine to form urea derivative, which reacts with benzylaniline to produce a red dye.

Primer tube:

$CCl_3NO_2 + I_2O_5 + H_2SO_4$ → $COCl_2$

Analyzer tube:

$COCl_2$ + Nitrobenzylpyridine → Urea derivative

Urea derivative + Benzylaniline → Red dye

54. No. 141 Ethyl Acetate

Ethyl acetate reduces bichromic ion of bichromic acid to trivalent chromous ion. Water content is eliminated in white precleaning layer of potassium carbonate.

$CH_3COOC_2H_5 + K_2Cr_2O_7 + H_2SO_4$ → $Cr_2(SO_4)_3$

55. No. 142 Butyl Acetate

Butyl acetate reduces bichromic ion of bichromic acid to trivalent chromous ion after water content is removed in white precleaning layer of potassium carbonate.

$CH_3COOC_4Hg + K_2Cr_2O_7 + H_2SO_4$ → $Cr_2(SO_4)_3$

Classification of Detecting Reactions in Gastec® Detector Tubes

56. No. 151 Acetone

Acetone reduces bichromic ion of bichromic acid to trivalent chromous ion. Water content is removed in white precleaning layer of potassium carbonate.

$(CH_3)_2CO + K_2Cr_2O_7 + H_2SO_4$ $\quad$ $Cr_2(SO_4)_3$

57. No. 152 Methyl Ethyl Ketone (MEK)

Methyl ethyl ketone reduces bichromic ion to trivalent chromous ion. Water content in sample is eliminated in white precleaning layer of potassium carbonate.

$CH_3COC_2H_5 + K_2Cr_2O_7 + H_2SO_4$ $\quad$ $Cr_2(SO_4)_3$

58. No. 153 Methyl Isobutyl Ketone (NIBK)

Methyl isobutyl ketone reduces bichromic ion of bichromic acid to trivalent chromous ion. Water content in sample is removed in white precleaning layer of potassium carbonate.

$CH_3COCH_2CH(CH_3)_2 + K_2Cr_2O_7 + H_2SO_4$ $\quad$ $Cr_2(SO_4)_3$

59. No. 161 Ethyl Ether (ether)

Ethyl ether reduces bichromic ion of bichromic acid to trivalent chromous ion. Water content in sample is removed in white precleaning layer.

$(C_2H_5)_2O + K_2Cr_2O_7 + H_2SO_4$ $\quad$ $Cr_2(SO_4)_3$

60. No. 171 Acetylene

Acetylene reduces iodine pentoxide to liberate iodine, which produces a greenish brown stain.

$C_2H_2 + I_2O_5 + H_2SO_4$ $\quad$ $I_2 + CO_2$

61. No. 172 Ethylene

Ethylene reacts with palladium sulfate to form an additive compound, which reacts with ammonium molybdate to yield molybdenum blue. Chemical formula of this reaction is still unknown.

$C_2H_2 + (NH_4)_2MoO_4 + PdSO_4$ $\quad$ Mo_3O_8

62. No. 191 Acrylonitrile

Acrylonitrile is decomposed by brown oxidizing agent consisting of chromic acid and sulfuric acid in the primer tube to liberate HCN. This HCN reacts with mercuric chloride to generate HCl, which changes the color of Hammett indicator (benzeneazodiphenylamine). Coexisting acidic gases are removed in white precleaning layer consisting of NaOH in the primer tube.

Primer tube:

$CH_2{=}CHCN + CrO_3 + H_2SO_4$ $\quad$ $HCN + Cr_2(SO_4)_3$

Analyzer tube:

$HCN + HgCl_2$ $\quad$ $HCl + Hg(CN)_2$

TABLE I.9

Paint and Varnish Manufacturing Solvent Inventory

Chemical compound	Usage: Major	Usage: Minor	TLV (ppm)	Catalog tube no.*	Remarks
Alphatic hydrocarbons (as *n*-hexane)		X	500	113	Mineral spirits, V.M.&P. naphtha
Methanol		X	200	119	
Ethanol		X	1,000	104A	Synasol® solvent
Isopropanol		X	400	150	
Butanols		X	100	–	C. S. solvent
2-Methoxyethanol		X	25	–	Methyl Cellosolve®
2-Ethoxyethanol		X	100	–	Cellosolve
2-Butoxyethanol		X	50	–	Butyl Cellosolve
Acetone		X	1,000	102A	
Methyl ethyl ketone (MEK)	X		200	139B	
Methyl *n*-butyl ketone		X	100	–	
Methyl isobutyl ketone		X	100	155	
Methyl isoamyl ketone		X	100	–	
n-Butyl acetate		X	150	138	
Isobutyl acetate		X	200	138	
sec-Butyl acetate		X	200	138	
Methyl amyl acetate		X	–	–	
Toluene	X		100	124	Limited exposure
Xylene		X	100	143	

*Unico®-Kitagawa® Model 400 and Unico gas detector tubes, 1956.[87]

were relatively minor compared to the "apparent" MEK results, which were well above the TLV limit (Table I.10), no attempt was made to correct the contribution of toluene to the reduction of the chromate reagent in the MEK tube.

Air was drawn through the detector tubes as specified in the instructions enclosed with each box of tubes.[87] Approximately 3 min was required to draw the 100-ml air sample through each tube. Leak tests were made on the pump, and leakage was found to be within acceptable limits (less than 5 ml/3 min). Stain lengths were determined in subdued daylight near outside windows and related to concentration on the charts supplied with the tubes.

Samples were taken in the breathing zone as close as possible to the operator's nose without interfering with his work assignments, or in the general room area at nose level. Doors and windows were closed, and ventilation was set to simulate cold weather operating conditions from about 10 a.m. to shift end.

Results – The results are summarized in Tables I.10 and I.11.

Basis for conclusions –

1. Any given operator would on the average spend the amount of time required to complete the operation at each of the stations sampled.
2. The operations observed represented the average daily activity in the building.
3. The ventilation was minimal as would be expected during cold weather operation.
4. Interferences in the determination of MEK were minimal.
5. The operator spent his entire 8 hr work shift in the manufacturing building.

This preliminary survey indicated that an integrated 8-hr average exposure could exceed the 1971 TLV by at least a factor of two, based on the CT value concept (ppm x hr). However, since considerable uncertainty is inherent in a preliminary survey of this kind, a more definitive study is required to determine whether or not the exposure potential for any given operator actually exceeds the 200 ppm TLV for MEK during an 8-hr work shift.

A personnel monitor survey would provide the

TABLE I.10

Air Analysis Results

Test number	Time	Temperature (°F)	Location	Occupancy	Material	Results (ppm)	Comments
	a.m.						
1	10:30	65	Paint dept. – 10 in. from rim of Hockmeyer mixer	5 min/day	MEK	100	
1A	10:35	62	As above	5 min/day	Toluene	50	
2	10:42	66	Stain dept. – pigment stain mixer	5 min/day	Naphtha	30	Hexane tube
3	10:50	63	Lacquer dept. – fill operator	30 min, 3 times/day, 2 times/month	Toluene	10	
4	10:55	64	Lacquer dept., center	30 min, 3 times/day, 2 times/month	MEK	750	
5	11:01	65	Lacquer dept. – top of tank #8 on platform	10–15 min, 4 times/day	Toluene	15	
6	11:06	65	Paint dept. – #2 mill – paste		Mineral Spirits	–	No odors
7	11:11	65	Paint dept. – 10 in. from rim of old Cowles® mixer	5 min/day	Toluene	120	
7A	11:15	65	As above – dumping N.C. into tank #8	5 min/day	MEK	1500	
8	11:20	66	Lacquer dept. – 40 in. from floor at opening	5 min, 12 times/day	Toluene	60	
9	11:26	63	1st floor, main aisle, south end – 20 ft from door	varied	Toluene	40	
10	11:30	63	As above	varied	MEK	50–100	
11	11:33	62	1st floor, main aisle, center	varied	Toluene	2	
12	11:35	61	As above	varied	MEK	50	
13	11:40	64	Men filtering batch of 30 gal of solvent and putting into 5-gal cans – adjacent to Cowles® mixer	10 min 10–15 times/day	Toluene	160	
14	11:43	64	As above	10 min, 10–15 times/day	MEK	100	
	p.m.						
15	2:40	65	Drum cleaning bldg., north room	240 min/day	MEK	1500	
16	2:53	65	1st floor, main aisle, north end – 20 ft from door	varied	Toluene	8	

TABLE I.10 (continued)

Air Analysis Results

Test number	Time	Temperature (°F)	Location	Occupancy	Material	Results (ppm)	Comments
17	2:55	64	Paint dept., center – at 55-gal drum mixer	15 min/day	Toluene	20	
18	2:58	64	As above	15 min/day	MEK	200	
19	3:08	64	Paint dept. – new Cowles mixer	10 min, 3 times/week	MEK	1200	
20	3:13	64	Lacquer dept. – on platform, 3 mixers running	15 min, 12 times/day	Toluene	5	
21	3:30	63	Paint dept. – old Cowles mixer changing mixer from one tank to another	5 min/day	Toluene	40	
22	3:37	63	1st floor, main aisle – center and area around mixers – paint dept. to east	varied	MEK	1000	

TABLE I.11

Exposure Summary for Methyl Ethyl Ketone (MEK) (2-Butanone)

TLV/1971 = 200 ppm
CT/8 hr = 8 x 200 = 1600 ppm hr

Test number	Results (ppm)	Length of exposure (min)	Exposures per day	CT value
1	100	5	1	10
4	750	30	3	1130
7	1500	5	1*	150
10	75		1	520
Av	65		1	520
12	50		1	520
14	100	10	12	240
18	200	15	1	50
19	1200	10	0.6	140
22	1000	1*	1*	1000
Total				3240

Major sources:
Test number 4 (Lacquer dept. – center at fill operation)
Test number 22 (1st floor, center, main aisle – around paint mixer perimeter)

Background, overall level:
Test numbers 10–12

*Estimated.

most reliable in-depth evaluation of exposure severity and would furnish the data necessary to establish conformance with the federal TLV.[88]

Briefly, this would require an operator to carry on his belt a small battery-operated pump that draws a determinable volume of air through a charcoal bed in a glass tube to adsorb the solvent vapor.[53] The tube is then capped and forwarded to a laboratory where, by gas chromatography, the weight of the individual solvents collected is determined. From the flow rate and sampling time, the volume, and then the concentration as ppm, are calculated. In order to average out daily variations in work activity and production schedules, this personnel monitoring should be carried on for at least 1 month (20 shifts). After a short training period, local supervision would be able to conduct the sampling. Gas chromatograph capability is available on location. To demonstrate continuing conformity with governmental regulations, periodic (for example, one 8-hr shift per month) monitoring may be required.

Conclusions –

1. This preliminary survey disclosed no potential health problems under minimum ventilation conditions employed to conserve heat during severe cold weather conditions.

2. Two categories of airborne contamination are involved: (a) hazardous vapors (toluene, xylene, cellosolve,® and methanol) and (b) nuisance vapors (ketones and acetic acid esters).

3. Based on the CT (concentration x time = ppm/hr) concept for the evaluation of TLV's derived from an integrated average for an 8-hr workday, exposure to toluene vapor was not significant. However, MEK vapor may present a nuisance problem. The other solvents used in comparatively minor quantities were not analyzed.

4. Personnel monitoring will be necessary to establish conformance with the TLV for MEK. The grab sampling survey indicated violations, but a more definitive sampling and analysis program is needed to establish the magnitude of the problem before corrective actions are considered.

5. If the gas chromatograph results confirm the assumption that the major fraction (75% or more) of the solvent vapor was indeed MEK, then recalibration of the MEK tube for 4-hr monitoring, as described for the CO detector tube,[53] should be intiated to minimize cost and provide quick results directly in the field.

From the foregoing example, one crystal clear conclusion can be drawn: The investigator must be able to identify the reagents in the detector tubes and understand the chemical reaction principles before he can evaluate the significance of his findings, or, indeed, decide whether available detector tubes could solve his problem. The manufacturer must supply at least a list of the components of the reagent system, if not the relative concentrations, to guide the user in choosing an appropriate analytical method and to avoid interferences. Without this information the investigator should not attempt a survey, as his conclusions can be entirely vitiated by coexistant components that may either mask or enhance the true result. Some manufacturers have been willing to provide this information, which is summarized in Tables I.8A to I.8D. If the reagent cannot be identified, the user is advised either to recalibrate with the interference present, if known, or better, to reject the system as a valid analytical tool. In any event, the preliminary survey carried out with detector tubes should be followed by a definitive, specific method such as gas chromatography if the apparent results fall in the 0.5 to 1.5 TLV region, as recommended in the paint and varnish plant report. This follow-up is critically important if a large expenditure for the correction of a non-existent hazard is to be avoided,[53] and if the validity of preliminary inspections based on detector tube surveys is to be successfully challenged. (See Chapter IV for a statistical approach to interference evaluation). In Tables I.8A to I.8D are listed a number of reported interferences, but the list is by no means complete and is offered only as a warning gleaned from the experience of others.

Some interference problems have been solved by incorporating a granular "guard" column in front of the reagent zone to trap the "impurities" quantitatively without removing a significant amount of the target component. In some cases the scavenger layer is packed in a separate tube and connected to the detector tube by a flexible collar[67] (Table I.12), and in other cases the guard is an integral part of the detector tube and separated from the reagent granules by a porous plug.[87] Generally this first granular layer is impregnated with a chemical such as sulfuric acid on silica gel, which reacts preferentially with the extraneous substance but retains only to a negligible degree the substance sought. Several examples illustrate this principle:

Passed to detector layer	Retained by guard layer
Carbon monoxide	Olefines (ethylene) by H_2SO_4, and nitrogen oxides
Carbon bisulfide	H_2S by lead acetate

TABLE I.12

Moisture Removal from Saturated (100% RH) Air Samples Before Analysis with Detector Tubes

Temperature range = 0 to 40°C
Guard layer (Kitagawa® tube No. 103BF) – H_2SO_4 on 40-60 mesh diatomaceous earth

Kitagawa detector tubes suitable for stack sampling:

No. 103a	Sulfur dioxide – high range
No. 103b	Sulfur dioxide – middle range
No. 103c	Sulfur dioxide – low range
No. 106a	Carbon monoxide
No. 106b	Carbon monoxide
No. 106c	Carbon monoxide
No. 106g	Carbon monoxide length-of-stain
No. 112a	Hydrogen cyanide – high range
No. 112b	Hydrogen cyanide – low range
No. 120a*	Hydrogen sulfide – high range
No. 120c	Hydrogen sulfide in coexistence of SO_2
No. 120hh*	Hydrogen sulfide – extra high range
No. 126a	Carbon dioxide – high range
No. 126b	Carbon dioxide – low range
No. 142	Mercury vapor
No. 159	Oxygen (not very sensitive to moisture)

Moisture-sensitive tubes that cannot be used with the desiccating guard tube:

No. 105a and 105b	Ammonia
No. 103d	Sulfur dioxide – D type
No. 109	Chlorine
No. 114	Bromine
No. 116	Chlorine dioxide
No. 117	Nitrogen dioxide
No. 120b	Hydrogen sulfide – low range

*Can be used without guard tube

Hydrogen cyanide	H_2S and HCl by lead compound, and NO_2, SO_2
Benzene	Toluene and xylene by acid and aldehyde
Phosphine and arsine	H_2S, mercaptans, NH_3, and HCl by copper compound
Sulfur dioxide	H_2S by copper compound

A number of detector reagents are sensitive to water vapor and must be protected by passing the air sample through a desiccant layer before the colorimetric reaction takes place. Some examples are listed in Table I.12. Limitations imparted by the use of sulfuric acid supported on silica gel should be carefully considered when applying this desiccant to other mixtures, especially those that contain reactive components of analytical interest. Ammonia, aliphatic amines, ethylene, etc., would be completely removed. Silica or alumina gel and anhydrous calcium sulfate (Drierite®) are effective desiccating agents in themselves, but are not specific for water and usually absorb a significant fraction of the substance under analysis. Calibration with the selected guards in place offers the only final answer relative to retention of the component for which an analysis is desired.

A rather limited number of indicator tubes require a preliminary reaction to convert the compound to a detectable derivative. The chlorinated hydrocarbons fall into this class. The first layer, which contains a strong oxidizing agent such as potassium permanganate or chromic acid, converts the chlorine atoms to elemental chlorine, which then reacts with the color-forming compound (dianisidine or tolidine). Carbon tetrachloride is converted to phosgene by fuming H_2SO_4, nitric oxide to nitrogen dioxide by chromic acid, and acrylonitrile to HCN by a chromate before passage into the detection zone in other examples.[67] While such multiple-layer tubes serve a useful purpose when properly constructed, they frequently have a shorter shelf life than single-layer tubes because the chemicals diffuse between layers and produce an accelerated deterioration.

Detector tube systems have been developed to eliminate the diffusion between layers in storage, and to improve the stability of the color-generating system, by incorporating one of the reactive components as a liquid or an impregnated gel in an integral fragile glass vial that is broken just prior to use. The Dräger methyl bromide detector, for example, retains the sulfur trioxide required for pre-oxidation sealed in the foresection of the glass detector tube. A plastic tube is shrunk on over this section to serve as a seal and to protect the operator after the tube is broken at this point.[67]

Many other complications can be expected when deviations from calibration conditions are introduced. Thus, for nitrogen dioxide the proportion of side reactions changes with flow rate. Altering sample volumes from calibration conditions cannot be assumed to be without error unless the detector reagent is known to be thoroughly free from interfering gases and humidity in the air.

The lack of specificity may be used to an advantage for determining substances other than those intended by the manufacturer. Thus, the trichloroethylene tube[67] can be applied also to the detection of chloroform, *o*-dichlorobenzene, dichloroethylene, ethylene chloride, methylene chloride, and perchloroethylene when present as sole chlorine-containing constituents in the atmosphere. The methyl bromide tube may be used for chlorobromomethane and methylchloroform. The chlorine tube has been used for bromine and chloride dioxide. Such conversion requires specific knowledge of the feasibility and proper corrections to calibration scales (recalibration with known mixtures of the desired compound).[62]

i. Observational Errors

When all of the foregoing precautions have been observed and effects of the variables compensated, the investigator arrives at the final stage, examination of the exposed tube, with a reasonable assurance that the best possible result is available. This final judgment on his part is to a great degree dependent upon his visual acuity. If he has carried out his own calibration, this variable can be minimized by practice. Some detectors are provided with color charts to be matched with the stain on the granules in the indicating portion of the tube. In others a prestained section is included in the detector tube.[67] Judgment depends not only upon the observer's color discrimination, but also on lighting conditions. This is the case when the reaction with test gas is relatively slow and color is produced throughout the length of the tube, since the gas is incompletely absorbed and the concentration at the exiting end is an appreciable fraction of that at the inlet. This type includes the Dräger tubes for benzene, nickel carbonyl, carbon bisulfide, cyanogen chloride, styrene, phosgene, and vinyl chloride.[67] Subtle color changes are difficult to detect, particularly under artificial light; and very often the standard color chart does not have the same tint or hue as the stain obtained. Subdued daylight through a north window provides the best comparison conditions. Exposure of the stain to direct sunlight should be avoided.

Other tubes such as arsine, stibine,[62] aniline, mercury, phenol,[67] ethylene oxide, and propane[65] require a variable volume of sample be drawn through until the first visible color is noted (gas titration). This technique requires the most acute judgment and should be made retrospectively by drawing additional sample through the system to deepen the stain and thereby confirm the validity of the first observation as being due to the gas sought and not to an artifact.

One manufacturer based his entire product line on this gas titration principle.[62,89] For reasons other than technical considerations, the producer of these gas detectors recently went out of business. However, a short description of the product is presented here to acquaint the reader with the possibilities for application to "instantaneous" alarm meters of the type described in Section I.A.2., "Inorganic Support Systems,"[4] as well as for handheld grab sampling. The air sample was drawn through the short (approximately 40 mm) detector tube at a manually regulated rate of 100 ml in 20 sec by a calibrated precision piston pump. The pump itself was quite similar in construction to those currently available models, but was not fitted with a flow-limiting orifice or a check valve in either the inlet or exhaust ports, which had to be manually regulated.[63,65,87] The volume of sample required to produce the first color change in the sensitive layer of the tube was related to the concentration of the test gas. Each detector tube was packed in a lead foil cylinder which was opened by bending the crimped portion sharply backward over the end of the detector tube.

The detector tube for phosgene serves as a good example of the application to field monitoring. The change from yellow to green indicated the titration end point, which was related to concentration by reference to the standardization chart:

Ml for first color change	350	100	50	25
Gas concentration (ppm)	0.1	1	5	10

The presence of 38 different gases and vapors could be determined with this detector system.[62]

To reduce the errors inherent in the variations between observers (see the study reported by Kusnetz, Saltzman, and Lanier for a statistical analysis of observer differences in estimating CO by the NBS detector tube[81]), most detector tubes have been converted from color shade matching to produce a variable length of stain on the indicator gel, although in a few tubes a variable volume of sample is collected until a standard length of stain is obtained (alcohol in breath samples[65]). This technique may be considered to be a variation of gas titration, as volume of "titrant" is related to quantity sought.[67] In most cases a fixed volume of sample is passed through the tube and the stain length measured against a calibration scale which may be printed directly on the tube[67] or on a chart.[65]

If the stain fronts are not well defined, or are not perpendicular to the axis of the tube, again a broad range in the interpretation of the results by different observers will be encountered. Experience in sampling known concentrations is indispensible in training the observer to decide whether to measure the length up to the beginning or end of the stain front and what portion of an irregularly shaped stain to use as the limit. Stains may change with time, and therefore the reading should not be delayed. Precision can be improved by taking several readings (four or five) and averaging the results, or better, by averaging the readings taken by two or more observers.

j. Repetitive Sampling

Repeated use of detector tubes as an economy measure may be commendable, but in the hands of the uninformed must be condemned as a habitual practice. Certain detector tubes may be used again after a negative test, but the investigator must bear in mind that some detector reagent systems are adversely affected by humidity, acid gases, alkaline gases, and other extraneous interferences likely to be present in the atmosphere sampled. For example, traces of ammonia will completely inhibit the color response of the mercury vapor detector tube even when used in a saturated atmosphere. Water vapor will have a depressing effect on stain formation in the benzene reagent mixture by diluting the sulfuric acid required in the condensation with formaldehyde (Table I.8). Other false negative reactions can be expected under high humidity conditions also (see Table I.12). Tubes that require a guard layer to scavange interferences are especially vulnerable to repetitive sampling. The palladium reduction for carbon monoxide detection will produce false positive results in the presence of ethylene if the guard layer is exhausted. Other examples should be consulted before decisions relative to reuse of any particular tube are made (Tables I.8A to I.8D).[65,87]

Information relative to the presence of interferences can be obtained by reversing a tube supplied with a guard layer after a negative result has been obtained from a sample drawn through in the designed direction. Color development in the unguarded end on reversal indicates the presence of a reactive component in the mixture other than the one sought. Such a color development with the carbon monoxide detector tube would indicate the presence of ethylene, acetylene, benzene, or H_2S.[67]

Symmetrical tubes – those filled with a homogeneous reagent composition from end to end – can be reversed and a second determination made. Results obtained from recalibration of the length of stain carbon monoxide detector tube are typical (Table I.13). At least one supplier has recommended this technique for those tubes that require guard layers.[87] The thrifty investigator will quickly note the cost savings in this recommendation.

However, the temptation either to reduce cost or to increase sensitivity by drawing additional aliquots of air through the same detector tube after a short, positive response has been noted can lead to significant errors. Calibration results indicate that at least two samples can be taken with the carbon monoxide (Table I.13) or the benzene detector tube[87] if the second aliquot is drawn immediately after the 3-min period required for the first sample (final result divided by 2). With the benzene tube, a time lapse of more than 10 or 15 min between aliquots produced errors that exceed 15%; and the wide deviation in the results (± 40% at 95% C.L.) renders a correction factor questionable at best. Even when taken in quick succession, a third sample through the same tube may be in error by as much as 20%. If recalibration for repetitive sampling is considered, deviations greater than those experienced with a single sample must be expected. Both graphic plotting and calculation of the slopes – $(\log Y) = b_0 + b_1 (\log X)$ – confirm a significant difference between the slopes of the second and third delayed aliquot curves and the slope of the initial

TABLE I.13

Recalibration of the Kitagawa® No. 100 Carbon Monoxide Length of Stain Detector Tube

Kitagawa tube No. 100
3-min stroke
Leak test: 0 ml/min

Standard CO (ppm)		Tube no.	Results (ppm) 1 Stroke	Results (ppm) 2 Strokes	Reversed ends (ppm) 1	Reversed ends (ppm) 2
25	In air	1	25	25	30	20
49	In air	1	50	45	–	–
		2	50	55	60	–
		3	60	50	–	–
99	In air	1	85	95	90	–
		2	90	–	–	–
		3	90	–	85	–
		4	90	85	90	75
50±5	In N_2	1	50	50	50	–
		2	50	–	50	–
100±10	In N_2	1	125	113	110	–
		2	110	100	90	–

sample curve (Figure I.26). Detailed results are summarized in Table I.14.

Suggestions have been offered to use two different tubes in series, such as passing the air sample first through a hydrogen sulfide tube and then through a phosgene tube to obtain two simultaneous determinations, or to remove interferences.[62,87] In selected cases this technique will resolve mixtures, but recalibration to establish the compatibility of the reagent systems and to compensate for the added resistance (Δ P) that will reduce sample flow rate must be carried out before even semiquantitative results can be obtained. In the foregoing example, serious losses of phosgene may occur in the hydrogen sulfide zone. Again, the identity of the reagent mixtures must be known in order to decide whether a given pair of different tubes can be expected to deliver the desired separation (Tables I.8A to I.8D). If a lead compound is incorporated in the gel to remove the hydrogen sulfide, then significant losses of phosgene in this zone would not be anticipated.[67] Also, the granular support may absorb the gas under investigation, i.e., the absorption of arsine on silica gel.[82a] Hydrogen sulfide is a common interference in the reaction of mercuric chloride with arsine and must be removed by a "cleansing" layer. The traditional scavanger – lead acetate on silica gel – also removes the arsine. Copper sulfate crystals (40-60 mesh) solve this problem by retaining the hydrogen sulfide without retaining the arsine fraction. Arsine produces no color in the lead acetate tube; therefore, simultaneous determinations can be carried out.[82a]

7. Concentration of Reagents

In general, as the concentration of color-forming reagent adsorbed on the substrate gel particles decreases, the length of the stained zone increases. However, the color becomes lighter and the stain front more diffuse. Therefore, a compromise between improved sensitivity gained by reduced reagent concentration and inaccuracies involved in determination of the zone boundary must be reached. The relationship for an arsine detecting system (mercuric chloride) is presented in Figure I.27.[82a]

8. Constant Rate and Continuous Sampling

Introduction of the Kitagawa gas detector in 1958 provided a wide range of significantly improved, granular reagent-filled detector tubes for quantitative analysis of the most frequently

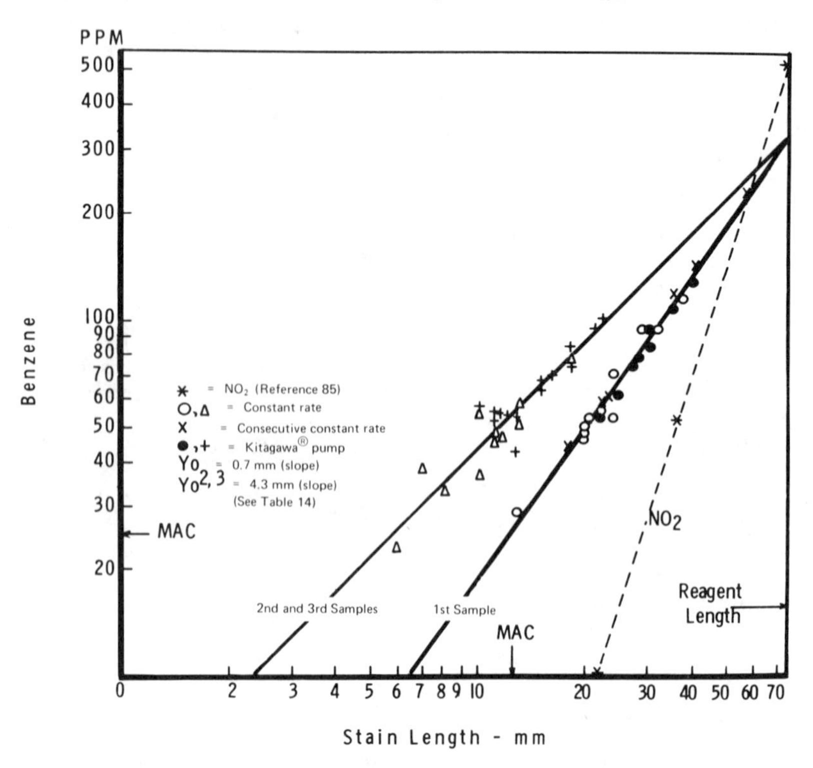

FIGURE I.26. Stain length vs. benzene concentration.

TABLE I.14

Multiple Sampling with the Kitagawa® Benzene Detector Tubes

Tube no.	Pressure drop (maximum) ($cm\ H_2O$)	1st 100 ml	2nd 100 ml		3rd 100 ml		Error[a] (% Decrease)	
		ppm	ppm	% Decrease	ppm	% Decrease	2nd	3rd
21[b]	–	107	105	2	106	1	1	0
15	–	84	66	21	57	32	11	18
9	–	16	14	12.5	16	0[e]	6	0[e]
7	72	67	56	16	52	22	7	13
12	–	79	69	13	57	28	6	14
17	–	78	64	18	58	26	9	14
19[b]	67	87	83	5	70	20	2	8
23[b]	71	128	117	9	–	–	4	–
25[b]	66	56	53	5	49	13	2	5
Average	69	–	–	16	–	27	8	15
Rapid succession[c]				5		11	2	4
S.D.[d]	1.2	–	–	3	–	4	4	2
Rapid succession[c]				3[b]				
S.D. at 95% C.L.	8.5	–	–	37	–	29	–	26

[a]Calculated error based on averaged results. Total benzene found was divided by 2nd or 3rd sample taken through the same tube.
[b]Tubes 19 to 25 used in rapid succession.
[c]These averages are from only a fraction of the total results.
[d]Standard deviation.
[e]Excluded from averages.

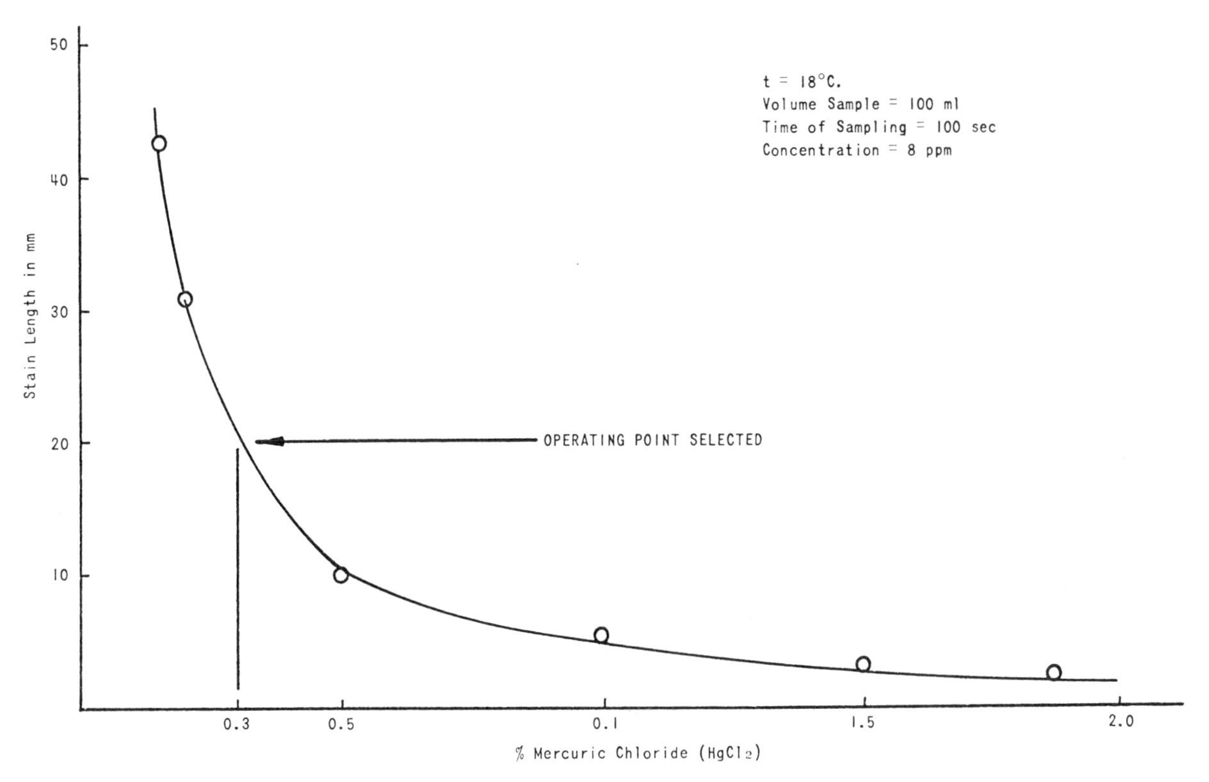

FIGURE I.27. Arsine detector tube – relationship of reagent concentration to stain length. (Data taken from Reference 82a.)

encountered airborne toxic gases and vapors by stain length measurement. Reproducible sample volumes and flow rates supplied by the 100-ml, single-stroke, manual piston pump fitted with a rate-limiting orifice assured volumetric precision. The increased utility attainable by extending the sampling period beyond the 3-min grab sample limit and by improving the sensitivity by increasing the sample volume soon became apparent. Consideration of the critical relationship between sample flow and stain development rates clearly indicated that recalibration or delineation of time-rate factors to produce results equivalent to existing calibration charts would be necessary.

Adaptation to a readily portable, self-powered pump unit from a manual operation offered the most practical approach to extended interval sampling. A sampler that derived its power by expansion of dichlorodifluoromethane or chlorotrifluoromethane through an aspirator without generating static spark potential was chosen on the basis of surge-free constant airflow at low sampling rates and suitability for use in explosion hazard areas.*[90-94]

Benzene, which had presented industrial hygiene exposure control problems, was selected on the basis of what appeared to be easily prepared standard air mixtures and because it is an example of a reagent system (formaldehyde and sulfuric acid) especially sensitive to sampling rate variations. The following procedure provides the reader with one approach to recalibration for extended periods at constant flow rates. The airflow characteristics of the Kitagawa hand pump (HP) have been described in Section I.A.2.d.6.c., "Packing Variations – Pressure Drop Effects," and the effect of variable flow rate on stain front velocity in Section I.A.2.d.6.d., "Sampling Rate." A low range rotameter was inserted in the vacuum line between the pump and the detector tube, and fine control was provided by a needle valve built into the rotameter body. The rotameter was calibrated by the soap bubble technique.[83a] In passing, a "trick of the trade" is worthy of note: The soap film can be stabilized by adding 10% glycerine and 3% gelatin to the soap solution. Quite durable bubbles can be produced from this mixture.

The mixing chamber and the sampling trains are shown in Figure I.28. Detector tubes were held in place by rubber tubing sleeves through which the detector tubes were inserted to a depth that provided an essentially glass-to-glass connection. The rubber stopper was covered completely with aluminum foil. Two midget impingers, fitted with glass ball joints packed with glass beads, were connected in series. The impingers then were immersed in a low temperature bath composed of a freezing mixture of liquid nitrogen and ethanol (–125°C).[91] In each freeze-out collection, 10 ml of isooctane was added to the cold impinger as soon as the run was finished to avoid benzene losses, and the optical density determined in 10 mm silica cells at 254 nm. The detector tubes were selected in pairs for identical reagent packing lengths.

A borosilicate glass carboy was conditioned before use by overnight equilibration with several hundred ppm of benzene. The excess benzene was removed by sweeping with fresh air for 20 min. Sampling ports and the agitator bearing were then sealed off and the carboy partially evacuated. The calculated amount of benzene in isooctane (302 mg benzene per ml isooctane) was introduced by a 0.02-ml pipette through a hypodermic needle into a sampling port. After equalizing to atmospheric pressure, the benzene-air mixture was agitated vigorously for 5 to 10 min.

Three samples were withdrawn simultaneously: one through a detector tube for the required 3 min by HP; a second through another tube for the time required for 100 ml, or a 3-min interval at constant rate; and the third 100-ml aliquot through the midget impinger train at a rate of 33 ml/min. Isooctane solutions were analyzed spectrophotometrically at 254 nm to detect benzene. The stain lengths were measured at 30-sec intervals.

A stain length equivalent to one stroke of the HP was obtained in 3 min at a constant sampling rate of 46 ml/min (Figure I.21 and Table I.15, column 8). Close control of aspirating rate, within ± 6% (± 2.5 ml/min), was required to maintain an overall precision of ± 10% (Figure I.24). Operation within these limits required recalibration of the rotameter to compensate for the pressure drop

*Since completion of the initial study of the adaptation of detector tubes to constant rate sampling, the manufacture of the Uni-Jet® constant rate sampler was discontinued in 1964. In its place a miniature battery-operated piston pump was developed and is currently available (see Chapter III).

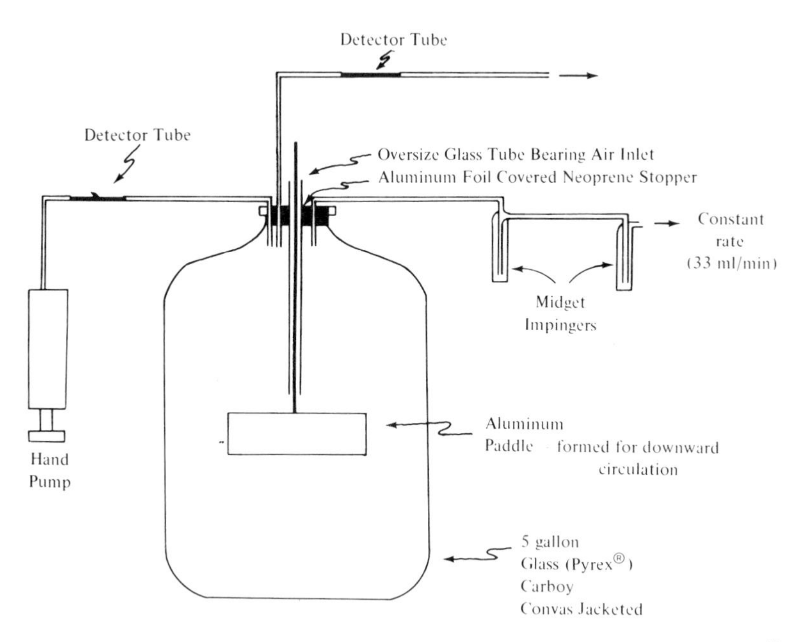

Figure I.28. Apparatus for calibration of detector tubes on the Uni-Jet® constant rate air sampler.

across the detector tube. The sample volume required at this constant rate increased to 138 ml. On an equivalent 100-ml HP volume basis, extrapolation of the calibration deviation curve (Table I.15, column 11) indicated that an equivalent stain length would be obtained in 1.2 min at a sampling rate of 85 ml/min (Figure I.24). This linear adjustment curve had a much more favorable slope than the 3-min time curve; therefore, considerably greater latitude in flow rate deviation could be tolerated. The estimated pressure drop (69 cm H_2O) would pose no problems. However, this alternative was not confirmed by laboratory trials due to curtailment of the experimental phases of this project.

When the constant sampling rate was reduced to 20 ml/min, benzene concentrations, after a 3-min sampling period, could be obtained by doubling the observed HP chart reading. However, no correction was required when the sampling time was extended to 9 min (Figure I.25). The ppm benzene readings on the right-hand ordinate axis are equivalent to the stain lengths taken from the HP calibration chart furnished with the detector tubes. Some sensitivity (mm stain length/ppm benzene) was lost, but for most purposes the results were satisfactory within a ± 10% range for time intervals of less than 5 min. For extended periods, 9 min or more, sensitivity actually improved. The stain front development rate passed through an inflection point at between 3 and 4 min to a significantly lower rate. Another tube exposed to the same benzene in air mixture at 34 ml/min did not exhibit such an inflection (Figure I.25). Results that required only 15% correction were obtained in 3 min at the flow rate of 38 ml/min (Figure I.25 and Table I.15).

Increased sensitivity was obtained by operating at a 46 ml/min constant rate for periods up to 10 min. Although the 9-min results averaged 15% less than three times the 3-min value, the much greater stain length compensated for the observational error in locating the color front; and at low benzene concentrations, the averaged value would be more reliable than a single 3-min or 100-ml analysis. However, additional data is needed to determine whether or not the correction factor developed for the 50 to 150 ppm range is directly applicable to the 10 to 40 ppm region. Better accuracy in the preparation of the standard benzene in air mixtures and more extensive calibration undoubtedly would reduce the uncertainty factor to within the desired ± 5% range. This technique certainly offers the most promise for increasing the reliability, sensitivity, and accuracy of granular reagent tubes for application to personnel monitoring.

Plots of the data as a log-log function of stain length vs. benzene concentration (Figure I.26) disclosed a significant difference between the slope

TABLE I.15

Hand Pump (HP) vs. Constant Rate (CR) Sampling

Hand pump (HP)				Constant rate (CR)			% Deviation on time basis (3 min/sample)[d]			% Deviation on volume basis (100 ml)[e]					
Tube no.	Stain (mm)	ppm	Error ±%[b]	Tube no.[c]	Flow rate (ml/min)	Pressure drop (cm H_2O)	1st	2nd	3rd	1st	2nd	3rd	Mass rate (μg benzene/min)[g]	Time (min)[a]	Vol (ml)[f]
21	35	107	5	22	20	14	-56	-65	-78	-28	-55[h]	–	6.8	8.3	166
15	30	84	5	16	32	–	-42	–	–	-24	–	–	8.6	4.3	138
7	25	67	6	8	36	20	-16	–	–	-22	-31	-45	7.7	3.4	122
12	28	79	5	11	41	31	-15	–	–	-20	-40	-52	10.3	3.6	148
25	22	56	7	26	46	34	+8	-21	-32	-18	-33	-55	8.2	2.5	115
17	27	78	5	18	46	–	+9	-17	-30	-19	-31	-45	11.6	2.5	115
19	30	87	5	20	46	31	-2	-18	-28	-18	-33	-46	12.8	3.2	147
23	40	128	3	24	46	33	-8	-47[h]	–	-21	-55	–	18.8	3.5	161
Average	–	–	–	–	46	–	+2	-19	-30	-19	-32	-49	–		
	–	–	–	–	All	–	–	–	–	-21	-34	-49	–		
S.D.[j]	–	–	–	–	46	–	9	2[i]	2[i]	2.0	1	–	–		
	–	–	–	–	All	–	–	–	–	3.4	4	4	–		
S.D. at 95% C.L.	–	–	–	–	46	–	–	11	7	11	3	–	–		
	–	–	–	–	All	–	–	–	–	16	12	8	–		

[a]To produce stain length equivalent to HP result
[b]Error – % difference vs. calculated ppm benzene in calibration mixture
[c]CR tubes 20 to 26 – continuous sampling
[d]% Difference in stain lengths at end of 3 min
[e]% Difference in CR vol vs. 100 ml (HP vol)
[f]Vol required to produce stain length equivalent to 100 ml by HP
[g]Mass rate $= \frac{\text{ppm x 0.00319 x 100 x ml/min}}{1000} = $ ppm x 0.00319 x ml/min
[h]Not used in averages
[i]46 ml/min rate
[j]S.D. = Standard deviation at C.L.

of the original Kitagawa calibration and the slope of the second and third aliquot results from the same detector tube after 5 or more min lapse between analyses. Since this slope was unity within reasonable limits, this reversion suggested complete sorption with a stoichiometric relation between quantity of sorbed gas and volume of the colored indicator that was independent of gas concentration (see Reference 86, Case 2). Results reported by Kinosian and Hubbard for nitrogen dioxide have been included for comparison with an entirely different system (Figures I.25 and I.26).[85]

Adsorption of benzene from the standard mixtures in air by the Pyrex® glass container walls occurred at a rate that reduced the benzene concentrations in the 100 to 200 ppm range by as much as 50% in less than an hour (Table I.15a). Only by equilibration overnight with mixtures that approximated the concentration required could reasonably stable conditions be attained. Simultaneous determinations with hand pump, constant rate, and control analysis were carried out to provide comparative results to minimize this variable.

A somewhat different approach was taken for recalibration of carbon monoxide detector tubes for personnel monitoring on a 4-hr basis. The report of this investigation, "Carbon Monoxide – Evaluation of Exposure Potential by Personnel Monitor Surveys" by A. L. Linch and H. V. Pfaff, is reproduced here by permission of the editor of the *American Industrial Hygiene Association Journal.*

INTRODUCTION

Downward revision of the Threshold Limit Value for carbon monoxide (CO) from 100 ppm to 50 ppm in 1967 by the American Conference of Governmental Industrial Hygienists (ACGIH)[95] and incorporation of the 1967 TLV's into Safety Regulation No. 3, issued December 1, 1967, by the State of New Jersey Commissioner of Labor and Industry,[97] focused attention on an area of potential exposure within the Consolidated Warehouse where a number of gasoline powered fork lift trucks are employed for drum storage within bays, and for loading box cars and truck trailers. Although no health problems had been encountered since completion of this very large building in 1955, reduction in ventilation capacity during cold weather when ventilators and the large loading ramp doors were closed to conserve heat could allow CO to accumulate beyond permissible levels. During January, 1968, fixed station CO alarms[52] set to trip at 50 ppm

TABLE I.15A

Benzene Loss by Adsorption on Container Walls

Tube no.	Benzene added (ppm)	Hours elapsed	Benzene found (ppm): Kitagawa®	Benzene found (ppm): Cold trap	% Loss	Remarks Temperature = (75±5°F)
6	100	0.5	56	–	44	
7	100	0.5	67	–	33	
9	100	16	16	–	84	
10	–	192	0	–	100	
11	200	16	100	–	50	
12	100	0.5	79	–	21	Flask conditioned overnight
15	100	0.5	84	80	16	Flask conditioned overnight – 100 ppm
17	–	1	78	80, 88	25	No additional benzene added
19	100	0.5	87	–	13	Flask conditioned overnight – 200 ppm
21	100	0.5	107	–	-7	Flask swept; 2nd 100 ppm added
23	200	0.5	128	–	36	100 ppm added after tube 21
25	50	0.5	56	–	-12	Flask swept; then 50 ppm benzene added
–	100	0.1	75	–	25	No conditioning for 30 days
		0.25	72	–	28	
		0.5	55	–	45	
		1	55	–	45	
		1.5	60	–	40	
		4.5	48	–	52	
		7.5	63	–	37	
		24	42	–	58	

were installed at the two centrally located "write-up" desks located in the general traffic area and in the fluorocarbon shipping areas where maximum exposure to the exhaust emissions from two to four continually circulating fork lift trucks would be expected and where at least one employee was in constant attendance while preparing invoices. Although these alarms did sound occasionally, the problem was not considered sufficiently acute to justify more extensive investigation while the more urgent concurrent study of tetraalkyl lead exposure in another area was in progress.[96]

The project was reopened in December, 1968. Frequent sounding of the alarms indicated concentrations of CO above the TLV at these locations for significant periods of time. Grab samples taken with length of stain indicator tubes[97a] whenever the alarm sounded disclosed concentrations in the range of 60 to 80 ppm for 26 out of 50 determinations taken during the period 8 a.m. to midnight over a ten consecutive day period. Only once did the indicated concentration reach 100 ppm. In one closely spaced series of five grab samples taken over a two hour period, the alarm tripped once in the 25–30 ppm concentration range, but not below 25 ppm. Since the alarm operates by electronic monitoring of the optical density of the stain which develops from CO exposure in a thin circular wafer of the same reagent used in the detector tube, this activation below the 50 ppm preset limit probably was derived from either an accumulation of stain as the concentration of CO slowly approached 50 ppm or from short term excursions of relatively high CO concentrations just preceding the grab sample. No correlation between traffic patterns, open or closed doors, nor operation of the ventilating fans could be established. However, on an averaged 8 hour work day basis, exposure to CO concentrations above the TLV was not indicated.

To further delineate the probable average concentration range and excursion limits, a recording non-dispersive infra-red (NDIR) analyzer was assembled in the Instrument Shop and installed at the general shipping area write-up desk. The chart tracing disclosed an extremely variable ambient atmosphere with respect to CO concentration. Not only did 30 minute averages differ by a factor of ten, but the short period excursions fluctuated as much as twenty-five ppm within the response time of the recorder (12 seconds). Three typical nine minute tracings taken from the eight hour night shift recording of May 20, 1969, illustrate this extreme variability in Figure I.29.

The recorder disclosed by "spike" count CO concentrations which exceeded 50 ppm during 16% of the time and 75 ppm for 2.5% of the same period. The range was 5 ppm to 158 ppm (single 12 second "spike").

Blood analyses by the pyrotanic acid colorimetric method[98] during this period disclosed less than 5% carboxyhemoglobin was present in the blood of personnel selected at random from the crew assigned to this building. The reliability of the colorimetric procedure had been confirmed earlier in our laboratory by comparison with the microgasometric procedure developed by Scholander and Roughton.[99] These results indicated the 8 hour average CO level must have been less than 25 ppm.[100,101]

By this time the conclusion that only by personnel monitoring could a true exposure level be determined was

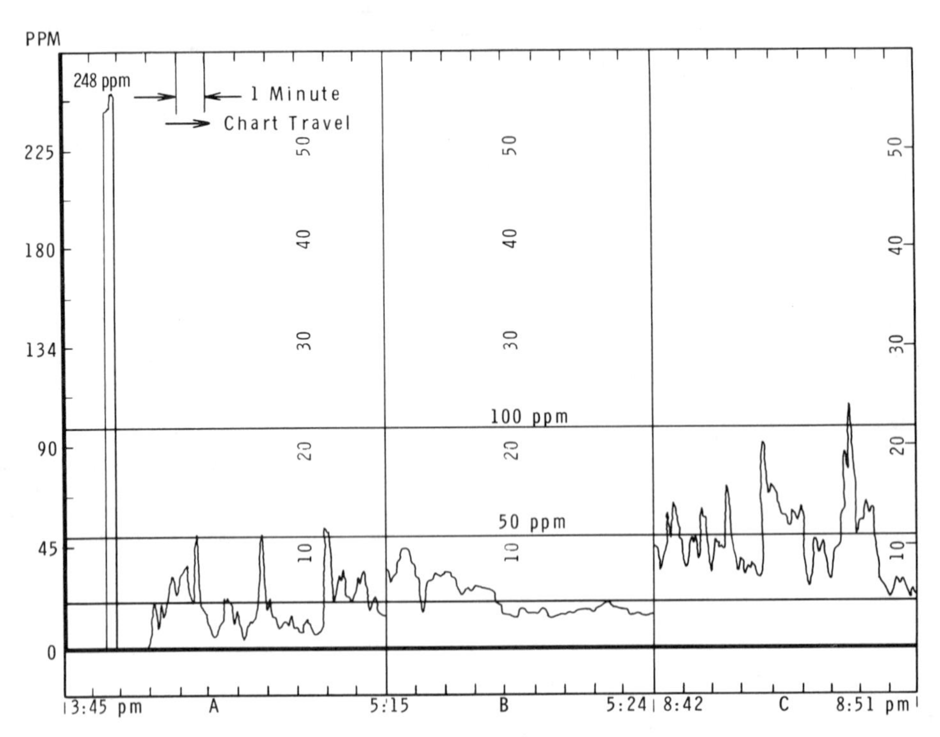

FIGURE I.29. Carbon monoxide in air by nondispersive infrared (NDIR) analysis. (From Linch, A. L. and Pfaff, H., *Am. Ind. Hyg. Assoc. J.*, 32, 745, 1971. With permission.)

obvious. This conclusion was reinforced by the results obtained from the concurrent tetraalkyl lead in air study which clearly demonstrated a complete absence of correlation between personnel and fixed station air sampling.[96] Further emphasis on the urgency relative to establishing the true exposure potential was provided by an engineering design estimate for $100,000 to provide additional ventilation required to maintain CO levels below 50 ppm.

Attempts to adopt either the colorimetric aqueous palladium chloride[102] or the silver para-sulfaminobenzoate[103] procedure to the microimpinger sampling technique met with complete failure due to the slow reaction kinetics. Other colorimetric methods were ruled out on one or more counts,[104] and the Hopcalite gravimetric procedure[105] required an impractically large sample. The highly accurate iodine pentoxide procedure required a reagent temperature of 150°C[104] which obviously cannot be considered for personnel sampling. Likewise, instrumental methods for measurement of physical parameters such as infrared absorption and combustion calorimetry have not been sufficiently miniaturized to permit use as personnel monitors.

A telephone call to Paul Toth at the Ford Motor Company plant in Dearborn, Michigan in early December, 1969 confirmed the soundness of our proposal to attempt recalibration of a length of stain CO detector tube for four-hour sampling based on our earlier experience with a benzene-in-air detector tube. The Dearborn application employed an MSA Catalog No. BY-47134 detector tube recalibrated for a one-hour sampling period at approximately one cubic foot per hour (470 ml/min) rate. Later in the month a reply from an earlier inquiry submitted to the manufacturer relative to the use of the Kitagawa No. 100 length of stain detector tube for personnel monitoring included a calibration chart for four-hour sampling at a 9.6 ml per minute rate.[97] A laboratory investigation was initiated immediately to confirm the calibration furnished by the manufacturer.

PROCEDURE

Calibration

A 10 liter (approximate) carbon steel cylinder which had been steamed and dried in accordance with specifications established for packaging refrigerant grade fluorocarbons was evacuated. The calculated quantity of carbon monoxide was injected from a glass syringe through a rubber septum connected to the inlet of the cylinder. The cylinder was then pressurized to 200 psi with purified air. The concentration of CO was confirmed by vapor phase chromatography on a column with the following specifications:

Column Packing	– 13 x molecular sieve
Length	– 4 meters
Temperature	– 40°C
Carrier Gas	– Helium
Gas Flow	– 110 ml/min
Sample Size	– 5.175 ml

The schematic drawing shown in Figure I.30 is largely self-explanatory. The valve on the calibration gas supply cylinder A is opened sufficiently to provide a definite gas overflow through trap B and to prevent introduction of outside ambient air into the system by the vacuum source C. After insertion of the detector tube D into the gum rubber tubing connections, the gas flow rate is adjusted to the desired rate by needle valve E as determined by the soap bubble flow meter F. (see References 76 and 79 for operational details). If secondary dilution is required, purified air is metered by a second soap bubble flow meter through a tee at G. The length of stain was read hourly and recorded in millimeters. The ambient temperature was 25 ± 2°C. Since the manufacturer disclaimed interference from humidity effects unless saturation conditions were approached, no humidity control was attempted.

A dynamic system assembled as shown in Figure I.30, except cylinder A contained purified air and pure CO was introduced at G from a syringe mounted in a constant speed drive which incorporated a wide selection of rates and volumes,[106] was abandoned before the criticality of concentration and flow rate relative to stain production rate was recognized. In retrospect, however, we have considered dynamic calibration as the better approach and will go back to this system when future calibration work becomes necessary.

The stain lengths were corrected to 60 mm and 80 mm reagent column lengths: corrected mm = $\frac{\text{mm found} \times 80}{\text{mm reagent length}}$. Calibration charts similar to those supplied for grab samples were constructed on mm cross section paper. A 100 mm (vertical) x 75 mm rectangle was laid out. Then a line was ruled from a point 60 mm above the 75 mm base on the left hand vertical edge to a point 80 mm

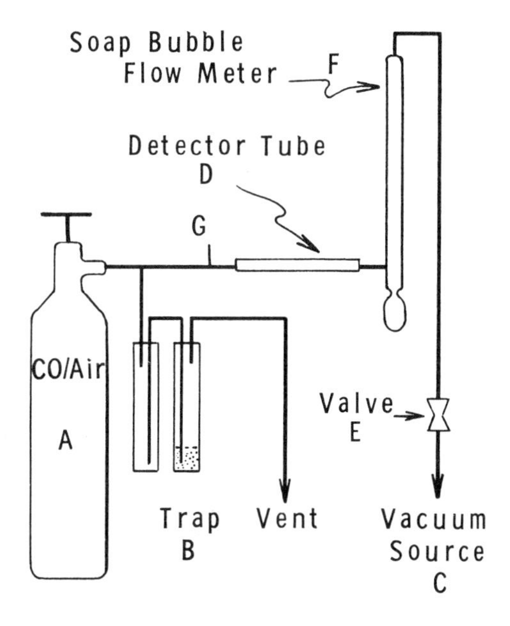

FIGURE I.30. Carbon monoxide in air calibration system. (From Linch, A. L. and Pfaff, H., *Am. Ind. Hyg. Assoc. J.*, 32, 745, 1971. With permission.)

above the base on the right hand side. The slope of this line then represented the total reagent column length variation and determined the position of an exposed tube for CO concentration readout. Calibration lines were constructed by plotting appropriate calibration stain lengths on the 60 mm and 80 mm axes.

Collection of the Sample

The personnel monitor employed in this study is shown in Figure I.31. The detector tube holder A was constructed from a section of polyacrylate plastic plate 135 mm long by 20 mm wide by 9 mm thick by milling a 5 mm semi-circular channel on the face. When fastened to the side of the miniature, portable, air sampler[107] by two brass machine screws with the channeled surface against the case, a secure, transparent holder was provided for the detector tube. If the connecting tube B was kept short as shown, the detector tube remained securely in place during field use.

In order to provide the constant, slow, air flow through the detector tube and yet allow the pump to operate at the 700 to 1200 ml/min optimum rate (rotameter C), a bypass capillary (0.737 mm inside diameter) was fitted to a tee inserted at D. The total air flow through the sampler was maintained at 1100 ± 25 ml/min to maintain the sampling rate within tolerable limits (8.0 ± 1.2 ml/min). Volumetric calibration was carried out with the soap bubble flow meter previously described[76,79] on a weekly basis. Performance is summarized in Table I.16.

After insertion of a freshly charged battery in the sampler, and connection of the detector tube located in the holder A to the rotameter C through tube B, the flow rate was adjusted to the calibration point. The assembly then was fastened to the employee's belt in a position which would least interfere with his work activities. The monitor was checked hourly, the pumping rate adjusted if necessary, and the detector tube examined for stain length progress. After four hours, a fresh detector tube was installed and the sampling continued for the balance of the 8-hour shift. A completely charged battery provided ample power for 8-hour operation without running to complete discharge. The CO concentration in ppm was read off the calibration chart per the manufacturer's instructions, recorded, and the tubes capped for future reference.

RESULTS

Calibration

Attempts to calibrate the detector tubes with 50 ppm CO in air at 2.5 ml/min flow rate or with 13 ppm CO at 9.6 ml/min in the dynamic system met with uniform failure. A stain 2 to 5 mm in length would develop in the first 20–30 minutes and then remain stationary for the balance of the 4-hour period. This failure was attributed to some undetected malfunction of the assembly and the system abandoned before operational range limitations had been recognized.

Carbon monoxide calibration mixtures in pressurized nitrogen were immediately available; therefore, the static system shown in Figure 30 was employed. A plot of the calibration stain lengths gave a linear curve entirely different from the one indicated by the data submitted by the manufacturer. Not only was the slope different, but the rate had to be reduced from 9.6 to 2.5 ml/min to remain "on chart."

Obviously, oxygen had a profound effect on the stain production rate; therefore, calibration mixtures were reconstituted in air. In order to include the 25 to 100 ppm CO range on a 4-hour sampling chart, the flow was

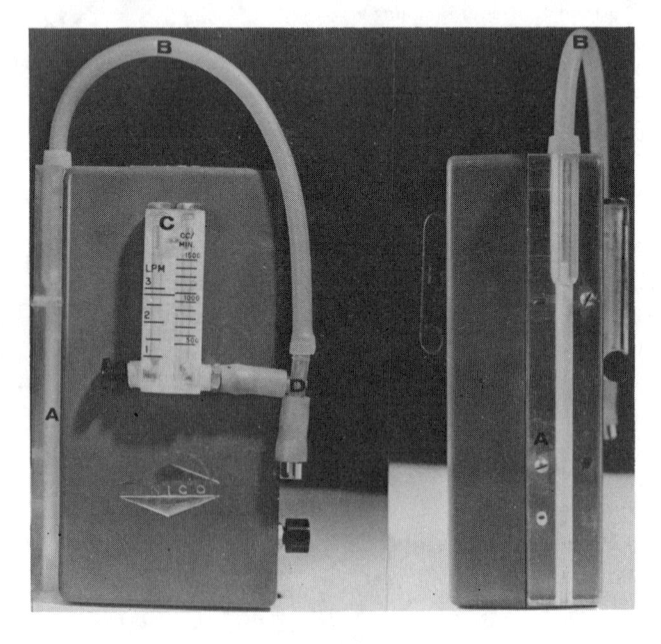

FIGURE I.31. The carbon monoxide personnel monitor. (From Linch, A. L. and Pfaff, H., *Am. Ind. Hyg. Assoc. J.*, 32, 745, 1971. With permission.)

TABLE I.16

Precision Limits for Calibration of Carbon Monoxide Length of Stain Detector Tubes

Variable	Number of observations	Average	Standard deviation at 95% C.L.
Reagent length (mm)	33	70	±5 mm
Flow rate through same tube for ±25 ml/min total flow rate (ratio = 0.017 ml/ml total)	10	8.0	±0.4 ml/min ±2.5 mm stain ±7 ppm
Total flow 1100±25 ml/min (0.737-mm orifice) through different tubes	24	8.0	±1.2 ml/min ±7 mm stain ±18 ppm
Stain length at 8.0 ml/min	5	28 ppm	±4 mm stain ±7 ppm

From Linch, A. L. and Pfaff, H., *Am. Ind. Hyg. Assoc. J.*, 32, 745, 1971. With permission.

reduced from 9.6 ml/min to 8.0 ml/min (Figure I.32). The effect of flow rate on the stain development rate was found to be critical in the 6.5 ml/min region (Figure I.33). Only the range 7 to 9 ml/min could be used for 4-hour sampling.

The criticality of concentration is clearly demonstrated by plotting stain length against concentration (Figure I.32), sampling rate (Figure I.33), or time (Figure I.34). The 30 ppm CO concentration region was found to be the lower operational limit in the construction of calibration charts in the range 0 to 100 ppm for 1-, 2-, 3- or 4-hour monitoring (Figure 32). Below 30 ppm the stain front progressed at a very slow rate and practically stopped after 2 hours (Figure I.34). Above 30 ppm the stain development velocity became linear and useable to the 100 ppm limit of the chart (Figures I.34 and I.35). The incremental increase in stain length vs. concentration (ppm/min = 2.8) was satisfactory for the short remaining range (25 mm) equivalent to 1/2 to 2 times the current TLV. Calibration charts prepared from the 1- and 2-hour calibration data indicated CO concentrations above 100 ppm could be accommodated, but this possibility was not explored further. Sharper delineation of the critical area between 20 and 30 ppm by additional calibration study might provide an extension of the useful region down to 20 ppm (dashed calibration line, Figure I.35).

Field Survey

A personnel sampling survey carried out in the Consolidated Warehouse on the 4–12 P.M. shift from September 11 to 22, 1970 when all ventilation sources were wide open disclosed no 4-hour periods during which the time-weighted average CO concentration exceeded 30 ppm. The "apparent" average was 7 ppm in a 5 to 11 ppm range. However, reliability in this region is equivocal. A similar survey over a 13-day period in early December with one exception also indicated less than 30 ppm CO in the breathing zone. One sample taken from 4 to 8 P.M. on December 8, 1970, by a fork lift operator's monitor while loading drums in a tractor trailer reached 50 ppm. The other 12 samples averaged 21 ppm in a range of 10 to 50 ppm apparent concentration. However, again the equivocal quality of all estimations below 30 ppm must be recognized.

DISCUSSION

Transitions in curves relating stain length to gas or vapor concentration had been reported previously.[85] Two entirely unrelated reagent systems for benzene and nitrogen dioxide are summarized in Figure I.25. Although the total time intervals are much shorter than four hours, the significance of these transitions for limiting the useful range of detector tubes for continuous monitoring is clearly implied. In each case these transition zones occurred beyond the usual limit for grab sampling – 5 minutes or less – imposed by the squeeze bulb or hand pump employed to draw the sample through the detector tube. Extension of the mathematical models developed by Saltzman[62] for short term sampling to periods beyond one hour undoubtedly will shed more light on the implications of the observations when more data become available.

Our failure to find conditions which would permit long-term monitoring for CO in concentrations below 25 ppm confirmed Ingram's earlier conclusion that currently available detector tubes for CO, SO_2, NO_2, and hexane do not possess sufficient sensitivity for use as personnel monitors in the range of 3 to 10 ppm under ambient air conditions.[72] However, Saltzman's calculations indicate possible success at higher flow rates since mass transfer rate appears to control these systems. Ingram's work indicated this rate must be greater than 12 ml/min.[72] Later studies at a much higher sampling rate (40 ml/min) for "a few hours" confirmed Ingram's conclusion that the state of the detector tube art does not support their use for the detection of air contaminants in the ambient air quality (non-industrial) concentration ranges, at least not for carbon monoxide and nitrogen dioxide.[108]

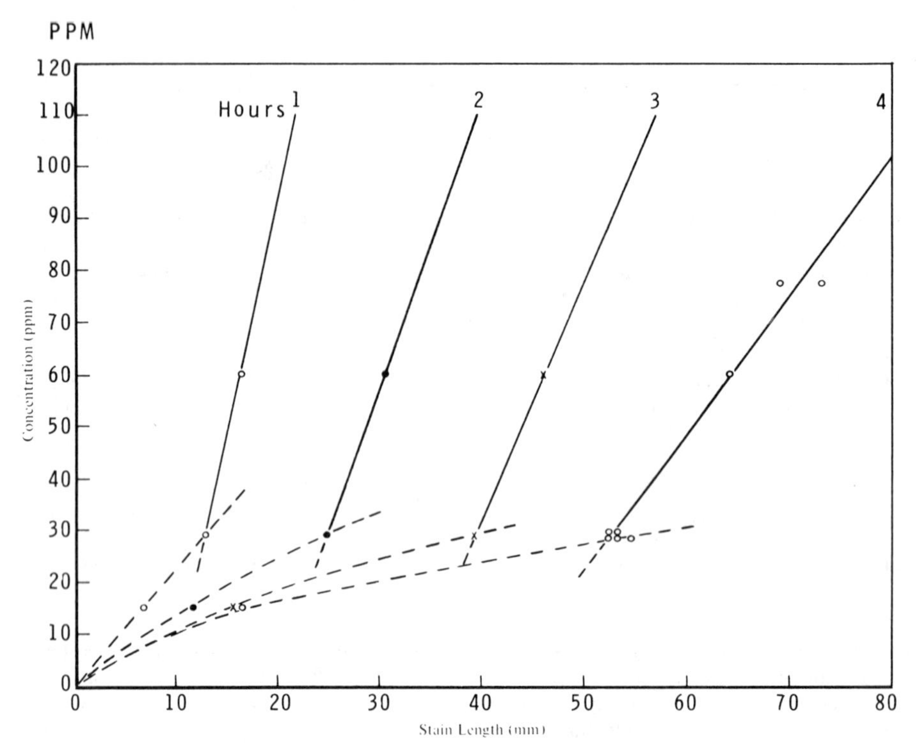

FIGURE I.32. The carbon monoxide in air calibration chart for personnel monitoring. (From Linch, A. L. and Pfaff, H., *Am. Ind. Hyg. Assoc. J.*, 32, 745, 1971. With permission.)

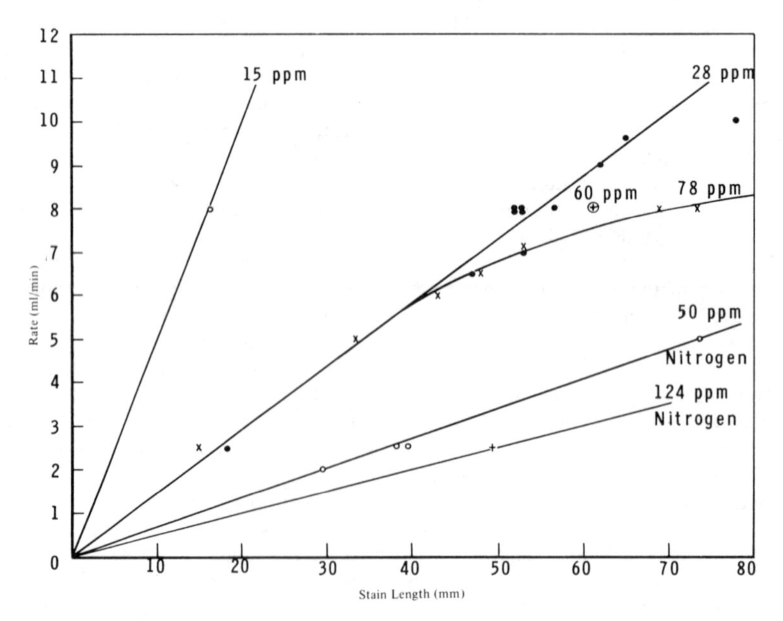

FIGURE I.33. Carbon monoxide in air – flow rate vs. stain length. (From Linch, A. L. and Pfaff, H., *Am. Ind. Hyg. Assoc. J.*, 32, 745, 1971. With permission.)

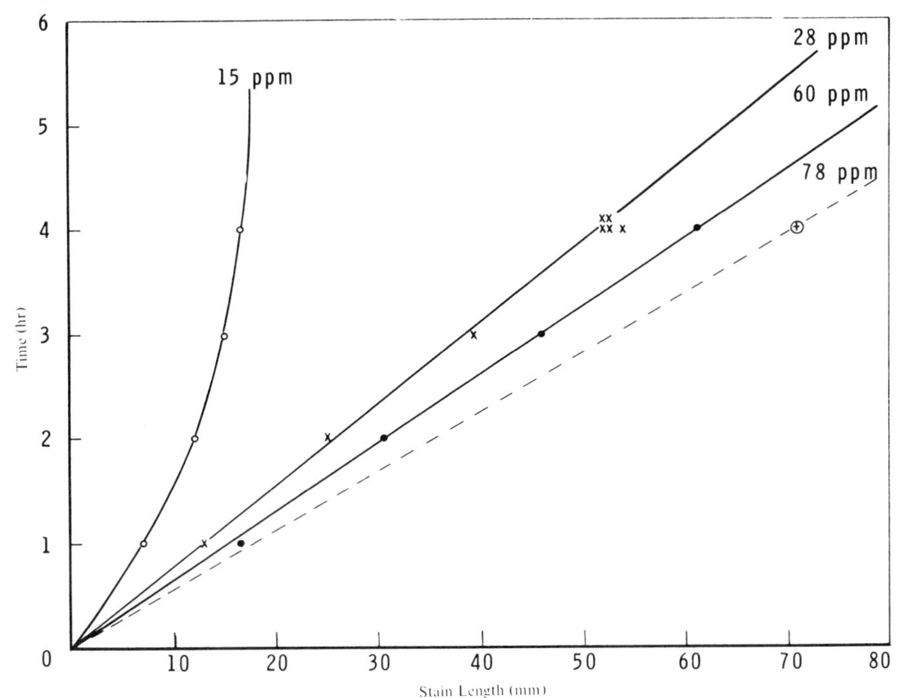

FIGURE I.34. Carbon monoxide in air – stain length vs. time. (From Linch, A. L. and Pfaff, H., *Am. Ind. Hyg. Assoc. J.*, 32, 745, 1971. With permission.)

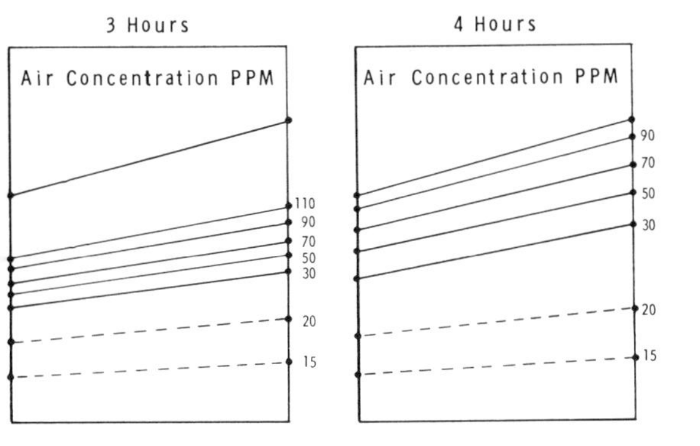

FIGURE I.35. Carbon monoxide in air calibration charts – Kitagawa® length of stain tube No. 100 (flow rate = 8.9 ml/min). (From Linch, A. L. and Pfaff, H., *Am. Ind. Hyg. Assoc. J.*, 32, 745, 1971. With permission.)

The black stain produced by CO reduction of the palladium sulfite complex to palladium metal on the silica gel support[50] can be bleached by atmospheric oxygen. This reversible property which provides the basis for recovery of the CO alarm detector tabs[52] may not be well recognized. Thus, yet another variable which is unique for CO is introduced into the length of stain detector tube system. If the two reaction rates converge, the overall effect would be catalytic oxidation to CO_2 without stain formation.

CONCLUSIONS

The length of stain carbon monoxide detector can be recalibrated for semi-quantitative application within restricted sampling limits as a personnel monitor for time intervals up to four hours. The flow rate must be maintained at 8.0 ± 1.2 ml/min to contain the standard deviation within a ± 18 ppm range. The useful range is limited to 30 to 100 ppm. However, this range does include the TLV within the range of one-half to twice the current 50 ppm limit.[96] With sampling rates in the 6.5 to 10 ml/min region, response of the reagent stain is not proportional to CO concentrations below 30 ppm.

Improved flow rate control would significantly reduce the range of analytical uncertainty. Part of this improvement must come from either reduction of pressure drop variability within the individual detector tubes, or introduction of a compensator within the air sampler system. Future studies should include simultaneous sampling and analysis in parallel with a standard method such as gravimetric Hopcalite[105] or the iodine pentoxide[104] procedure for "in place" calibration at a fixed station site in the highly variable warehouse atmosphere. For the first time the capability for at least semi-quantitative correlation of CO body burden with industrial environmental exposure in the TLV range has become available. Therefore, in future surveys, CO analysis of breath samples collected at the end of the 4-hour sampling period for correlation with the personnel monitor results should be included.

Surveys in a large Consolidated Warehouse served by

TABLE I.17

Carbon Monoxide Detector Tube Flow Resistance Calibration – Reagent Length vs. Flow Rate

By-pass capillary size (I.D.) = 0.737 mm
Values ( ) = Calibrated flow rate in mm/min vs. nominal calibration scale

Tube no.	Column length (mm)			Flow rate through pump (rotameter ml/min)			Flowmeter rate for 8 ml/min through tube	
	Reagent	Retainer	Total	(510) 700	(900) 1100	(1160) 1400	Nominal	Actual
1	68	9	77	4.0	9.6	13.8	1000	800
2	70	10	80	3.0	7.0	9.6	1200	980
3	74	10	84	–	6.1	8.6	1325	1100
4	72	14	86	–	6.3	8.6	1325	1100
5	68	10	78	3.7	9.6	13.8	1000	800
6	65.5	9.5	75	3.2	8.1	10.9	1125	920
7	76.5	9.5	85	3.6	8.6	11.9	1075	870
8	66	16	82	3.5	8.6	11.9	1075	870
9	69	12	81	3.5	8.6	11.3	1100	900
10	75	12	87	3.0	7.8	10.4	1175	960
Average	70.5	11	81.5	3.4	8.0	11.1	1140	930
Variance	±3.8				±1.2			

gasoline-powered fork lift trucks under restricted ventilation conditions imposed by cold weather confirmed CO levels which were in conformance with the current 50 ppm TLV.[96] A recording NDIR instrument centrally located in the general traffic area indicated transient concentrations equivalent to one and one-half hours continuous CO exposure above the TLV per 8-hour shift. Again, the necessity for personnel air sampling or biological monitoring for evaluation of personnel exposure in highly variable industrial environments has been vindicated.*

After the foregoing paper had been published, the study was reopened in an effort to improve the reliability of these carbon monoxide detector tubes for field use. First to be considered was the relationship between reagent length and pumping rate (total rate indicated by the pump rotameter with by-pass capillary in place – Figure I.31) to determine whether a chart could be developed for control of the detector tube flow rate at 8.0 ± 0.4 ml/min (± 5% volumetric error) based on reagent length. The results were disappointing. Reference to Table I.17 and Figure I.36 indicated an intolerable scatter of data. Obviously other factors, such as packing density and uniformity of particle size distribution, were overriding the granular column length as the prime source of variations in flow resistance. Pressure drop (differential) data are presented in Table I.18. Estimates of the magnitude of permissible variable tolerances are summarized in Table I.19.

The final solution to the sampling rate control problem was suggested by the development of a miniature precision glass rotameter designed for the range 3 to 30 ml/min airflow rate.[84] The original model was modified to fit into the carbon monoxide monitor previously described (Figure I.37).[153a] This innovation provided sample flow rate control within the ± 5% to 8.0 ± 0.4 ml/min goal specification.

The previous descriptions of constant rate sampling through detector tubes over prolonged periods of time have been limited to systems using battery power driven pumps to draw the air sample through the tubes. In at least one instance an evacuated rubber squeeze bulb was employed to create the vacuum needed to draw the air sample. The slow sampling apparatus which used a 120-ml deflated thick-walled rubber bulb was designed primarily to draw air through a trichloroethylene indicator tube at a constant rate over a period of 8 hr. The method depends upon the "reasonably" constant pressure exerted by the rubber wall irrespective of the degrees of deflation. Against a suction head of "about" 120 mm of mercury, a sample flow of 15 ml/hr was obtained through a 20-mm long x 0.05-mm bore (I.D.) glass

*Reproduced from Linch, A. L. and Pfaff, H., *Am. Ind. Hyg. Assoc. J.*, 32, 745, 1971. With permission.

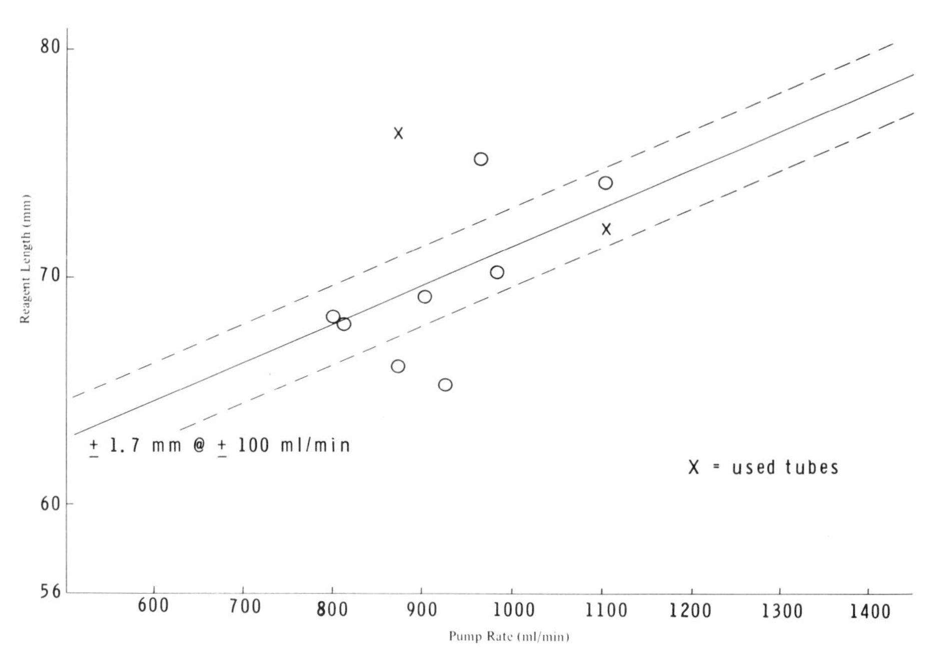

FIGURE I.36. Pump rotameter rate (calibrated) vs. reagent length. (By-pass capillary size (I.D.) = 0.737. mm; rate = 8 ml/min.)

TABLE I.18

Pump Rotameter Calibration

Bypass capillary size = 0.737 mm.

Nominal	Rate (ml/min) actual	Pressure drop – (cm H_2O)
1400	1160	34
1100	900	36
700	510	39

tube (polarography capillary). The capillaries were protected from dust by porous glass disks fitted to either end. The authors also found that longitudinal air "leakage" through a 100-mm length of 13-mm diameter electrode carbon rod sealed in a glass tube furnished a suitable sampling rate.[110]

9. Reagent Systems

In the foregoing sections only a few of the better known colorimetric reagents that have been developed to detect trace materials that could be present in the industrial atmosphere have been presented. Selection was based on availability, familiarity, extent of field application, published performance evaluations, and consensus standards. However, many potentially valuable detectors have been described in the trace chemistry literature. Campbell and Miller published a comprehensive review of this literature in the form of annotated abstracts.[109] Although this publication has long been out of print, the contents have been made available by the commercial press.[45]

A description of the preparation, calibration, and use of detector tubes for phenol, formaldehyde, SO_2, and NH_3 that might not ordinarily come to the attention of the field inspector would add two additional (phenol and formaldehyde) capabilities to his surveys.

The tubes are suitable for measurements up to 40°C, so that waste gases above this temperature must be cooled before being tested. Phenol, formaldehyde, sulfur dioxide, and ammonia are the substances of principal interest in connection with oven drying. Test tubes for phenol show a blue discoloration and have a measuring range up to 5 ppm. For formaldehyde, the discoloration is reddish, and the measuring range of the tubes is 2 to 40 ppm. The results obtained with this method in a plant for sheet metal packing materials coated with varnish, where oven drying and catalytic afterburning are included in the operation, are reported by Merz.[172]

10. Testing Gas and Vapor Detector Pumps

The critical factors governing the response of

TABLE I.19

Flow Rate Tolerances for Carbon Monoxide Detector Tube No. 100

Bypass capillary size (I.D.) = 0.737 mm
Tube flow rate = 8.0 ml/min

Variable	Magnitude	Result		Reference
		Measured	Magnitude	
Pump rate	±100 ml/min	Tube flow	±0.7 ml/min	Table 17 (Tubes 3 and 4)
			±1.5 ml/min	Table 17 (Tubes 1 and 5)
	±25 ml/min	Tube flow	±1.2 ml/min	53
Reagent length	±1.7 mm	Pump rate	±100 ml/min	Figure 36
Stain length	±2 mm	Concentration	±6 ppm	Figure 32 (at 50 ppm)
Stain length	±2 mm	Tube flow	±0.6 ml/min	Figure 33
Tube flow	±1.2 ml/min	Concentration	±12 ppm at 50 ppm	

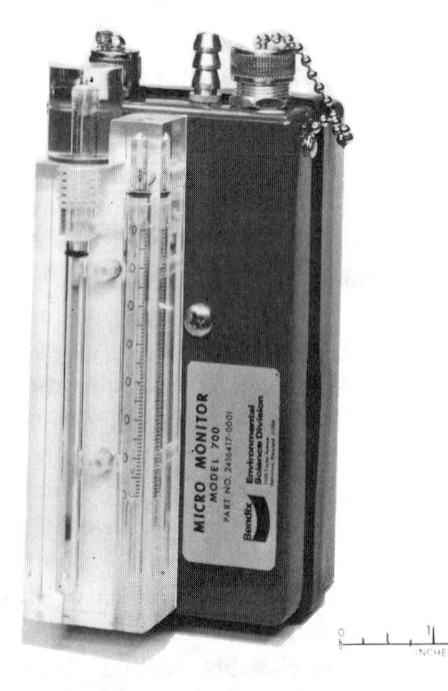

FIGURE I.37. Micro Monitor, Model 700. This model utilizes a Bendix-Gastec® reagent type gas and vapor detector tube which is inserted in a transparent holder mounted on a Micronair® sampling pump. Total exposure to a contaminant is clearly and instantly indicated by the length of color stain in the detector tube. More than 80 different gas detector tubes are readily available to monitor the TLV of most common industrial contaminants. This monitor also functions as a universal collector by sampling with a silica gel or charcoal-filled tube to collect a sample for subsequent analysis by gas chromatography or other suitable means. The micro monitor indicates accumulated exposure and features a built-in flow-meter and flow-adjustment valve. (Courtesy of Bendix Environmental Science Division, Baltimore.)

chemical detector tubes are (1) the volume of air sampled (contaminant mass, or concentration) and (2) the airflow rate (contact time with the detecting reagent). Therefore, the detector pump must be calibrated when procured and periodically thereafter. Three common types of pumps are available: (1) squeeze bulbs with or without limiting orifices,[62] (2) bellows-type pumps with limiting orifices (Figures I.12 and I.13), and (3) precision piston pumps with limiting orifices (Figures I.10, I.11, I.15, and I.16). Only types (2) and (3) are recognized as permissible by OSHA.

Two procedures are presented: (1) testing for leaks and (2) determining airflow characteristics by measuring the volume of air sampled and the flow rate. In leak testing, the manufacturer usually describes the recommended procedure in his instructions. In general, to test for leakage, a sealed detector tube is placed in the pump inlet port and the time to reinflate the bulb or bellows, or to reach atmospheric pressure with a piston pump, is determined.

The tentative "Detector Tube Certification Schedule" proposed by the National Institute for Occupational Safety and Health delineates pump performance as follows:

> (1) Aspirating pumps for drawing the air to be sampled with detector tubes shall be free from leakage which would result in erroneously low tube readings, and shall be calibrated by the manufacturer to sample at whatever flow rate is deemed appropriate by the manufacturer in order to assure accurate measurements using the respective manufacturer's detector tubes.

After the pump has been evacuated, leakage shall not be more per minute than 2% of the pump's volume capacity in a single stroke, if a sealed detector tube is used to plug the pump inlet.

The flow rates which are set by the tube manufacturer are regulated by limiting orifices or other restrictions of flow at the pump inlet. These flow control devices shall regulate the flow rate to within ± 5% of the rate stated by the manufacturer for each flow rate control device. The choice of the flow rate into the pump is the manufacturer's prerogative.

Leakage tests and flow rate tests shall be made at room conditions of temperature, pressure, and humidity.

Pumps shall be calibrated by the manufacturer to insure that they are capable of sampling accurately the stated volume at the stated flow rate. Subsequent to a check of proper flow rate and volume, the pump shall be capable of drawing 100 full-capacity strokes of air without deviating more than ± 5% from the calibration flow rate.

(2) The pump shall be capable of drawing the volume stated by the manufacturer to within ± 5% of said volume. The pump shall be able to continue to draw this volume throughout its normal working lifetime, with minor repairs only.

(3) The pump shall be capable of drawing ten consecutive strokes using one detector tube with no more than ± 5% error from the theoretical sampling volume.

(4) The manufacturer shall provide printed instructions with each detector tube pump so that users may verify by a simple field test the absence of leakage and capability for accurate flow rate sampling. The manufacturer shall also provide instructions for making simple adjustments to the pump and replacing minor components.*

In the field the test may be carried out rapidly after inverting an unopened detector tube in the inlet port of the piston pump by pulling the handle back completely, locking in position by rotation of the handle, and then 3 min later releasing the piston cautiously. The piston should return to within 5% of its original totally advanced position, i.e., out no more than 5% of original withdrawn position. In a 100-ml pump this would be equivalent to 5 ml leakage in 3 min. If the residual volume exceeds 5 ml, replacement of check valves, tightening connections (especially the inlet port-limiting orifice assembly), replacement of tube connections, greasing piston, or advancing the piston expansion nut (Kitagawa pump) is indicated. This test should be performed prior to each use of the device.

In normal use, the pump should be taken apart, cleaned and re-lubricated after about 100 to 200 samples have been drawn. To disassemble, unscrew the cylinder end cover and slowly pull the pump handle until the piston leaves the cylinder. Remove the inlet and pressure-relief valve from the front end of the pump, and clean all the bearing surfaces carefully using non-abrasive wipers and a good degreasing solvent such as trichloroethylene or perchloroethylene. Lubricate all surfaces which make metal-to-metal contact with the special silicone grease supplied with the pump. Reassemble the pump carefully, making sure that abrasive particles which could score the walls do not get inside the pump.

If excessive leakage is found, it usually takes place either at the pump inlet, or between the piston and cylinder walls. The latter source of leakage is much less likely than the former, and can usually be eliminated simply by cleaning and relubrication. Leakage at the inlet may result from a poor seal between the detector tube and the rubber inlet tip, or between the flange of the rubber inlet and the pump body. To check for leakage at the tip, reposition the sealed detector tube or replace it with another one, and repeat . . . If the leakage persists, it is probably at the second location, and simply tightening down the inlet clamping nut may suffice to eliminate it. If not, remove the rubber inlet and examine it for cracks and/or foreign matter on its bearing surface. If it is clean and undamaged, replace it in the pump inlet, and repeat the leak test. If necessary, replace the worn rubber inlet with a fresh one.**

In some instances, the manufacturer's instructions include measurement of flow rate and the volume of air sampled by their equipment. These procedures may furnish only a rough estimate of performance. A more accurate method of measuring these parameters, a simple procedure employing a soap bubble flowmeter, follows. This calibration should be performed when the equipment is procured and at frequent intervals during use thereafter. When experience has been gained with a particular detector kit, the user will recognize changes in performance that indicate a need for rechecking sampling characteristics.

Basic equipment required consists of a 100-ml laboratory burette, several short pieces of rubber and glass tubing, a solution of synthetic liquid detergent in water, and a stopwatch. Since this

*From Roper, P., Detector Tube Certification Schedule, first draft, NIOSH, Health Services and Mental Health Admin., Public Health Service, Department of Health, Education, and Welfare, Cincinnati, O., June 1971; *Fed. Register,* 37(No. 84), 19643, September 21, 1972; 42CFR, part 84.

**From Kohn, R. J., Unico® Carbon Monoxide Detector Kit (Model 400 Precision Gas Detector Pump with Model 100 Carbon Monoxide Detector Tubes), National Environmental Instruments Corp., Box 590, Pilgrim Station, Warwick, R.I., 1971. With permission.

apparatus will be used, in some cases, to measure air volumes slightly greater than 100 ml, it is necessary to increase the scale on the burette by 6 to 8 ml in increments of 1 ml. This is done by filling the burette with water to the 100-ml mark, adding successively 1 ml of water from a 1-ml volumetric pipette, and marking the burette after the addition of each 1-ml portion.

A solution of synthetic liquid detergent and water is prepared, and the foam is allowed to subside. Soap films may be stabilized by adding 10% glycerine and 3% gelatin to the soap (1%) or synthetic detergent. The solution is poured into the burette carefully to prevent additional foaming, and then poured out in a manner that will coat the inner surface of the burette with a thin film of the solution. The open end of the burette is placed momentarily in contact with the surface of the detergent solution. When the burette is withdrawn, a soap film will adhere to the mouth of the burette. Upon the application of suction to the tip of the burette, the soap bubble can be drawn through the length of the burette. In the calibration procedure, start the bubble at a reference point, draw air through the burette by means of the detector pump being tested, and note the total volume after the bubble has come to rest. The filling time of the pump can be determined also by measuring with a stopwatch the time it takes the bubble to move from the reference point to the rest position. Further details have been described in the technical literature;[83a] and several commercial sources supply both the glassware and complete kits in size ranging from 2 ml to 2,000 ml (Figures I.38 and I.38a)[83b-83f]

It can be determined from the manufacturer's instructions whether or not the detector pump should be calibrated with an open detector tube in place. In general, pumps equipped with limiting orifices may be calibrated by the soap bubble technique without a detector tube. Where the rate of airflow is governed only by the resistance to flow of the detector tube, the device should be calibrated with the tube in place. With some pumps, it will be necessary to use short pieces of glass and/or rubber tubing to connect the air intake of the device to the burette. In some devices the major resistance to the airflow is in the chemical packing of the tube (Figures I.12 and I.13). Therefore, each batch should be checked before use. Error in the flow rate indicates a

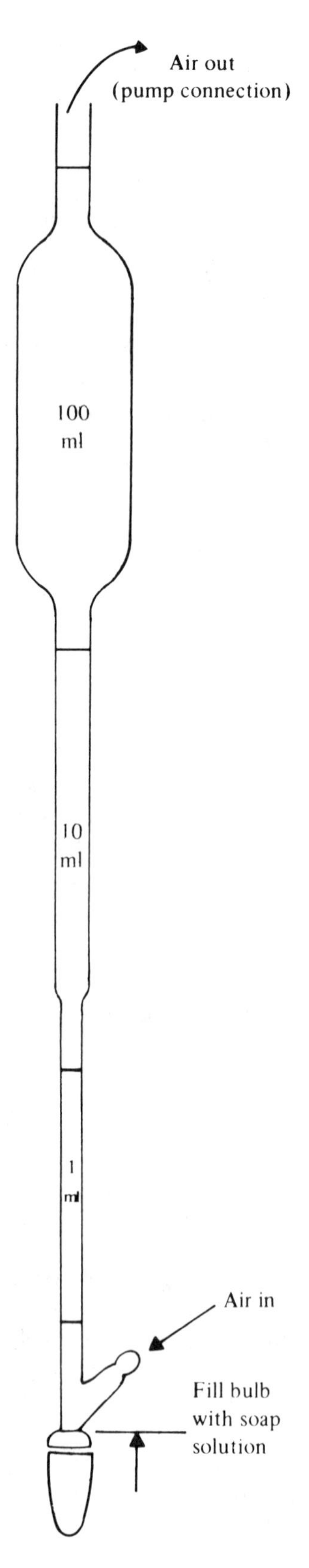

FIGURE I.38. Soap film flowmeter – representative design.[83e]

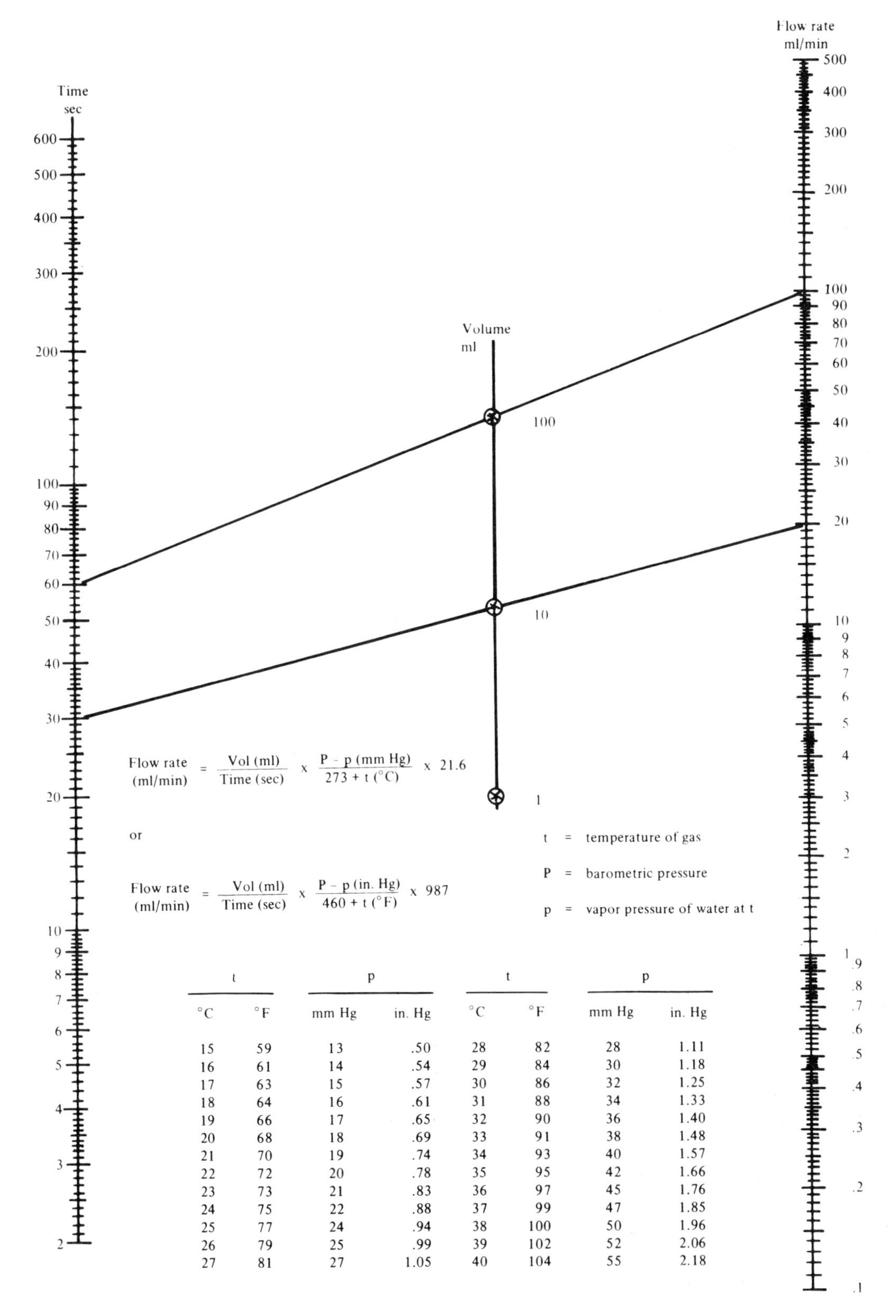

$$\text{Flow rate (ml/min)} = \frac{\text{Vol (ml)}}{\text{Time (sec)}} \times \frac{P - p\ (\text{mm Hg})}{273 + t\ (^\circ C)} \times 21.6$$

or

$$\text{Flow rate (ml/min)} = \frac{\text{Vol (ml)}}{\text{Time (sec)}} \times \frac{P - p\ (\text{in. Hg})}{460 + t\ (^\circ F)} \times 987$$

t		p		t		p	
°C	°F	mm Hg	in. Hg	°C	°F	mm Hg	in. Hg
15	59	13	.50	28	82	28	1.11
16	61	14	.54	29	84	30	1.18
17	63	15	.57	30	86	32	1.25
18	64	16	.61	31	88	34	1.33
19	66	17	.65	32	90	36	1.40
20	68	18	.69	33	91	38	1.48
21	70	19	.74	34	93	40	1.57
22	72	20	.78	35	95	42	1.66
23	73	21	.83	36	97	45	1.76
24	75	22	.88	37	99	47	1.85
25	77	24	.94	38	100	50	1.96
26	79	25	.99	39	102	52	2.06
27	81	27	1.05	40	104	55	2.18

FIGURE I.38.a. Nomograph for computing flow was using soap film flowmeter. Nomograph gives flow rates at ambient temperature and pressure, uncorrected for the vapor pressure of water. To correct observed values to standard conditions of temperature (0°C, 32°F) and pressure (760 mm Hg, 29,92 in. Hg), use either of the equations and the table of vapor pressures.

partially clogged strainer or orifice which should be cleaned or replaced.

The flow rate for bellows and squeeze bulb type pumps can be checked expediently in the field simply by timing the period for complete expansion. The rate is quickly calculated by dividing the known volume by the time observed. Alternatively, a used detector tube may be mounted in the holder and connected by a short piece of rubber tubing to a low flow, precision rotameter, and the air sample then drawn through the assembly. The rate than is read from the calibration chart furnished with the flowmeter. One particular model (Visitrol®) is provided with 6 interchangeable tubes to cover the range 5 to 140,000 ml/min with stainless steel, or borosilicate glass floats. The unit may be rotated 180° to permit reversal of flow through the system. Precision has been rated at ± 5%.[84]

With the piston-type pump, a field check for flow rate can be made as follows:

a) With no detector tube in the pump inlet, pull the handle all the way out and lock it.
b) Wait exactly 30 seconds. Release the pump handle.
When the handle is released, it should spring back part way toward its initial position. It will come to rest at an intermediate position at which the pressure on both sides of the piston will be equal. If the pump is operating properly, and the orifice at the inlet is free of obstructions, approximately 35–40cc of air should enter the pump in 30 seconds. The volume that entered can be read from the markings on the pump handle. If less than 35cc was drawn in, it will be most likely due to the orifice being blocked by a stray particle or fiber. It will usually be possible to dislodge this foreign matter simply by flushing the pump several times with air. If the obstruction remains, the orifice must be taken out and be cleaned or replaced with a clean one.
The leakage and flow-rate tests just described should be performed whenever the pump is used after an extended idle period, and in routine use should be repeated after every 10–12 samples to insure accuracy and reliability.*

Restriction in either the orifice or the detector should be checked. The first can be determined by retracting the piston and locking without a detector tube in place. When the handle is rotated off of locking position, no residual vacuum should remain, i.e., the piston will not voluntarily move back into the pump (less than 100 ml sample indicated on the shaft) after 3 min. A defective indicator tube which has high internal pressure drop will exhibit the same residual vacuum after 3 min. For example, a CO tube under test indicated that only 90 ml of a 49 ppm CO in air standard mixture had been drawn through the tube. The indicated concentration was 30 and 35 ppm instead of the 45 to 50 ppm found with other tubes of the same lot.

Pumps should be considered defective if either of the following conditions is found: leakage that cannot be corrected by the user, following the simple procedures prescribed by the manufacturer; or deviation of the airflow characteristics by more than ± 5% from the manufacturer's specifications. New pumps that are found defective should be returned to the manufacturer in accordance with the terms of the warranty.

The use of factors to adjust observed air volumes or rates of airflow to the values stated by the manufacturer should not be considered. This manipulation assumes a linear relationship between air volume and/or rate of airflow and the response of the detector tube. This assumption is either not valid at all, or is applicable only over a narrow range of operating conditions that must be determined by recalibration.

B. COLLECTION OF CONTAMINANT FOR LABORATORY ANALYSIS

Direct reading visual comparators such as indicator papers, chemical dosimeters, and detector tubes provide an unquestionably valuable service in the hands of the qualified industrial hygienist for obtaining immediate semiquantitative measurements from screening surveys. However, as previously emphasized (see Section I.A.2.d.6.h, "Specificity and Interferences"), the preliminary survey should be followed by a quantitative evaluation, especially when results fall within the one half to one and one half TLV range, or when the results will provide the basis for assigning a TLV to a hazardous substance. If a mixture of similar compounds is known to be present, or interferences are anticipated, separation procedures must be carried out in the laboratory to minimize errors inherent in nonspecific methods. Furthermore, by reserving a portion of each field

*From Kohn, R. J., Unico® Carbon Monoxide Detector Kit (Model 400 Precision Gas Detector Pump with Model 100 Carbon Monoxide Detector Tubes), National Environmental Instruments Corp., Box 590, Pilgrim Station, Warwick, R.I., 1971. With permission.

sample for later analysis, constituents other than those initially investigated can be determined without returning to the location to secure additional samples. Also, the validity of the study can be improved by replicate analysis by the same or different methods (see Chapter IV). Laboratory analysis of collected samples minimizes the loss of valuable data missed when the direct reading instrument or spot detector responds erroneously or not at all. Most analytical laboratory methods quantitate color densities in a spectrophotometer, thereby eliminating subjective interpretation, a major source of analytical error. Sensitivity can be greatly increased through concentration of the target component, or by application of sophisticated instruments that are not amenable to field application (gas chromatography, thin layer chromatography, ultraviolet or infrared spectroscopy, atomic absorption spectroscopy, etc.).

The most obvious solution to the complete analysis problem would be to transport a sample of the atmosphere itself back to the laboratory. However, as applied to personnel monitoring, this technique is at best impractical due to the large volumes required for most analyses. In some few cases where sufficient sensitivity is available to permit the analysis of small sample volumes, conceivably a sample could be collected over a period of hours by a fine capillary "bleed" into a small evacuated rigid vessel (glass tube) or by slowly evacuating a rigid container in which a plastic bag (bladder) is sealed by means of a bulkhead capillary fitting. The sample would be drawn as the vacuum created within the rigid container expanded the bag.

With the exception of a rubber squeeze bulb restricted by a limiting orifice to draw an air sample through a trichloroethylene detector tube at a constant 15 ml/min rate over an 8-hr period,[110] equipment of this type is not commercially available, nor has it been described in the literature. Furthermore, the relatively large surface to volume ratio greatly increases the errors due to adsorption and chemical interactions of the air contaminants on the container walls, and losses from leaks.

1. Impinger Systems – Liquid Reagents

a. The Midget Impinger

The alternative to transporting the entire air sample is to collect the contaminants from a known volume of the atmosphere in a small volume of liquid absorbant or on a solid adsorbant. One of the most widely used and thoroughly studied collection devices suitable for application to personnel monitoring is the midget impinger. The essential specifications and design features of a currently available model are shown in Figures I.39 and I.40. The original model (1938) employed a rubber stopper to retain the impinger tube,[112] and an all-glass counterpart in which a ring seal replaced the rubber stopper, and was introduced commercially shortly thereafter (1944).[113] A Teflon® (Registered U.S. Patent for Du Pont's fluorocarbon resins and film) sleeve

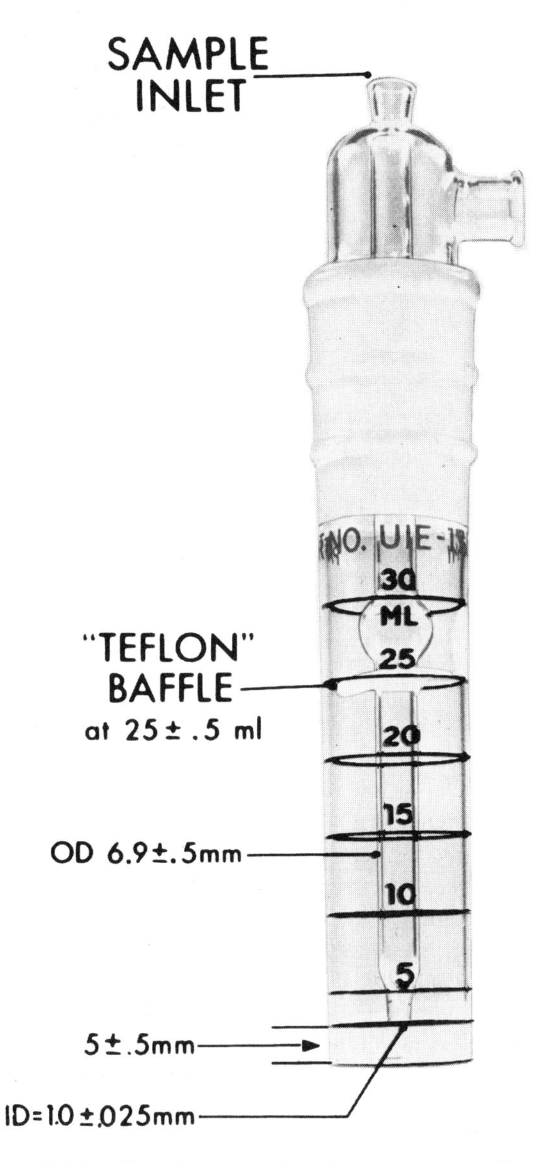

FIGURE I.39. The standard midget impinger specifications. (From Linch, A. L., Wiest, E. G., and Carter, M. D., *Am. Ind. Hyg. Assoc. J.*, 31, 170, 1970. With permission.)

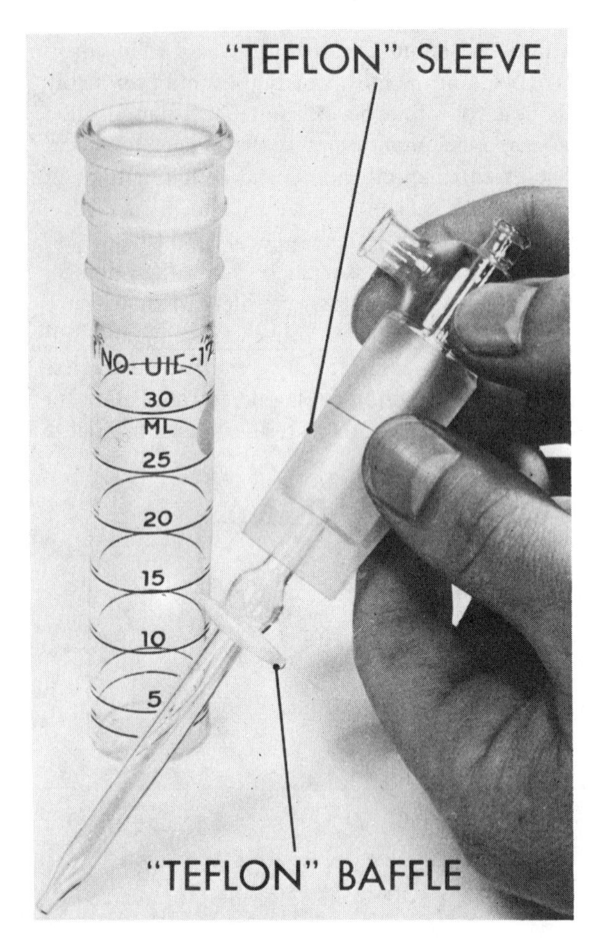

FIGURE I.40. The standard midget impinger – exploded view. (From Linch, A. L., Wiest, E. G., and Carter, M. D., *Am. Ind. Hyg. Assoc. J.*, 31, 170, 1970. With permission.)

fitted to the inner member of the joint as shown in Figure I.40.[93] The elimination of contamination derived from reagent contact with either the rubber stopper or the joint lubricant is especially critical if the analytical procedure is based on ultraviolet or infrared spectrometry.

Although the midget impinger was designed primarily for aerosol (dust and liquid) collection, its application to gases and vapors has been well documented.[115,116] Collection efficiency in the range of 85 to 95% is usually attained, and may reach 95 to 99% if the gas or vapor to be collected is highly soluble in the collecting liquid or if the gas reacts rapidly with a component of the liquid reagent, for example, ammonia collected in dilute sulfuric acid. The major factors that influence this system are[117]

a. Solubility of the contaminant in the collecting liquid.

b. Rate of diffusion of contaminant into the liquid.

c. Vapor pressure of contaminant at the sampling temperature.

d. Chemical reactivity of the contaminant with the reagent (including hydrogen bonding as in collecting acetone in chloroform).

e. Contact area (size of air bubble) and time of contact (flow rate) of the contaminated air through the liquid reagent.

f. Volatility of the collecting liquid.

g. Analytical techniques to be applied after the sample has been collected.

Without a reagent to react with the gas in the liquid medium, the rate of absorption may be slower, and a portion of the gas that does dissolve may be swept out by subsequent aeration (escape factor). The ratio of the air sample volume to the volume of collecting liquid is a controlling variable in collection efficiency, and beyond the critical limit of this ratio the efficiency declines significantly. For acetone in water, the ratio would be approximately 51 ml air per 20 ml water distributed between two "bubblers" connected in series. Therefore, if an attempt were made to collect a 20-l air sample in this system at room temperature, the collection efficiency would be very poor, in fact, unacceptable. Methylethylketone (MEK) would be even more poorly absorbed than acetone in water.[118] In some cases this escape factor can be calculated.[119]

The close-tolerance specifications (Figure I.39) required to maintain a reproducible flow rate of 2.8 l/min at 305 mm pressure drop (Δ P) across the impinger nozzle and an impinging velocity of 60 m/sec intended for dust collection are not critical for gas and vapor collection. In fact, excessive liquid entrainment at the 2.8 l/min rate may reduce the collection efficiency.[120] Considerable latitude in flow rate (0.56 to 5.6 l/min) without sacrificing more than 5% efficiency is permissible[116] unless the limiting orifice effect of the nozzle is required to maintain a reproducible, constant flow rate on which to base sample size. In systems provided with a calibrated rotameter and a timing device to calculate the total volume collected, the impinger nozzle must be matched with the pump assembly or the geometric tolerances held within limits consistent with the established limits of sampling error.[93,114] In any event, the distance from the orifice to the cylinder base is not at all critical for gas or vapor collection.

Two impingers connected in series will increase the overall collection efficiency when the rate of absorption is slow, or higher sampling rates are desired. In developing a collection method, series collection always should be employed to establish the average and range of collection efficiency. The amount collected in the first midget impinger divided by the total collected multiplied by 100 equals % efficiency. If the amount collected in the second impinger is greater than 10%, then the performance cannot be considered satisfactory. The added Δ P introduced by the second impinger nozzle must be taken into consideration when calculating the sample volume. The Δ P effects are not directly additive; therefore, the most prudent approach is to calibrate volumetrically with all components of the system in place.[121] The following section is included here by permission of the copyright holder.

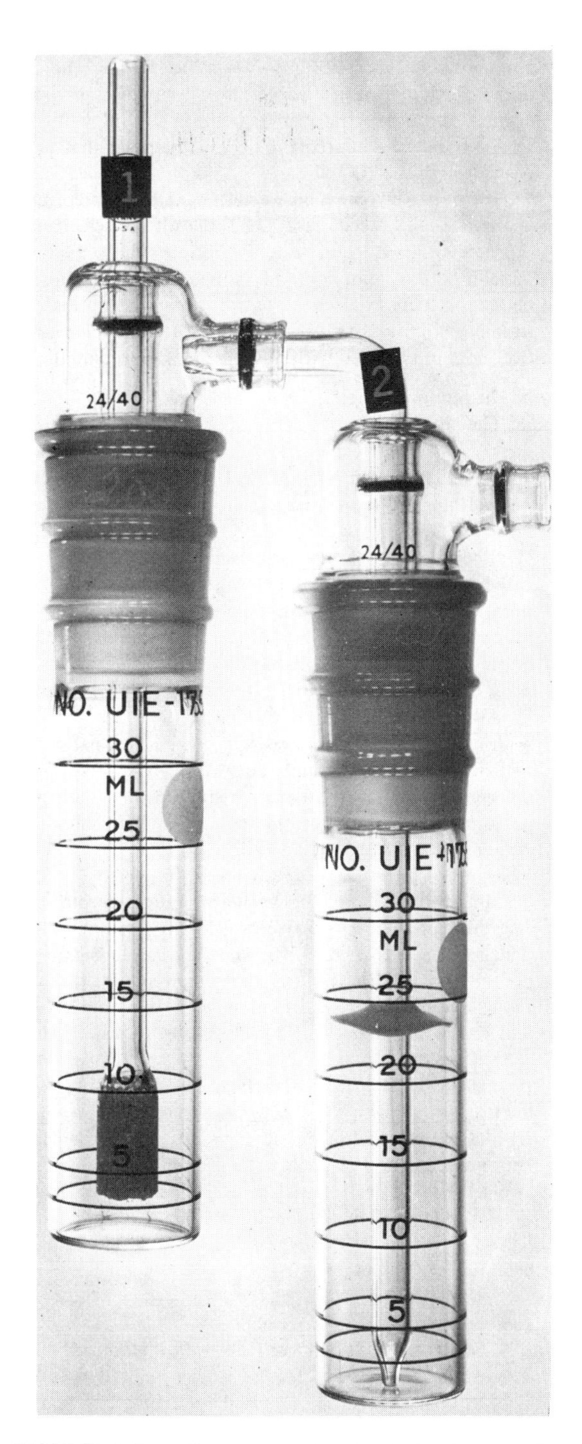

FIGURE I.41. The improved midget impinger and bubbler in series. (From Linch, A. L. and Corn, M., *Am. Ind. Hyg. Assoc. J.*, 26, 601, 1965. WIth permission.)

Replacement of the rubber stopper in the original midget impinger by an all-glass ring seal to support the impinger tube did eliminate undesirable contact of reagent with rubber, reduced spray carryover and provided more convenience in assembling, but at the cost of flexibility in choice of orifice size or replacement of damaged nozzles. The introduction of elastic "O" ring seals for support of glass mercurial thermometers in standard-taper holders offered an alternative design which would retain the advantages of an all-glass system with the flexibility of the original model.

The fragile glass ring seal in the midget impinger head was replaced with a pair of "O" ring seals in an indented tube to provide a flexible support for the impinger tube (Figure I.41). The distance between the orifice and the cylinder base is adjustable by sliding action, but lateral movement of the orifice is prevented by spacing the "O" rings, which limits motion to rotation and longitudinal movement. Tubing to be connected to these seals is slightly tapered and conservatively lubricated – usually water is sufficient – and then inserted with a rotating motion to prevent displacement of the "O" ring from its channel. The optional porous glass bubbler (Figure I.41) illustrates one of many possible alternative inlet tube forms which can be substituted for the easily removed impinger tube. Because the distance between orifice and cylinder base was adjustable, and impinger orifices were ground within tolerances that permit interchange without preliminary calibration, the matching of midget impinger components was not necessary for reproducible performance. The rapid replacement of damaged inlet tubes or changeover to a different orifice size not only reduced cost but also facilitated field sampling. Tolerances required for proper fit in the inlet tube "O" ring seal (tube outside diameter, 6.4 to 6.5 mm; "O" ring retainer inside diameter, 7.0 to 7.1 mm) did not pose a problem for competent glass shops.

The addition of a third "O" ring in the outlet tube provided essentially glass-to-glass connections for series assembly or for direct coupling to other collection components (Figure I.41), without sacrificing the freedom associated with flexible tubing connections. However, addition of a right-angle bend in the inlet tube for series connection limited the tip to a diameter that would pass through the "O" ring.

The relatively inexpensive impinger tubes were interchanged or replaced easily and permitted selection of orifices within Δ P-flow rate tolerances which made the matching of assemblies and individual calibrations unnecessary. Tubes fabricated with a variety of orifice sizes, porous glass bubblers, spiral construction or other modifications could be used with the same head and cylinder.

b. The Microimpinger

The recent development of a self-powered aspirator[93,114] and battery-powered pumps[96,107] small enough to be worn as personnel monitors by workmen during their routine job assignments created an immediate need for further miniaturization of the midget impinger. Consideration of several proposed designs[122,123] indicated that consolidation of desirable features and incorporation of "O" ring seals could provide a versatile microimpinger one third the size of the standard midget impingers, without sacrificing the performance characteristics of the original impinger design.

Although every effort was made to scale down the midget impinger on the basis of critical design parameters – that is, air velocity and pressure drop – this was not considered sufficient basis for assumption of equivalent performance. Therefore, the collection efficiencies for aniline vapor, SO_2 gas, and silica dust were selected as measures of probable field utility.

Miniaturization of the standard midget impinger to one third of the original size resulted in a microimpinger which operated at one-fifth of the sampling rate and consumed one-fifth of the reagent required with the larger model (Figure I.42). Since all physical parameters were scaled down by one-fifth, it is important to note that it is not necessary to recalibrate or to revise analytical procedures established for the standard midget impinger.

In our design, "O" ring seals provided the same flexibility and interchangeability features described for the revised midget impinger.[113] Replacement of the conical ground orifice with a short length of precision-bore capillary tubing eliminated the need for adjustment of the orifice to provide matching impinger assemblies.[107]

Calibration

The diameters of the orifices were measured with a microscope equipped with a low-power objective (8x) and a 10x eyepiece with a calibrated graticule. The body of the microscope was tilted to the horizontal position, and the substage condenser was removed to permit insertion of a one-hole rubber stopper to permit the operator to focus on the orifice plane. The measured capillary diameters agreed with the manufacturer's values to within the error of measurement. From these results, air velocity in the nozzle was calculated for flow rates of 0.01, 0.02, and 0.03 cfm:

where

$$V = 0.0283\, q/60A$$

V = air velocity, meters per second (m/sec).
q = air flow rate, cubic feet per minute (cfm).
A = orifice area, square meters (m^2).

Relationships of pressure drop (Δ P) to flow rate for orifices of different diameters were determined with an open-end water manometer inserted between the vacuum source control valve and the vent of the impinger. A dry test meter was connected to the impinger tube (American Meter Company, Model 2-40) to measure air volume. All runs were made with 10 ml of water in the midget impinger or 2 ml of water in the microimpinger.

Figure I.43 illustrates the nozzle velocities obtained with some commercially available capillary tips. Figure I.44 summarizes the measured pressure drops across the microimpinger with three different capillary tips. The reproducibility of flow characteristics with different impingers made it possible to operate simultaneously two or more units in parallel without the need for individual flow meters, control valves, or volume correction. As many as five microimpingers have been operated in parallel without increasing the vacuum capacity required for a single midget impinger.

At rates as high as 0.04 cfm, physical entrainment of the absorber liquid did not create a problem if care was taken to exclude detergents, greases, and other foam-generating materials. Consistently good results were obtained by soaking units in 1% aqueous trisodium phosphate in water for several hours and following this with 1:1 nitric acid and distilled water rinse.

EFFICIENCY EVALUATION

Aniline Vapor

Sampling. Two to three drops of reagent-grade aniline were dropped into a 5-gallon "Pyrex" glass bottle fitted with a rubber stopper covered with aluminum foil. Two sampling lines and a glass sleeve for the shaft of a "Teflon" bladed agitator were inserted through the stopper. The air in the bottles was agitated until the aniline was completely vaporized and then for an additional hour to ensure thorough mixing. Two impingers connected in series, each containing 10 ml of spectrograde ethanol and chilled in an ice pack, were used for sampling.

Recovery of aniline in alcohol was more consistent both in the laboratory and in the field than was recovery in dilute (1% or 10%) aqueous hydrochloric acid. Therefore, alcohol was adopted as the collecting reagent of choice for both aromatic amino and nitro compound collection from air samples. Air flow through the midget impinger was held at the 0.1-cfm rate which Roberts and McKee found to be the mid-point of maximum collection efficiency.[116] The contents of the impingers were analyzed separately to determine efficiency of collection by the first impinger. Aniline was determined as follows:

REAGENTS

Hydrochloric Acid, 10%. Dilute 200 ml of analytical reagent grade 37% hydrochloric acid with 540 ml of distilled water.

Chicago Acid. Dissolve 0.5 gm of 8-amino-1-naphthol-5, 7-disulfonic acid in 1 ml of 37% hydrochloric acid and 45 ml of distilled water, and dilute to 50 ml with distilled

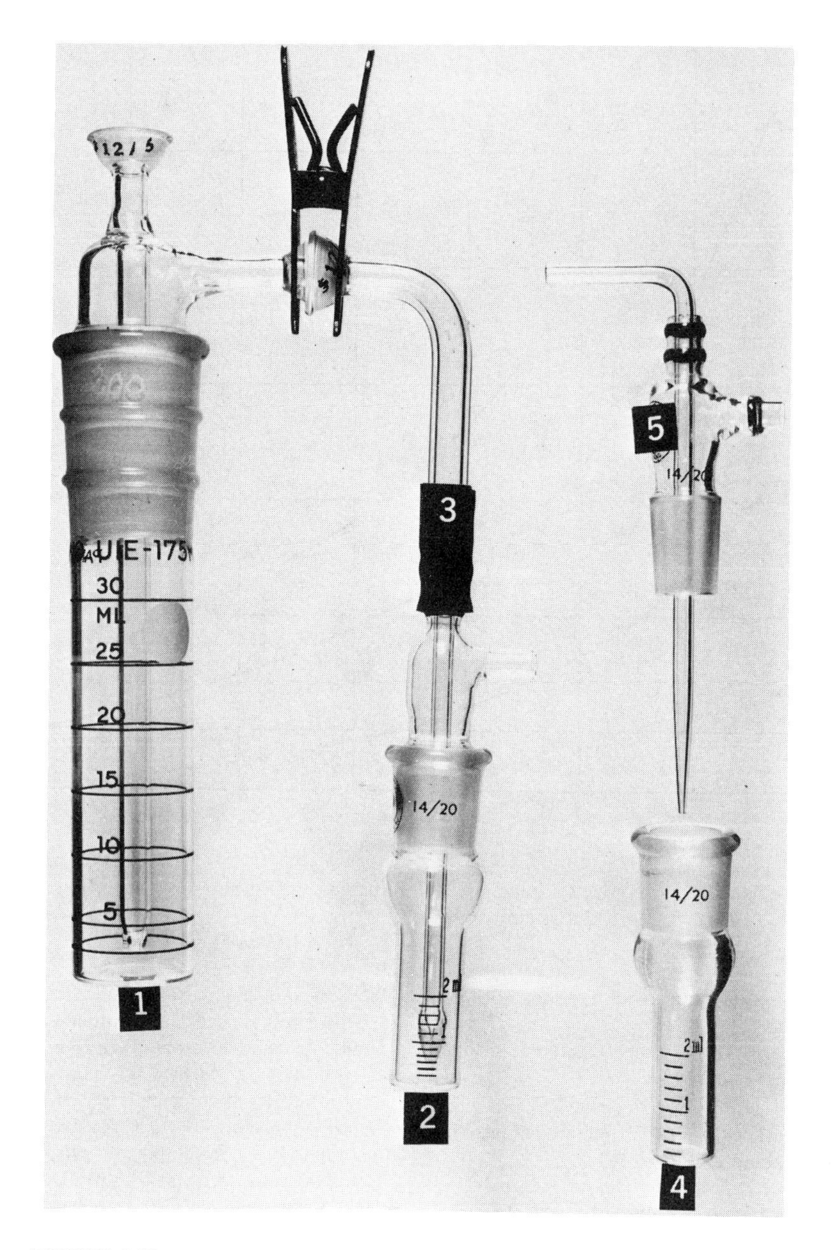

FIGURE I.42. The standard midget impinger vaporizer coupled to the micro-impinger gas absorber. (From Linch, A. L. and Corn, M., *Am. Ind. Hyg. Assoc. J.*, 26, 601, 1965. With permission.)

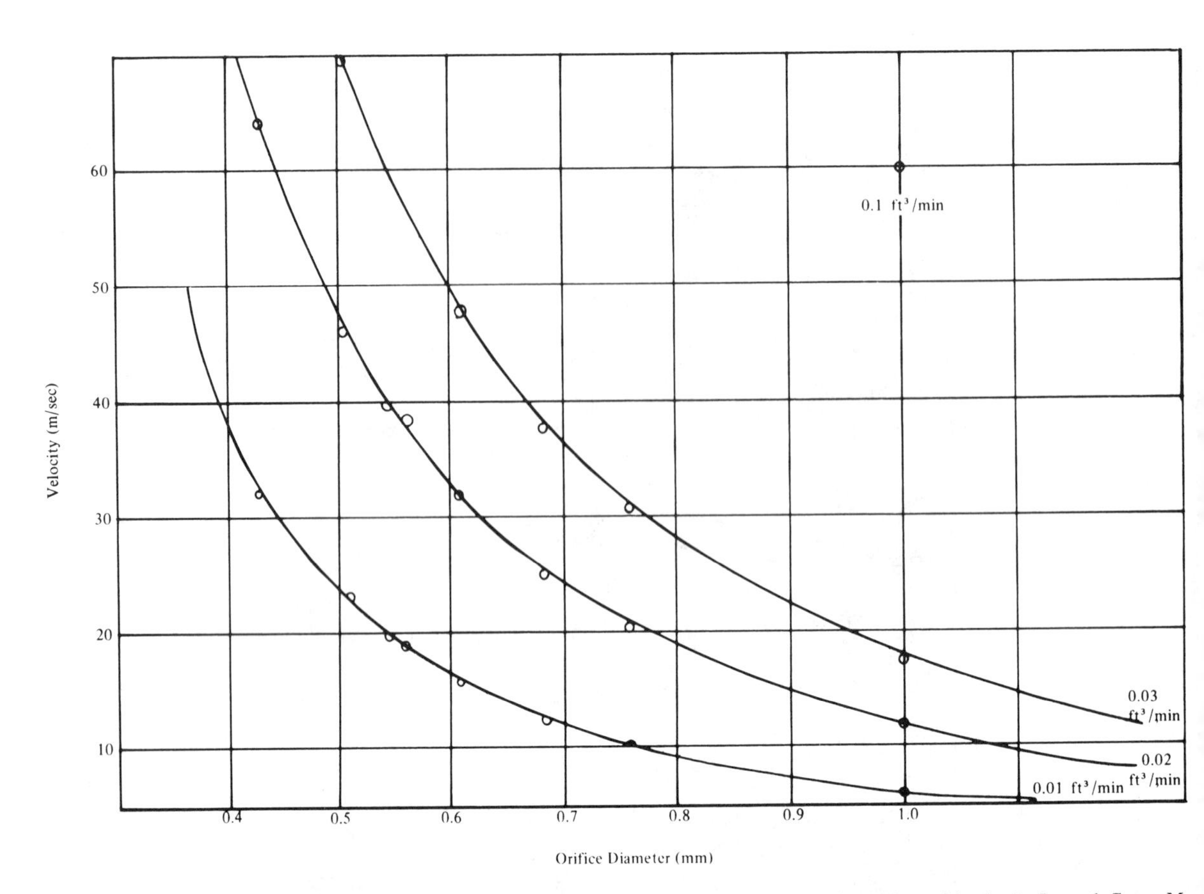

FIGURE I.43. Calculated air velocities for commercially available capillary nozzles. (From Linch, A. L. and Corn, M., *Am. Ind. Hyg. Assoc. J.*, 26, 601, 1965. With permission.)

water. Store in refrigerator and avoid exposure to light. Discard after the third day.

Sulfamic Acid. Dissolve 10 gm of 99% sulfamic acid in distilled water and dilute to 100 ml.

Sodium Nitrite, 3% Aqueous Solution. Discard when color appears any darker than a faint lemon-yellow. This reagent is stable for about 2 weeks.

Sodium Acetate Buffer. Dissolve 320 gm of analytical reagent grade sodium acetate trihydrate in 1000 ml of distilled water.

Procedure – After collection with 10 ml of ethanol in the midget impinger, transfer the contents of the midget impinger cylinder to a 50-ml beaker, rinse into the beaker with 2 to 3 ml of distilled water, and add 1.5 ml of 10% hydrochloric acid and a boiling stone. Slowly heat on a hot plate until the alcohol has evaporated completely (temperature of residue reaches 100°C).

Transfer the 2 to 3 ml of residue to a 50-ml volumetric flask, and dilute to volume with distilled water. Mix thoroughly, and pipet 5 ml of this solution into each of two 25-ml volumetric flasks. Chill to 0° to 5°C in an ice bath, add 10 ml of distilled water, and diazotize by adding 1 ml of sodium nitrite reagent to each flask. Mix by swirling, and after 1 minute destroy excess nitrous acid by adding 1 ml of sulfamic acid reagent. Mix thoroughly, let stand at 0°C for 10 to 15 minutes, and test for nitrous acid by spotting on cadmium iodide-starch test paper. If a positive test appears, add another 1 ml aliquot of sulfamic acid, and let stand for another 20 minutes.

Add 0.5 ml of Chicago Acid reagent to one flask only, and 5 ml of sodium acetate reagent to each flask. Mix by swirling, heat to 50° ± 5°C for 5 to 10 minutes to ensure complete color development, and dilute to volume with distilled water. Within 4 hours after coupling, determine light transmission at 520 mμ. The micrograms of aniline are read from a calibration chart (semilogarithmic graph paper) based on known quantities of aniline in aliquots taken from a standard solution (100 μg/ml) in 1% aqueous hydrochloric acid.

$$\text{ppm Aniline} = \frac{10 \times \mu g \times 263}{1000 \times 28.3 \times \text{vol. (ft}^3)} = \frac{\mu g \times 0.093}{\text{vol.}}$$

The sensitivity for trace concentrations can be increased by repeating the analysis with a 20-ml aliquot from the initial 50 ml of dilution mixture, and revising the volumetric correction in the final calculation (factor 2.5 instead of 10).

Analysis of the 2 ml of alcohol used to collect aniline in the microimpinger is identical except that all volumes are divided by 5, and final color dilution is made in 5-ml graduated cylinders calibrated to 0.1-ml intervals.

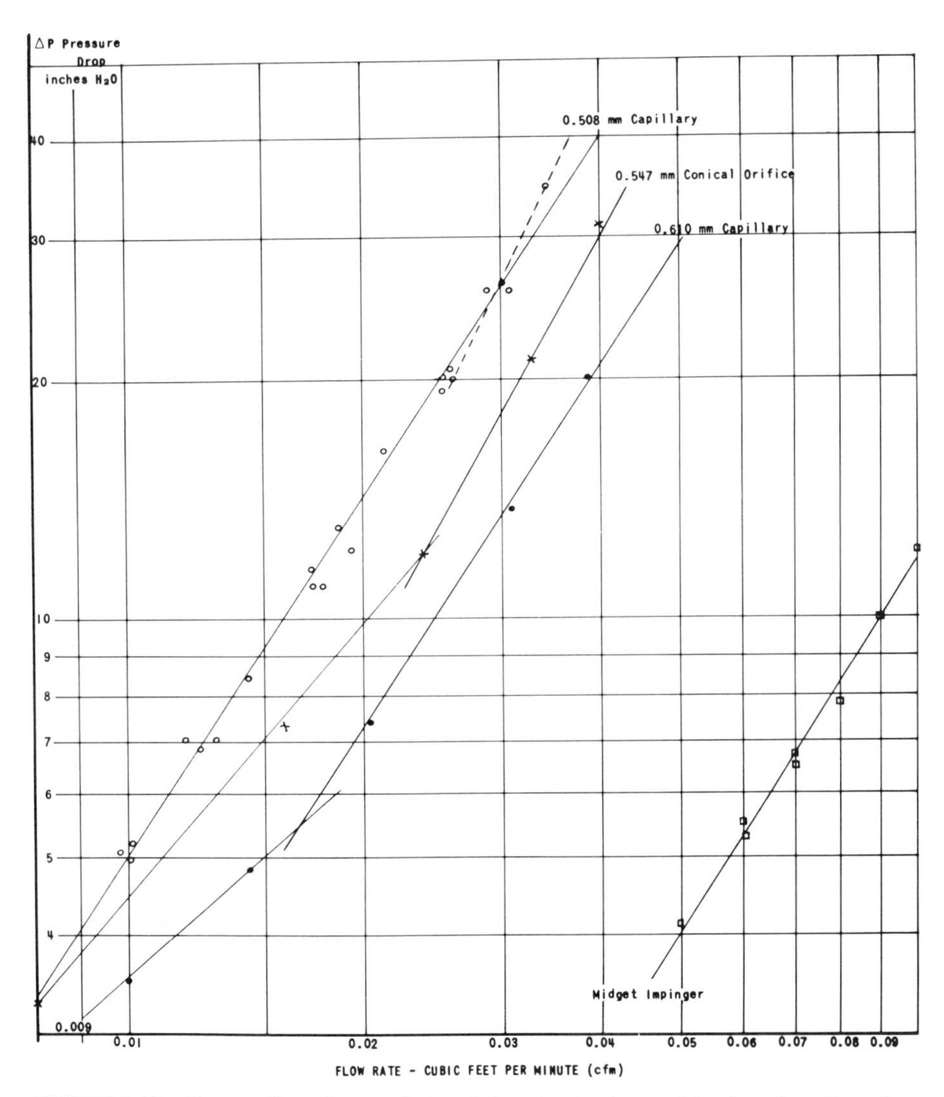

FIGURE I.44. Pressure-flow characteristics of the microimpinger with selected capillary tips. (From Linch, A. L. and Corn, M., *Am. Ind. Hyg. Assoc. J.*, 26, 601, 1965. With permission.)

When significant losses of aniline by absorption in gum rubber tubing connections were confirmed (see Table I.20), the essentially all-glass system shown in Figures I.41 and I.42 was adopted for efficiency evaluation. In this case, "clean" air was drawn through the system at 0.1 cfm for the midget impinger or 0.02 cfm for the microimpinger. A known amount of aniline in the range of 10 to 100 μg was then pipetted into the impinger tube (part 1, Figure I.42) and allowed to vaporize into the collecting impinger. The last traces of aniline were vaporized, probably from absorption on the glass, by heating the vaporizing impinger (part 1, Figure I.42) for 5 minutes in a 60°C water bath. A solution of 100 μg/ml in alcohol was used.

Results – Serious aniline losses were encountered when the contaminated air sample was drawn through gum rubber tubing. Furthermore, this loss increased as the age and number of uses increased (Table I.20). Even 3-inch connections introduced significant losses, but the loss was not entirely proportional to hose length. With an all-glass system (ball joints or "O" ring seals), 100% recoveries could be attained. However, at room temperature some adsorption on the glass surfaces of the vaporizer was encountered which could be cleared by heating to 60°C.

Based both on sampling from the prepared air mixture and use of the midget impinger as a vaporizer the efficiency of the microimpinger operated at 0.02 cgm was equivalent or superior to the performance of the midget impinger at 0.1 cfm (Tables I.21 and I.22). Near quantitative recoveries (98 to 100%) were attained with either impinger when all-glass systems were employed. Increasing the microimpinger flow rate from 0.02 to 0.04 cfm did not significantly reduce collection efficiency, but increasing the midget impinger rate from 0.1 to 1.5 cfm lowered the efficiency by more than 15%.[116]

TABLE I.20

Aniline Losses in Gum Rubber Tubing Connected to the Midget Impinger Vaporizer

Collected in 95% ethanol in midget impinger
Vaporization finished at 60°C for 5 min

Exposed tubing length (in.)	Age of tubing (days)	Aniline recovery (%)
0	–	100
2.5	–	91
9.5	0	83
9.5	1	76
9.5	7	65
9.5	8	57
19.5	0	80
19.5	7	58
19.5	8	50
29.0	10	41

From Linch, A. L. and Corn, M., *Am. Ind. Hyg. Assoc. J.*, 26, 601, 1965. With permission.

TABLE I.21

Recovery of Aniline Vapor From Mixtures in Air by Two Impingers in Series

Sampling time = 10 min
Absorbed in 95% ethanol

Pressure drop (in. H_2O)	Sampling rate (ft^3/m)	Efficiency of first impinger (%)	Aniline (ppm)
Microimpinger			
12.3	0.02	90.0	11.9
22.3	0.02	93.7	11.8
16.4	0.02	96.0	9.6
19.0	0.02	95.8	5.5
19.6	0.02	88.0	3.9
23.7	0.03	89.5	1.8
25.0	0.04	91.7	1.1
Average	–	92.6	–
19.5	0.02	83.3*	1.4
Midget Impinger			
12	0.10	89.7	11.8
12	0.10	93.8	3.2
12	0.08	95.5	3.2
–	0.15	85.2	1.4
–	0.16	81.0	1.2
Average	–	89.0	–
12	0.10	94.5*	1.0

*Absorbed in 10% aqueous HCl

From Linch, A. L. and Corn, M., *Am. Ind. Hyg. Assoc. J.*, 26, 601, 1965. With permission.

TABLE I.22

Aniline Recovery in the Midget Impinger Vaporizer

Absorbed in 95% ethanol

Aniline added (μg)	Vaporization time (min) 20°C	60°C	Aniline recovered (%) Micro-impinger	Midget impinger
Rubber tubing connection (3 in.) from vaporization to absorber				
110	10	–	42.2*	65.0*
92	10	–	61.0	85.6
103	20	–	80.8	85.3
103	10	5	95.0	91.0
Ball joint or O-ring seals				
100	5	5	–	101.0
105	5	5	98.1	96.0
11	5	5	109.0	100.0
13	5	5	96.2	108.0
1150	5	5	102.6	102.2

*Absorbed in 10% aqueous HCl

From Linch, A. L. and Corn, M., *Am. Ind. Hyg. Assoc. J.*, 26, 601, 1965. With permission.

TABLE I.23

Comparison of Midget and Microimpinger Efficiencies for SO_2 in Air

Midget impinger collection efficiency = 100%
Midget impinger reagent volume = 10 ml
Microimpinger reagent volume = 2 ml
Midget impinger sampling rate = 0.099 ± 0.001 ft^3/m (25°C and 760 mm Hg)
Microimpinger sampling rate = 0.0194 ± 0.004 ft^3/m (25°C and 760 mm Hg)
Two samples at each rate

SO_2 Concentration by midget impinger collection (ppm)	Microimpinger performance: Sampling time (min)	Collection efficiency (%)
0.069	30*	99.5
0.10	10	100.0
0.20	10	99.1
0.65	10	99.8
1.0	10	100.0

*Final volume = 1.8 ml

From Linch, A. L. and Corn, M., *Am. Ind. Hyg. Assoc. J.*, 26, 601, 1965. With permission.

Sulfur Dioxide.

Procedure – Standard atmospheres containing sulfur dioxide in the concentration range 0.5 to 5 ppm were prepared by means of a dynamic dilution system modeled after the one described by Kusnetz et al.[81] Compressed gas in a stainless-steel tank at a nominal SO_2 concentration of 300 ppm in nitrogen was mixed with purified dilution air to achieve the above concentrations of SO_2. The midget impinger and microimpinger, each containing 0.1 M disodium tetrachloromercurate(II) scrubbing solution, were run simultaneously and used to sample standard concentrations from a flask at the final stage of the dilution system. The scrubbing solution and method of analysis were those of West and Gaeke.[124] As reported by Roberts and McKee,[116] collection of 1 ppm of SO_2 in air in disodium tetrachloromercurate is essentially complete with the midget impinger at a flow rate of 0.1 cfm. Therefore, in this evaluation the microimpinger was compared to the midget impinger; that is, collection efficiency for the midget impinger was considered to be 100%.

Results – Table I.23 summarizes results of tests with the micro and midget impingers at 735 to 760 mm Hg and 20° to 25°C with different concentrations of SO_2 and at a sampling rate of approximately 0.02 cfm. Collection of SO_2 at concentrations equal to or less than 1 ppm was essentially complete, with deviations from 100% efficiency being attributed to experimental errors.

SUMMARY

A threefold reduction in the dimensions of this "O" ring model midget impinger gave a microimpinger which performed efficiently at one-fifth the flow rate and with only one-fifth of the sample and reagent volumes required for the larger counterpart. Results for aniline vapor, sulfur dioxide, and dust collection confirmed performance equivalent or superior to that of the standard midget impinger. Replacement of the conical ground orifice with a short section of precision-bore capillary permitted fabrication precision consistently within tolerances which permitted interchangeability without calibration to ensure matched performance. As many as 5 units could be operated in parallel with no increase in aspirating capacity over that required for a single midget impinger, or need for individual flow control meters. Combined with a recently available battery-operated miniature pump, the microimpinger provides a compact personnel monitoring instrument which delivers standard midget impinger results.[107]

An all-glass vaporizer assembly was used to determine impinger performance after significant losses of aniline by adsorption on gum rubber connecting hoses were confirmed. Collection efficiency in alcohol from a static mixture through impingers in series varied from 81% to 96% for the midget model in a concentration range of 1 to 12 ppm. Under similar conditions the microimpinger gave 88% to 96% recoveries. Both impingers gave essentially quantitative recoveries (96% to 100%) when assembled with the all-glass vaporizer system. Microimpinger performance was not significantly altered by increasing the sampling rate from 0.02 to 0.04 cfm, but a

50% rate increase (0.10 to 0.15 cfm) through the midget impinger decreased the efficiency by more than 15%.

Collection of sulfur dioxide from air in a concentration range of 0.07 to 1.0 ppm was essentially complete for both impingers when operated at the recommended flow rates.*

However, when application to the personnel monitoring for which the unit was designed was attempted, a major design deficiency was discovered. The reagent contained in the cylinder (2) (see Figure I.42) would drain out through the side tube (3) when the unit was rotated beyond 80° from the upright position. Not only was the absorbing solution lost, but, unless an effective trap was installed, the sampling rate meter and the pump or aspirator would be badly damaged. Since the employee is expected to perform his normal work activities while wearing the monitor, rotation beyond the critical angle cannot be avoided.

To correct this design imperfection, the cylinder (4) length was increased slightly, and the upper part was expanded sufficiently to accommodate 3 ml of liquid when the cylinder was placed in a horizontal position. Then, to form a trap for 180° rotation, the length of the tapered section of the head (5) was extended by using a standard-taper 14/35 joint inner member in the original standard-taper 14/20 joint outer member on the cylinder (Figure I.45).

A second trap is formed by extending a glass tube through the side arm until the open end nearly touches the impinger tube. As long as air flows through the impinger, no liquid can be lost through the inlet. For storage or shipping, a rubber cap can be used to seal this opening. In the field, to avoid accidental expulsion of a drop of corrosive reagent through the impinger the tube may be recessed by shortening down to the first O ring in the head, and a cylinder of porous Teflon or polyethylene pressed in place on top. This filter also provides removal of dust which might otherwise plug the impinger orifice. Replacement of the impinger tube with a porous glass diffusion tube will also minimize this possibility.**

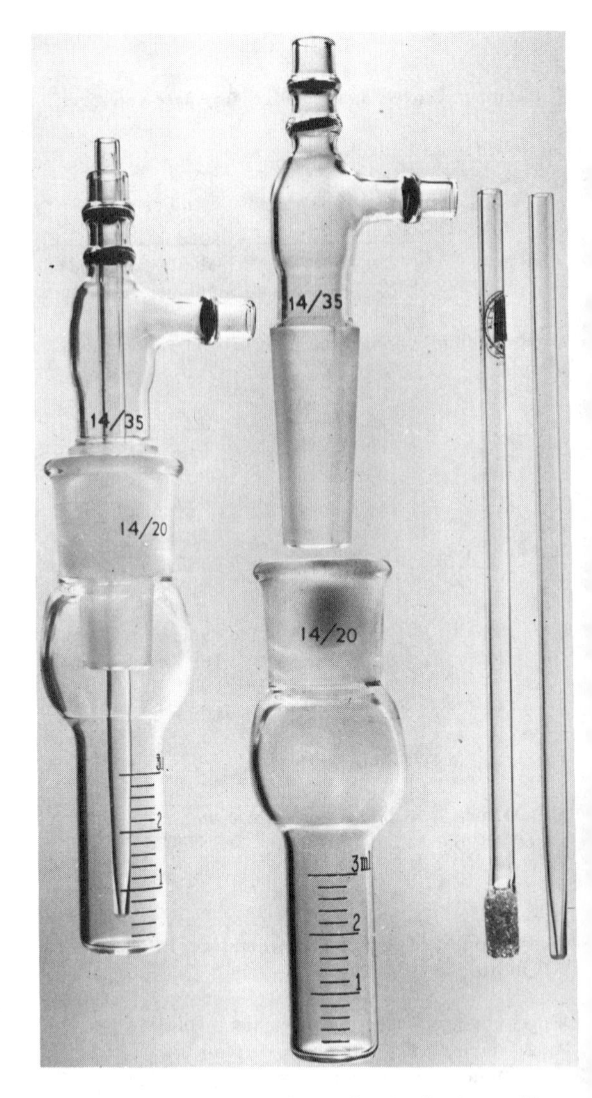

FIGURE I.45. The "spill-proof" microimpinger. (From Linch, A. L., *Am. Ind. Hyg. Assoc. J.*, 28, 497, 1967. With permission.)

This is the model that has been used extensively for personnel monitoring for organic and inorganic lead[96] and methylene-*o*-chloroaniline[126] exposure control. Late in 1971 the design was further miniaturized (Figure I.46) to permit reduction in the overall dimensions of the personnel monitor assembly (Figure I.47). The head height of the microimpinger (Figure I.46) was decreased by locating the exhaust port on the side of the ground joint and drilling a hole through the side tube seal to engage a matching hole in the head (6). This has the further advantage in providing a closure by rotating the head which is fitted with a Teflon sleeve as a sealing "lubricant." Part 2 is the impinger tube fitted with a Teflon baffle and precision bore nozzle (3). Part 5 is a rigid Teflon gasket which retains an O-ring against the cap (4) to produce a compression seal when the cap is screwed down tightly (7).

No liquid entrainment, or carry-over was found when the impinger was operated at flow rates up to 600 ml/min with 2 ml of either water, ethanol, or 0.01 *N* sodium hydroxide in the cylinder (1). The spill proof feature was confirmed by rotating the assembly through 180° without any loss of the liquid contents. Insertion of the impinger tube

*From Linch, A. L. and Corn, M., *Am. Ind. Hyg. Assoc. J.*, 26, 601, 1965. With permission.
**From Linch, A. L., *Am. Ind. Hyg. Assoc. J.*, 28, 497, 1967. With permission.

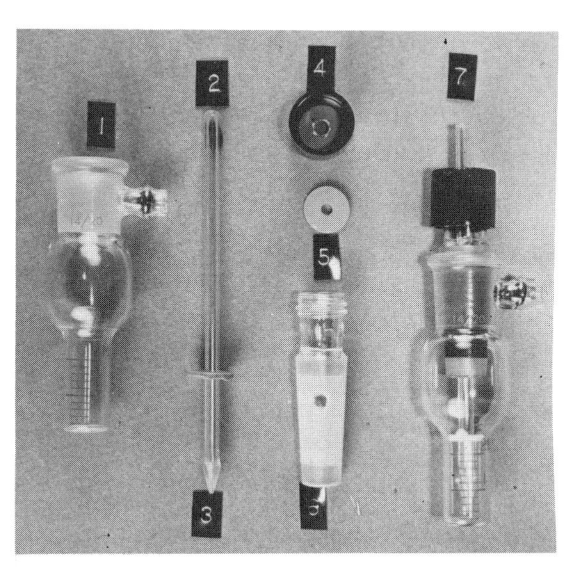

FIGURE I.46. The improved "spill-proof" micro-impinger.

through the head (6) also has been simplified as compared to the sometimes difficult manipulation through the two O-rings of its predecessor (Figure I.45).

C. The Wilson Impinger

This modification collects at a 5 l/min flow rate under a 107-cm Δ P. The nozzle velocity is approximately 100 m/sec, which is equivalent to the velocity attained by the larger (Greenburg-Smith) counterpart. This higher nozzle velocity provides a greater collection efficiency for fine particulate matter. The greater height, which would be undesirable for personnel monitoring, lessens the possibility of loss through liquid entrainment. Collecting liquid volume is the same as for the midget impinger – 10 ml. This design is commercially available.[84]

d. Porous Glass Bubblers

The following description is included by permission of the copyright holder.

Fritted (sintered) bubblers designed for sampling gases and vapors provide fine, dispersed streams of bubbles capable of increasing the diffusion so that contaminants, sparingly soluble in the collecting medium, are retained with acceptable efficiency. In general, the fritted glass bubbler improves efficiency by increasing the contact area between the air stream and the absorbing medium.

Frits come in various grades, usually designated as fine, medium, coarse and extra coarse. Although fine bubbles are desirable, the fine frits have no advantage because a high pressure drop is required. The effective gas bubble size is practically independent of pore size when the bubble population is high. If a useful flow rate is maintained through a fine frit, numerous bubbles will coalesce near the surface of the frit, making the effective gas bubbles essentially the same size as those from a coarse frit.[119]

The contact time of the bubbles can be increased either by increasing the height of the liquid or increasing

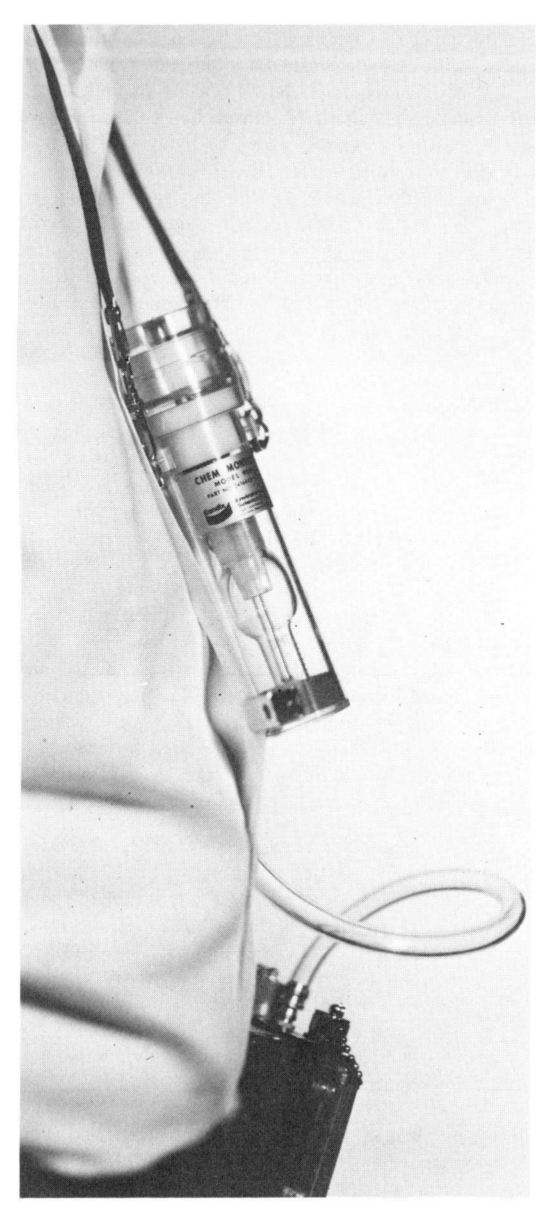

FIGURE I.47. The Chem Monitor, Model 900. This personal monitor has a 37-mm membrane filter for the collection of aerosols and a spill-proof microimpinger for the collection of gases and vapors. (Size = 40 x 146 mm; weight = 156 g.) (Courtesy of Bendix Environmental Science Division, Baltimore.)

the diameter of the sampling flask holding the liquid. An increase in the diameter of the flask lowers the velocity of the air passing through the liquid because of the greater cross-sectional area. For a 22 mm diameter flask (midget impinger flask) a flow rate of about 2 lpm is satisfactory; the contact time drops off rapidly at flow rates above this.

A coarse frit is usually best if the gas or vapor is appreciably soluble or reactive. A medium porosity frit may be used for gases and vapors difficult to collect, but the sampling rate must be adjusted to maintain a flow of discrete bubbles (fairly low bubble population). This necessitates an increase in sampling time.

If fritted bubblers are used, the gas or vapor sample must not form a precipitate when it reacts with the collecting medium, and the air sample must contain very little dust. If the air being sampled contains particulate matter which might clog the pores or interfere with the analysis, a non-reactive and non-absorbing pre- filter, such as dry fibrous glass, porous plastic, cellulosic paper or asbestos[117] should be used.

Calibration of fritted bubblers is neither so convenient or reliable as that of single orifice systems. There is considerable variation among bubblers caused mainly by the multiple orifices in the frit which are not of a consistent size. Each bubbler, containing the same volume of the chosen medium required for the field sampling procedure, must be calibrated individually. Air flow rates can be measured directly with a spirometer or wet test meter or, indirectly, with any of the following devices, previously calibrated with a reliable primary standard:

1. A manometer or vacuum gauge to measure pressure drop across the fritted bubbler(s).[128]
2. A limiting orifice between the bubbler(s) and the pump (this requires a high pressure drop).
3. A rotameter between the bubbler(s) and the pump. It may be necessary to dry the air entering the rotameter to prevent condensation of moisture therein.
4. A rotameter at the inlet of the bubbler(s) is not recommended.

Strongly alkaline media will attack glass frits and alter the calibration, but the effect is not serious unless the alkali remains in contact with the frit for extended periods. Alkaline solutions should be poured into the bubbler just prior to sampling and removed promptly after collection. Glass fritted bubblers, used regularly in alkaline media, should be recalibrated frequently.*

In most cases the bubbler has little to offer as the collection efficiency of the impinger exceeds 90%. A typical example is illustrated by the results from the collection of tetraethyl lead vapor. Three midget impingers, each of which contained 10 ml of 1 *N* aqueous iodine and two of which were packed with glass wool, were operated side by side at the standard 2.8 l/min rate for 40 min in an atmosphere known to contain organic lead vapors. Results were as follows:

1. Without glass wool – 14 μg Pb
2. With glass wool – 13 μg Pb
3. With glass wool – 15 μg Pb

   Average – 14 μg Pb

The glass wool was used primarily to reduce frothing, but the results indicate no increased lead absorption from increasing the glass surface area.

The exception to this generalization would be collection of contaminants that are dependent upon surface reaction with the reagent for collection. In this case an extended glass surface area is desirable. The collection of nitrogen dioxide in Saltzman's reagent serves as an example of this condition. As applied to the microimpinger, even the coarse frits cannot be sealed to the impinger tube without incorporating an excessive degree of airflow resistance which requires an excessive ΔP for operation even at low flow rates.

Packing the impinger with glass beads or spheres offers a better solution to the extended surface area problem.[91] Three-mm glass beads, packed to the 25-ml graduation in a standard midget impinger, will increase the collection efficiency significantly. The beads contribute an extended surface area for absorption, act as baffles to extend the air bubble liquid contact time, and increase the liquid turbulence to promote mixing. A smaller volume of liquid absorbent is needed when beads are used. Micro-glass spheres are available for packing the microimpinger.[129] This type of packing also greatly reduces liquid entrainment at high flow rates, or at maximum liquid fill level.

e. Traps

Even for water, a protective trap should be inserted between the pump and the liquid absorbant. Even though the absorbant may be inert insofar as the aspirating train is concerned, a liquid film can foul the rate meter (rotameter floats especially), reduce the pump's efficiency, or interfere with valve action. Activated charcoal is probably the best all-purpose scavenger for solvent and corrosive vapors, and silica gel serves well for water vapor. A mixture of the two would be suitable for certain aqueous systems. A polyethylene tube 110 mm (75-mm packed section) x 75 mm diameter packed with 8/10 mesh activated

*From Monkman, J. L., Analytical Guides, AIHA, Akron, O., 1965. With permission.

charcoal provided protection for 40 hr of operation at 420 ml/min when used with the iodine monochloride reagent to collect tetraalkyl lead vapors.[96]

Solid Adsorbents

a. Silica Gel

The following is taken from "Analytical Guides" by permission of the copyright owners.

Advantages of Silica Gel

Silica gel is a valuable medium for the collection of gases and vapors from air. While silica gel does not have as great an affinity for organic materials as does activated charcoal, it has a decided advantage as an analytical tool in that adsorbed materials can be removed easily, consistently and quantitatively. It is a clean, white, granular solid and it is easy to work with. Silica gel generally surpasses liquid substrates on solid supports in its ability to hold materials and thus has gained wide acceptance as the preferred adsorbent or carrier for the direct indicating gels.

For the industrial hygienist interested in quantitative air analysis adsorption on silica gel offers a variety of advantages. This technique is, primarily, one of concentrating the air contaminant of interest. Secondly, only the concentration step needs to be carried out in the field, leaving the delicate manipulations of analysis for the laboratory. With the use of a granular adsorbent such as silica gel quantitative transfers from the sampling vessel to a sample bottle, in the field are simple and rapid. Finally, because the phenomenon of adsorption is quite nonspecific, a single air sample taken on silica gel may be used to determine quantitatively several different materials coexisting in the sampled air.

Mode of Adsorption on Silica Gel

When a gas or a vaporized liquid is brought into contact with a solid, there is a tendency for the formation of a solution of the gas or vapor in the solid, as well as a tendency for the vapors to collect on the surface of the solid. Although the first tendency is usually slight because of the low solubility of vapors and gases in most solids, the second tendency of adsorbing vapors and gases at the surface is exceedingly high for a number of solids.[129] *

Many theories have been proposed to account for the adsorption of vapors on materials such as silica gel. At present, none of these theories appears to explain the phenomenon of adsorption in such a manner that accurate predictions can be made without trial runs and experimental data. However, despite the lack of a good theoretical base, it is known that silica gel does adsorb many materials quantitatively under the proper conditions and will release these adsorbed materials quantitatively under the proper desorption conditions.

1. Factors Affecting Adsorption

Factors that affect the dynamic adsorption of materials from air on silica gel include the following:

Previous history of the gel
Size of the gel particles
Size range of the gel particles
Mass of gel used
Depth of gel used
Superficial velocity of the airstream through the bed of gel
Temperature of the gel bed
The molecular species being adsorbed
Concentration of the air contaminant
Humidity of air contacting the gel bed
Duration of sampling

These factors vary in importance, and some appear to interact with others. Until more information is obtained, use of silica gel for air sampling must be based on empirical data.

a. Previous History of the Gel

Experience has shown that silica gel must be used in a standardized manner if quantitative results are to be obtained. The first consideration is the particular gel to use. Silica gel is not difficult to make, but making it in exactly the same manner each time over a period of years is a problem. Therefore, most users purchase their gel to rather exacting specification. The Davison Chemical Company (Baltimore) manufactures gels well-suited for analytical methods. Good results have been obtained from "part No. 05-08-05-215, 6-16 mesh, type 3," "PA100, 14-28 mesh," and "PA400, 14-28 mesh."[130,131]

As purchased, these gels are almost completely free of contaminants but may contain small amounts of ammonia. In addition, they may contain more moisture than desirable for use of the gel as a quantitative adsorbent. Therefore, all purchased gel should undergo a standardized pretreatment before use.[132,133]

One method for removing all ammonia and water is baking the gel in an oven overnight at 350 to 400°C in a dry nitrogen atmosphere. Enough nitrogen is allowed to flow through the oven to

*From Monkman, J. L., Ed., *Analytical Guides,* AIHA, Akron, O., 1965. With permission.

maintain a slight positive pressure and to sweep out any contaminants driven off the gel. After the baking has been completed, the gel is allowed to cool in uncontaminated room air until warm to the touch. It is then stored in sealed containers. This pretreatment is necessary only once: If the gel is used for studies involving thermal desorption, each successive desorption reactivates the gel; if the gel is used for studies involving leaching, it is discarded.

b. Size of Gel Particles

Investigations on the use of silica gel as a quantitative adsorbent have been carried out with gel ranging in size from 200 mesh to about 6 mesh. In general, the larger mesh sizes have proved easier to handle, but they have a somewhat reduced adsorption capacity. The reduction in capacity is not serious, however, and gel in the size range of 6 to 28 mesh performs quite satisfactorily.

c. Size Range of Gel Particles

For best results the range of particle sizes of gel used for quantitative adsorption should be kept small. Investigators have used 6-16 mesh and 14-28 mesh gel with success,[130,131] but wider ranges than these should be avoided. Gel that has been used repeatedly will gradually produce fines. As soon as fines accumulate, the gel should be discarded.

d. Mass of Gel

With other variables remaining the same, doubling the mass of gel used will increase the breakthrough time by a factor of about four. The breakthrough time is the time required before the first traces of the air contaminant can be detected downstream from the gel bed. It is the duration of time that adsorption is quantitative.

e. Depth of Gel Bed

Preliminary data have shown that the depth of the silica gel bed has an effect on the breakthrough time that is somewhat independent of the mass of gel used. With all other conditions kept constant, increasing the bed depth increases the breakthrough time and the total amount of contaminant adsorbed on the gel. The ideal container for air sampling is a relatively long tube with a relatively small diameter.

To avoid channeling, the minimum diameter of the adsorbing column probably should be several times the diameter of the largest gel particles.

Limited bed depth studies have indicated that channeling, wall effects, and nonuniform adsorption become important as the bed depth for 6-16 mesh gel is decreased below 2 cm. Less bed depth is satisfactory for smaller gel sizes.

f. Superficial Velocity of Airstream Through Gel Bed

The superficial velocity is the velocity that would be observed if the gel were not present in the adsorber. It is determined by dividing the airflow rate (cm^3/min) by the cross- sectional area of the adsorber (cm^2). Doubling the superficial velocity will result in a decrease of the breakthrough time by at least a factor of four. This relationship, however, is not linear and is dependent also upon the concentration of the contaminant in the air. Superficial velocities normally encountered in sampling operations range from 50 to 1000 cm/min.

g. Temperature of Gel Bed

Few, if any, experimental data are available relating bed temperature to retention or breakthrough time. Although most of the work with silica gel has been done at room temperature, 20 to 35°C, some investigators have used an ice bath when sampling for extremely volatile materials.[134] However, halogenated hydrocarbons, with boiling points as low as −28°C, have been quantitatively adsorbed on silica gel at room temperature.[130]

During the sampling period the temperature of the gel bed is elevated from the heat of adsorption contributed both by the contaminant and by water vapor in the airstream. A "front" can easily be detected passing through the bed of gel simply be feeling the adsorber. This lack of uniformity of the bed temperature has undoubtedly complicated the investigation of that variable.

h. The Molecular Species Being Adsorbed

The affinity of silica gel for water and for water vapor has long been known; normally, it is used as a desiccant. Silica gel apparently will adsorb water in preference to any other material. Some investigations have shown, however, that its affinity for ammonia and a few other materials may be almost as great as for water. Both because of structure and known affinity for water, silica gel should best adsorb materials possessing a hydroxyl group.

Most of the commonly encountered halogenated hydrocarbons, benzene, many other hydrocarbons, and other organic polar and nonpolar materials have been determined quantitatively by collection on silica gel.

Materials with high boiling points appear to adsorb somewhat better on silica gel than do those with low boiling points. Attempts to determine methyl chloride in the air by silica gel adsorption at room temperature have been unsuccessful; methyl bromide, on the other hand, adsorbs quite readily. Variables affecting the adsorption of these materials have not been examined thoroughly.

i. Concentration of Air Contaminant

Limited investigation has shown that the concentration of the contaminant in air may have very little effect on its adsorption on silica gel below a certain critical concentration. For carbon tetrachloride the critical concentration appears to be about 400 ppm. That this level varies appreciably for different materials has been established, but a thorough study has not been conducted. With other conditions held constant, an increase in concentration above the critical level causes a decrease in the breakthrough time.

j. Humidity of Air Contacting Gel Bed

Because of its affinity for water, the usefulness of silica gel for air-sampling could be quite limited when the absolute humidity of the air is high. Most materials other than water adsorb very poorly, or not at all, on drying agents such as magnesium perchlorate or Drierite. Silica gel in a sampling train may be preceded by a drying agent to remove the water.[20,135] This allows the silica gel to be used to full capacity for the air contaminant of interest. This technique should be used with caution. Under normal circumstances, silica gel has been used without a drying agent to sample air for most of the halogenated hydrocarbons.[130,136]

k. Duration of Sampling

The duration of sampling should, in all cases, be somewhat less than the breakthrough time under the conditions encountered. The most practical solution to a particular problem is to sample known concentrations under conditions similar to those anticipated in field sampling. If the time necessary to obtain sufficient contaminant for analysis is greater than the breakthrough limit, the other variables will have to be altered. Breakthrough times ranging up to several hours are attained by proper manipulation of the sampling variables.

2. Practical Air-sampling Details

a. U-Tubes

Several different types of adsorbing vessels may be used with silica gel. A common type is the U-tube with silica gel in both arms of the tube. A train of two or more U-tubes has been used,[20] but experience has shown that one such tube is usually sufficient.[131] A single U-tube with silica gel in one arm and a desiccant in the other may also be employed.[135] U-tubes should hold at least 10 g of gel with a bed depth of more than 2 cm. Add the silica gel to the U-tube slowly with a slight tapping to obtain uniform distribution which limits airflow channeling through the gel bed. Vessels constructed of 16-mm tubing, 150 mm long, have worked satisfactorily when airflow rates have not exceeded 2 lpm.[136]

b. Sintered Fiberglass Adsorbers

Straight tubes with a sintered glass support for the gel bed have also been used. Usually, the sintered glass is of a medium or coarse porosity to ensure efficient distribution of air throughout the bed without causing an excessive pressure drop. Adsorber diameters are commonly 30 or 40 mm. With 20 g or more of silica gel in such an adsorber, flow rates up to 3 lpm may be used.[130]

c. Other Methods

A technique for using the adsorption properties of silica gel, but avoiding the difficulties of controlling the variables encountered in a dynamic flow system, has been developed. It might be called a "static adsorption system." In this method, 10 g of silica gel is added to a 3- or 5-l saran bag that is subsequently filled with air and sampled by means of a glass hypodermic syringe pump equipped with two glass check valves. The bag is shaken occasionally, and a few hours later the gel will have absorbed the air contaminant quantitatively.

3. Desorption of Vapors and Gases

Very little adsorbed vapor is lost if the gel is stored at room temperature in a closed container (provided that colder temperatures were not found necessary for sampling). Air samples collected on silica gel can be held for at least 3 to 10 days without loss of contaminant. This retentive property of silica gel is very useful when considerable time must elapse between sampling and analysis.

Adsorbed materials can be determined on the gel or after removal. Determination on the gel is the basis for the detector tube technology. Usually the detecting reagents are adsorbed onto the gel before exposure to the contaminant, but the reverse technique also has been used. Many factors render these tubes acceptable only for semi-quantitative analysis. For quantitative analysis the adsorbed material is removed from the gel.

Removal of adsorbed vapors or gases from silica gel is not difficult when the proper conditions are selected. Desorption is accomplished thermally or by solvent leaching. The method of choice is determined by the analytical technique selected and by the efficiency of removal.

a. Thermal Desorption Methods

All thermal desorption methods require elevated temperatures and a carrier gas to collect the desorbed material. Usually the gel is transferred to a tube which, in turn, is placed in a tube furnace. As the gel temperature rises, desorbed material is swept along with the carrier gas into a collecting device. The carrier gas instead of the gel may be heated, for example, superheated steam. Heating the gel alone, without a flowing carrier gas, produces a hysteresis effect that prevents good recoveries.[129a]

Desorption temperatures and the carrier gas composition have been varied widely with good results. Using helium, Altshuller[134] found that acetylene could be desorbed at 50°C; low molecular weight sulfur compounds (H_2S, SO_2, CH_3SH) at 100°C; hydrocarbons such as *n*-butane, isobutane, isopentane, propane, and propylene at 150°C. Jacobs reports benzene can be freed from silica gel at 100°C with air.[137] Peterson et al. in experiments with trichloroethylene, found that the time required for complete desorption varied with temperature but that, at 350°C with a flow rate of 2 lpm of dry air, complete desorption was assured.[130] Halogenated hydrocarbons with boiling points ranging from -28°C to 190°C were recovered quantitatively under these conditions. Erley used a temperature of 400°C and about 7 l dry nitrogen at 0.5 lpm for the desorption of several materials.[132]

After the material has been desorbed from the gel as a vapor into the carrier gas, the concentration is greater than in the originally sampled air. If the carrier gas is superheated steam, the analysis proceeds after condensation as if the material had been trapped in water. When the carrier gas is noncondensable at ordinary temperatures, the mixture may be fed into a second furnace at a higher temperature for direct combustion, or may be collected in a saran or polyethylene bag for analysis by various instrumental techniques. Both gas chromatography and long-path infrared spectrometry have been employed with excellent results, especially when more than one molecular species has been adsorbed on the gel. Methods of analyzing the concentrated vapor are limited only to the ingenuity of the analyst.

b. Solvent Leaching Methods

Some analytical techniques require the contaminant to be in solution, such as UV absorption methods, gas chromatography techniques, or wet chemical determinations. The collected contaminant may be removed quantitatively by immersing the gel in a solvent. If interferences are introduced by the gel, or if removal is not quantitative, the contaminant should be collected in a solvent and not on silica gel.

When removing the contaminant, the silica gel should be added to the solvent slowly, almost grain by grain, to avoid a build-up of heat in the solution. If the solvent is added to the gel, the heat of reaction may be sufficient to drive off some contaminants; moreover, since the gel is also an active catalyst, a polymerization or a breakdown of some substances could occur. Decrepitation of the gel may produce fine fragments that will cause interferences in UV absorption analysis.

A convenient dispenser, shown in Figure I.48, is attached to a 50-ml round-bottom flask containing the absorbing solution. The flask with dispenser is placed in an ice bath on a laboratory shaker, and as the flask and dispenser are shaken the glass rod with the slots swirls around in the dispenser allowing a few grains of silica gel at a time to drop into the absorbing solution.

Much of the information on the desorption of micro amounts of contaminants from silica gel by leaching with solvents is unpublished. Members of the AIHA Analytical Chemistry Committee have

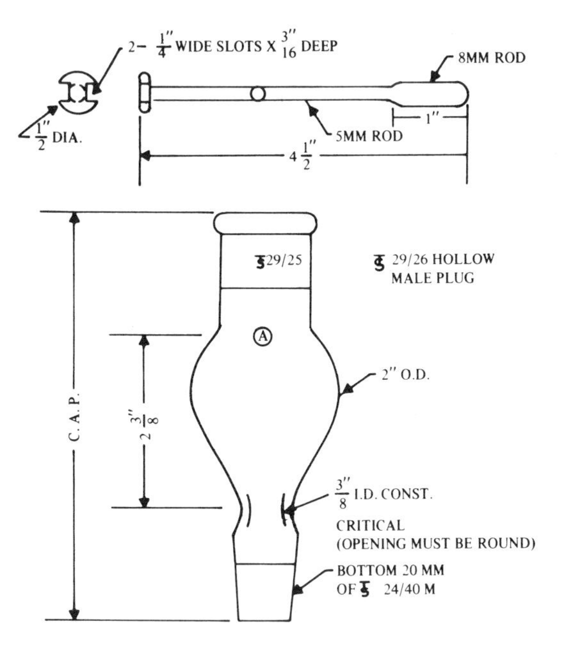

FIGURE I.48. Silica gel dispenser. (From Monkman, J. L., Analytical Guides, AIHA, 1965. With permission.)

reported that the following materials can be leached quantitatively by the agent indicated:

Materials	Leaching agent
Benzene Divinyl benzene Alpha methyl styrene Isopropyl benzene Toluene	Isooctane plus water
Allyl alcohol Allyl chloride Propylene dichloride	*n*-Hexanol
3,4-Dichloronitrobenzene Nitrobenzene	Methanol
Aniline Benzene Isophorone Nitrobenzene Styrene Toluene	Ethanol
Dinitrophenol Ethyl benzene	Isopropanol

When the materials listed above are sampled separately and leached with the appropriate solvent, satisfactory results can be expected. Difficulties have been encountered, however, when sampling mixtures with poor recovery of one or all of the components are present. Recoveries of solvents in mixtures must be checked in the laboratory before attempting to use this technique for atmospheric sampling.

Likewise, recoveries of some individual materials were found to be low and erratic; for example, recovery of acetone leached from silica gel with ethanol was about 50%; the same solvent used for methylethylketone resulted in recoveries varying from 60 to 80%. In any case, when definite information is not available, recoveries must be verified with known concentrations before field sampling is undertaken.

In UV absorption methods, particulate matter released from the gel can interfere; these particulates are not removed by centrifuging. Marked interference has occurred with grade 05 silica gel. It is advisable to use either the PA100 or PA400 grade gels mentioned earlier: PA400 gives the least difficulty. Interference may be serious at wavelengths below 250 nm. Silica gel is not recommended when the absorption peak is below 250 nm.

Available information on solvents indicates that the greatest interference occurs with ethanol. Relative interference, in descending order, was found to be: ethanol, methanol, water, isopropanol, isooctane. Generally, isooctane is a very satisfactory solvent, even at low wavelengths. However, a small amount of water to obtain satisfactory release of the adsorbed contaminants is necessary.

Additional factors that contribute to interference are leaching time and temperature. Interferences increase as time and temperature increase.

In 1966 Campbell and Ide contributed one of the most significant developments for the use of silica gel in collecting aromatic hydrocarbons from the atmosphere.[138]

A column of silica gel, divided into four sections, was held in a glass tube 3 mm I.D. by 10 cm long. The gel was in four 2 cm lengths separated by glass wool, each section contained approximately 125 mg of silica gel. Air contaminated with an aromatic hydrocarbon was drawn through the tube at 60 ml per minute with a commercial, hand syringe-type pump (Figure I.49.a).

After the air sample was taken, the tube was cut into portions representing the first through fourth sections, and the silica gel from each section was transferred to a separate 15 ml centrifuge tube. The silica gel then was eluted with 5 ml ethanol for the 30 min. standing time (isopropanol may be substituted), the mixture centrifuged at 18,000 rpm for 5 min. and the alcohol extract analyzed

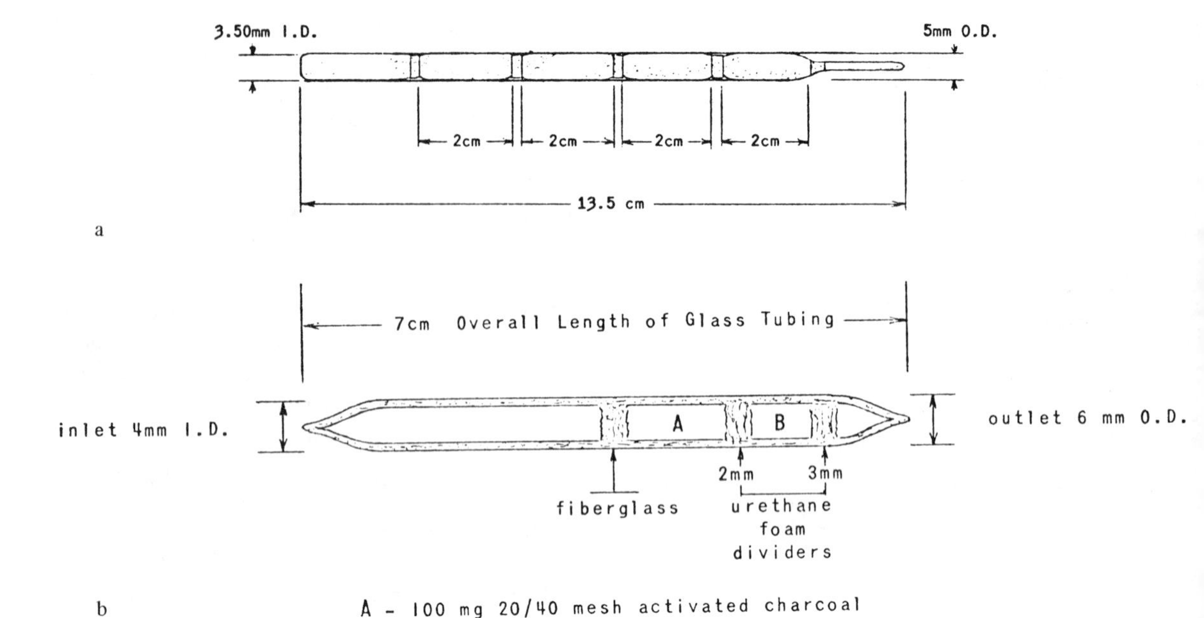

FIGURE I.49. Silica gel sampling tube. The ends of the glass tube are flame sealed to ensure circular breakage of 2-mm diameter section and to restrict contamination of the charcoal prior to sampling. a. Silica gel sampling tube. (From Campbell, E. E. and Ide, H. M., *Am. Ind. Hyg. Assoc. J.*, 27, 323, 1966. With permission.) b. The charcoal sampling tube. (From Kupel, R. E. and White, L. D., *Am. Ind. Hyg. Assoc. J.*, 32, 456, 1971. With permission.)

by UV spectrophotometry. The alcohol interferences noted previously were not encountered by these investigators.

The resulting OD-wavelength plot was converted to amount of aromatic hydrocarbon, and the efficiency of the tube determined. The series of silica gel sections readily compensated for factors which affect retention of the contaminant, such as ambient temperature at the time of sampling, relative humidity, and interfering substances.

Polar substances cause aromatic hydrocarbons to migrate through the sampling tubes by displacement. If c is the concentration of the substance in the sample, and c_1, c_2, c_3, and c_4 are the amount of substance in each section, experiment shows that Equation 1, which is the sum of a standard geometrical progression, gives the true concentration of the substances in the air sample.

$$c = c_1 + [c_2(k^n-1)/k-1)] \qquad (1)$$

The amount of the substance in section one, c_1, was found always higher than was expected from the amounts in the remaining sections, probably because of impingement, condensation, and other factors influencing equilibrium. The constant k is defined as

$$k = c_{n+1}/c_n \text{ and } c = c_1 + c_2(K^{n-1})/(k^{-1})$$

where n realistically should be less than 10 (a theoretical number of sections).

If the concentration in section four, c_4, is zero, the amount of substance in the air sample is the arithmetical sum of the sections. If section four, c_4, is equal to or nearly approaches section three, c_3, no calculation can be made. An estimation is possible, however, if the difference is less than 30% and if similar sampling conditions can be duplicated in the laboratory for comparison.*

The background optical absorbance was corrected by base line calculations. The UV absorbance curve of the ethanol solution between 210 and 340 nm was plotted and a line drawn tangent to the minimum points. The absorbance on this base line at the wavelength of maximum total absorbance was then subtracted from the total absorbance and the difference related to a calibration curve constructed in a similar fashion. Each section of the recommended tube design contained 115 to 120 mg of silica gel. Therefore, the background optical absorption due to the fine silica gel fragments that arise from the shattering of the gel during desorption can be so controlled that the simple centrifugation and base line construction minimizes this problem. Only *o*-dichlorobenzene, nitrobenzene, methylaniline, and aniline failed to desorb completely in ethanol during the 30-min extraction period.

Pretreatments of the silica gel did not affect its retention characteristics. The four-section tube

*From Campbell, E. E. and Ide, H. M., *Am. Ind. Hyg. Assoc. J.*, 27, 323, 1966. With permission.

eliminated most extrinsic (relative humidity, temperature, sampling rate variations, etc.) variability as total recovery was the same under all conditions studied. The first section (Figure I.49.a) did not completely remove nonpolar materials such as benzene, toluene, cyclohexane, or most aliphatic hydrocarbons. For example, 88% of the benzene in a 100-ml air sample was retained on the first section and the balance (12%) on the second section. From these values, a silica gel saturation of 0.045 mg benzene per mg silica gel was calculated. Polar compounds – aniline, pyridine, nitrobenzene – were retained completely on the first section.

The silica gel tubes were prepared as follows:

> The silica gel is poured into the tube through a small funnel or from a filling jar through a glass tube in a one-hole rubber stopper, then packed uniformly by tapping the tube on a hard surface. Glass cloth or glass wool pledgets secure the ends of the column and separate the sections. The first section is filled to a level 0.25 cm above the 2-cm mark, tapped to pack the gel just to the 2-cm level, and secured with a glass wool pledget. The second 2-cm mark is measured from the surface of the pledget, and the second section is gel filled to 0.25 cm above the mark and compacted, as before, The third and fourth sections are prepared similarly. Routinely, the tubes are left open and stored over silica gel, but for field use they may be capped with rubber (tip of rubber policeman) or with Parafilm® or other suitable covering.*

The pressure drop (Δ P) across these tubes packed with 42-80 mesh silica that had been activated by heating to 150°C and had an acid equivalent titer of 9.1×10^{-6} in Eq of H^+ per mg at various flow rates were

30 ml/min – 22 cm H_2O or 8.6 in. H_2O
60 ml/min – 46 cm H_2O or 18.2 in. H_2O
90 ml/min – 70 cm H_2O or 27.5 in. H_2O
120 ml/min – 94 cm H_2O or 37 in. H_2O

J = D dx/dc = diffusion flux in mol/cm^2

> The tubes may be used to collect air samples for vapor chromatography analysis. The silica gel may be placed in special sample holders and the vapor removed by thermal displacement, or the tube itself may be placed in line with the carrier gas and the contaminant removed by polar-solvent vapor displacement. Gas chromatographic analysis is more sensitive than the method described here, but is more time-consuming. The silica gel tubes may well serve as reactant beds for field or laboratory procedures similar to those described by the Chemical Warfare Service.[139] The silica gel tube approximates a critical orifice; therefore, the tube may be carried into the field on routine surveys and used with aspirators, house vacuum systems, or any other convenient vacuum source capable of overcoming a pressure drop of 34 mm of mercury, to sample atmospheres for qualitative or semiquantitative evaluation in the laboratory. This convenience permits the industrial hygienist to analyze any air contaminant without having to carry a large number of specific sampling media.*

Loss of nonpolar contaminants from the silica gel was found to be very nearly constant. For example, an average 13.8% of the benzene retained on the first section desorbed into each 100-ml aliquot of clean, dry air passed through the tube after the initial 100 ml of nearly saturated benzene vapor sample collection. However, under "normal" sampling conditions, all test compounds, including benzene, were quantitatively retained at concentrations up to 500 ppm even when one liter of clean air was drawn through the tube after sample collection. The microcolumns may be shipped without sample loss and may be stored for some time with only slight migration of the aromatic hydrocarbons to other sections of the column.

The sensitivity limits that determine the parameters for sample size and optimum concentrations for 11 aromatic derivatives in the eluting solvent are presented in Table I.24. Although the system was designed for grab sampling with a 100-ml hand-operated syringe pump, direct adaptation to personnel monitoring would require only the addition of a small electrically driven pump and accessories to determine flow rate and sample size.

Silica gel columns also have been used to determine the accuracy of length of stain gas detector tubes.[140] Grab samples were collected simultaneously in the field on silica gel and detector tubes for benzene, toluene, and xylene. The aromatic hydrocarbons were desorbed from the silica gel in solvent and analyzed by UV spectroscopy as in the Campbell reference.[138] The collection efficiency was 93%. The three sources of detector tubes read ± 60% of the true value in the range 15 to 40 ppm. Isooctane in a midget impinger did not collect quantitatively. Sherwood used silica gel cartridges to collect benzene vapor from the breathing zone with a battery-operated pump (personnel monitor). Although the sampling period extended only over a 15-min interval, the

*From Campbell, E. E. and Ide, H. M., *Am. Ind. Hyg. Assoc. J.*, 27, 323, 1966. With permission.

TABLE I.24

Sensitivity of Silica Gel Sorption and Subsequent Spectral Determination

	TLV (760 mm Hg)					Minimum air volume needed to determine ml	
Substance	ppm in air	mg/m³ in air	Spectral sensitivity (μg in 5 ml EtOH)	Optimum concentration (μg/ml in 5 ml EtOH)	Wavelength (maximum)	TLV (ppm in 5 ml EtOH)	Optimum concentration (ppm in 5 ml EtOH)
Aniline	5	19	0.4	3.5	233.	100	920
Benzene	25	80	10	175	253.9	625	11,000
Chlorobenzene	75	350	15	175	263.3	214	2,500
m-*p*-Cresol	5	22	2	35	273.5	454	8,000
o-Dichlorobenzene	50	300	17	175	268.8	283	2,920
Ethylbenzene	100	870	13	177	260.7	150	1,018
Methylaniline	∿2	∿9	0.3	3.5	244	167	1,950
Nitrobenzene	1	5	0.5	8.75	258	500	8,750
Styrene	100	420	0.2	3.5	247	2.5	0.42
Toluene	100	750	10	175	261.2	134	1,170
Xylene	200	870	1	175	264.5	6	1,000

From Campbell, E. E. and Ide, H. M., *Am. Ind. Hyg. Assoc. J.*, 27, 323, 1966. With permission.

application to 8-hr time-weighted sampling was obvious. Gas chromatographic analysis of an ethanol elute was elected for analysis.[141]

Peterson and co-workers developed procedures for the analysis of airborne halogenated hydrocarbon contaminants by collection on silica gel and combustion of the collected material to yield a mixture of halogen and halogen acid, which is then titrated by the micro Volhard procedure.[142] Desorption is accomplished by passing air at a 1 to 3 l/min rate through the collection tube in a furnace at 350° to 450°C. for 10 min. The gel may be reused without loss in absorption capacity. The technique was applied to 14 halogenated hydrocarbons including CCl_4, chlorotrifluoroethylene, dichlorobenzene, methyl bromide, and vinyl chloride with 100 ± 10% recovery. The anomalous results encountered with methyl chloride suggested that silica gel at room temperature should not be used for halogenated hydrocarbons boiling below 0°C. The pyrolysis gases were absorbed in aqueous mixture of 1% Na_2CO_3 and 1% sodium formate to reduce the evolved halogen to halide ion. Other useful source material can be found in References 132 and 142 to 145.

5A molecular sieves have been installed in front of the silica gel bed to remove water vapor from the air sample before sample adsorption takes place.[145] However, the possibility that the target contaminant may be partially if not completely absorbed in the surface of the molecular sieve, which usually is organic in nature, must be considered when attempts are made to avoid the desorbing effect of water vapor. Whitman and Johnston reported on an investigation in which contaminated air was drawn at 1 lpm through an 8 mm I.D. x 245 mm tube containing 5 g 5A molecular sieve and 5 g 12-28 mesh silica divided by a pledget of glass wool. The contaminants were displaced by adding first 1 ml isopropyl benzene to the silica gel followed by 50 ml cold water. The lighter solvent layer was decanted and analyzed by gas chromatography (GC).

Recoveries of benzene, toluene, and xylene were in the range 95 to 100%. Concentrations were confirmed by drawing a simultaneous sample through an isopropanol scrubber at 0.5 lpm and analysis by UV spectroscopy.

b. Alumina Gel

Although alumina gel is an excellent adsorbant for water vapor, very little information relative to collection of air contaminants is available. Application to the collection of alkaline components such as methylamines, pyridine, morpholine, piperidine, etc., should be worth investigating, especially in those applications where the acidic nature of silica gel interferes with desorption. Tests of the efficiency of 80-200 mesh adsorption alumina indicated that columns of alumina not only absorbed a higher percentage of parathion and gamma benzene hexachloride insecticides than glass wool, Florisil,® or Dry Ice®-acetone, but also desorbed in acetone or 95% ethanol more readily.[145a]

C. Carbon

The use of activated carbon or charcoal to adsorb undesirable components from the atmosphere was established prior to World War I and employed during and after that period in the canisters of gas masks to remove toxic chemical vapors from inhaled air to protect the respiratory tract. Application to air analysis as well as to air purification has been widely employed, and no attempt will be made to review the subject here. The most recent developments will serve to illustrate the utility of activated charcoal for the collection of air contaminants for subsequent analysis. The following summary quoted from the methods for air sampling prepared by the Intersociety Committee serves as an excellent introduction to the subject.

Activated carbon is electrically non-polar and is consequently capable of adsorbing organic gases and vapors in preference to atmospheric moisture. In fact, even activated carbon with previously adsorbed moisture will lose such moisture by displacement with organic gases and vapors from contaminated atmospheres. Such displacement of moisture is not exhibited to any significant extent by siliceous adsorbents.

The following comparison between activated carbon and silica gel will further illustrate the differences between the two types of adsorbents for purposes of sampling atmospheric gases and vapors:

a. Activated carbon has a greater over-all capacity for adsorbing and retaining atmospheric gaseous and vaporous contaminants, but silica gel exhibits a greater selectivity among them.

b. Activated carbon is more suitable for collecting a conglomerate mixture of gases and vapors, especially when these are present in very small concentrations and must, therefore, be collected over relatively long periods of time.

c. Desorption of gases and vapors from silica gel is easier than from activated carbon, and it is in fact feasible to obtain some separation of gases and vapors by selective desorption from silica gel.

d. Each adsorbent can be suitably impregnated to render it more suitable for chemisorption of specific gases or vapors.

To determine which gases and vapors may be sampled by physical adsorption using activated carbon, the following approximate criteria, based on available information, may be used: True gases, usually having critical temperatures below −50 C and boiling points below −150 C are virtually nonadsorbable at ordinary temperatures by physical means. These include, for example, hydrogen, nitrogen, oxygen, carbon monoxide, and methane. Low boiling vapors having critical temperatures between approximately 0 and 150 C and boiling points between −100 and 0 C are moderately adsorbable. The adsorption efficiency at ordinary temperatures and concentrations, however, is not quantitative. Such vapors can be concentrated from the atmosphere by using a "thick carbon bed" preferably refrigerated. These vapors include ammonia, ethylene, formaldehyde, hydrogen chloride, and hydrogen sulfide. Chemisorption methods may also be used for such low boiling vapors. The heavier vapors (boiling above 0 C) are readily adsorbed and retained by activated carbon at ordinary temperatures; these include most of the odorous organic and inorganic substances.

Adsorbed gases and vapors may be desorbed from activated carbon by (*a*) displacement of the adsorbed material by superheated steam and (*b*) heating the carbon under vacuum and distilling the desorbed material into cold traps.

For steam displacement, the saturated carbon is flushed with superheated steam at 300 C or higher. The effluent steam and displaced vapors and condensate are held for analysis. Flushing is continued until the condensate is substantially odorless. Upon standing, the condensate may separate into an oily and an aqueous layer. Both oily and aqueous fractions may then be subjected to qualitative chemical or spectrometric analysis for detection of inorganic ions and organic functional groups. The steam displacement method, although very effective as a means of desorption, suffers from the disadvantage that steam will hydrolyze some organic compounds, such as esters, thereby causing qualitative changes. This effect can be circumvented by using the vacuum method described in the following paragraph.

For vacuum desorption, the saturated carbon is connected to a train of three traps in series, which are immersed in ice-salt, Dry Ice-Cellosolve and liquid nitrogen, respectively. The entire system is evacuated and pumping is continued at a pressure of about 1 μ until distillation practically ceases. During the pumping, the carbon sample is held at a temperature of 200 to 250 C.

At the end of the pumping period, the system is slowly returned to atmospheric pressure by bleeding in air or nitrogen. The traps are then removed and their contents examined. Generally, most of the water and some nonaqueous matter will be found in the ice-salt trap, most other atmospheric vapors and some water in the Dry Ice trap, and some of the lighter gases in the liquid nitrogen trap. These materials, with or without further fractionation, may be subjected to chemical or spectrometric qualitative analysis.*

*From Katz, M., Ed., Tentative method for preparation of carbon monoxide standard mixtures (42101-01-69T), *Lab. Health Sci.*, Suppl. 7, 72, 1970. With permission.

Some of the disadvantages expressed by Campbell and Ide include:

> Sorption media such as three forms of charcoal, two molecular sieves and three silica gels were compared, using several types of aromatic hydrocarbons from the air stream, the hydrocarbons could not be removed from the charcoals with any degree of reliability. The charcoal fines from mechanical breakage were lost during transfer from the sampling tube to the elution tube, many of the particles remaining on the glass wool pledgets and on the walls of the sampling tube. The molecular sieves did not remove the aromatic hydrocarbons from the air stream with any degree of efficiency; the background spectrum, however, was somewhat reduced. The passage of the hydrocarbons to successive (4) sections in the sampling tube was greatest with the molecular sieves, while silica gel and charcoal essentially retained the hydrocarbons in the first section.*

The collection of organic solvent vapors on activated carbon for analysis by gas chromatograph and the delineation of instrumental operating parameters were rather completely reported by Franst and Hermann in 1966.[146] Variations in recovery efficiency of methyl ethyl ketone, toluene, trichloroethylene, butyl acetate, 2-methylcyclohexanol, and styrene and desorption into carbon bisulfide were evaluated in the TLV to five times TLV concentration range. The experimental collection tubes were 6 mm I.D. x 83 mm long and were filled with ordinary water purification charcoal that had been size graded by sieving. A 50-mm fill required approximately 0.7 g charcoal. Desorption was carried out in 10 ml of CS_2 and allowed to stand for approximately 1 hr. The following conclusions were drawn from the experimental results:

1. The optimum carbon particle size was found to be that of those passing through 10 mesh and retained on a 30 mesh sieve.
2. In general the optimum sample rate was 100 ml/min for the sampling tube fabricated to the above dimensions. Although most reproducible at this rate, the efficiency of recovery was not always at a maximum.
3. When sampling at 100 ml/min the recovery efficiency was relatively independent of the volume of vapor collected.
4. Higher sampling rates did not always provide the same results as higher concentrations, which indicated that the amount of vapor collected might not be the controlling factor.

An investigation reported later in 1968 by Reid and Halpin confirmed the reliability and flexibility of charcoal as a collecting medium.[146a] The conditions employed were quite similar to those reported by Franst and Hermann, and later recommended by NIOSH (Table I.25). Excellent recoveries were obtained with relatively small sample volumes in the one half to two times TLV range (Table I.26). Trichloroethylene recoveries ranged from 88 to 100% in the concentration range 50 to 200 ppm, sampling rate range 1.5 to 2.0 lpm and sample size 3 to 10 l. The GC parameters are summarized in Table I.26. This technique is ideally suited for direct application to personnel monitoring, expecially for chlorinated hydrocarbons, aromatic hydrocarbons, ketones, alcohols, and esters with a minimum of time and effort to calibrate a GC system selected from those already established (Table I.26).

The collection of mercury vapor on charcoal has been shown to be both rapid and complete.[147] Glass tubes 4 mm I.D. x 152 mm long were packed with two 25-mm sections (180 mg each) of 20-40 mesh impregnated charcoal activated by heating at 600 to 800°C for 1 hr, which were separated and retained by fiberglass plugs. The tubes gave quantitative recoveries of mercury from air or vaporized from urine, tissue, and water. Analysis of the sections separately indicated complete absorption in the first section in every case. The mercury was quantitated directly from the absorbant in a tantalum sample boat that was introduced into an atomic absorption spectrometer oxidizing air-acetylene flame. The minimum detectable quantity of mercury was 0.02, μg.

In order to minimize the variability of results obtained with charcoal sampling, NIOSH has recommended tentative standard dimensions for the charcoal tube, which are reproduced as follows:

> The charcoal sampling tube used by the National Institute of Occupational Safety and Health, Cincinnati, Ohio, for the collection and concentration of solvent vapors has been modified to meet the requirements for a personal sampler. The packing size and configuration of

*From Campbell, E. E. and Ide, H. M., *Am. Ind. Hyg. Assoc. J.*, 27, 323, 1966. With permission.

TABLE I.25

Operational Parameters for Charcoal Absorption Tubes

Parameter	Franst and Herman[146]	Reid and Halpin[146a]	NIOSH Recommendations[148,151]
Tube diameter (I.D.) (mm)	6	6	4
Tube length (mm)	83	127	70
Charcoal sieve size	10/30	12/20	20/40
Charcoal type	Water purified	Darco®	–
Charcoal weight (gm)	0.7	0.6	0.15 (2 sections)
Pressure drop adjustments (mm Hg)	–	35–40/2 lpm	25/1 lpm
Optimum sampling rate (lpm)	0.100	1–2	1
Desorption liquid	CS_2	CS_2	CS_2
Concentration range (TLV's)	1–5	0.5–2	0.5–2
Recoveries (%)	30–100	88–100	100
GC column packing	3% SF96/Celite®	20% Carbowax® 1500/ 60-80 mesh GC-22	10% FFAP/80-100 mesh acid-washed DMCS Chemosorb® W
Sample size (liters)	–	–	10

TABLE I.26

Parameters for Determination of Several Halogenated and Aromatic Hydrocarbons[a]

Compound	Chromatographic settings: Oven temperature (°C)	Detector temperature (°C)	Injection block temperature (°C)	Helium flow rate (ml/min)	Attenuation	Minimum air samples (liters)	Efficiency[b] (%)	Retention time (minutes)	Conversion factor (to ppm)
Benzene	50	50	100	175	x 2	10	89	5.8	313
Carbon tetrachloride	50	50	100	175	x 1	10	95	4.1	158.8
Ethylene dichloride	75	75	150	156	x 1	4	96	6.3	247
Methyl chloroform	50	50	100	175	x 2	4	100	4.2	183.3
Methylene chloride	50	50	100	175	x 2	4	97	4.8	288
Perchloroethylene	75	75	150	156	x 1	4	95	4.6	147.4
Toluene	75	75	150	156	x 2	4	93	5.7	266
Trichlorotrifluoroethane	50	50	100	175	x 4 & x 16	2	92	0.8	130.5
Trichloroethylene	75	75	150	156	x 1	4	95	4.2	186
Xylene	100	100	200	140	x 1	4	95	5.4	230.7

[a]In all cases the column inlet pressure is 24 psi, and the detector is set at 8 volts.
[b]Efficiences are based on one half to twice TLV concentrations and sampling rates of 1 to 2 l/min.

From Reid, F. H. and Halpin, W. R., *Am. Ind. Hyg. Assoc. J.*, 29, 390, 1968. With permission.

TABLE I.27

Results Obtained with Other Compounds*

Compound	Chamber concentration (ppm)	Air volume (liters)	Range of recoveries (ppm)	Average recoveries (ppm)	Average recoveries (%)
Perchloroethylene	50	4	45–47	46	92
	100	4	95–96	95	95
	200	4	192–193	193	97
Methyl chloroform	175	4	168	168	96
	350	4	355–363	358	102
	700	4	682–721	697	100
Ethylene dichloride	25	4	24	24	96
	50	4	45–49	47	94
	100	4	96–100	98	98
Carbon tetrachloride	5	10	4	80	80
	10	10	12	12	120
	25	10	20–25	23	92
Methylene chloride	250	3.8	234–235	234	94
	500	3.8	484	484	97
	1000	3.8	950–1010	988	99
Benzene	12.5	10	12	12	96
	25	10	21	21	84
	50	10	46–48	47	94
Toluene	100	3.8	90	90	90
	200	3.8	180–189	186	93
	400	3.8	382–391	386	96
Xylene	50	3.8	48–49	48	98
	100	3.8	90–92	91	91
	200	3.8	194	194	97
Trichlorotrifluoroethane	25	0.5	26–29	27	108
	50	2	49–51	50	100
	500	2	476–509	491	98
	1000	2	900–959	939	94
	2000	2	1756–1835	1797	90

*In each case three samples were collected at a sampling rate of 2 l/min.

From Reid, F. H. and Halpin, W. R., *Am. Ind. Hyg. Assoc. J.*, 29, 390, 1968. With permission.

the tube assure that the pressure drop across the tube will not exceed one inch of mercury when sampling is accomplished at a flow rate of one liter per minute. This feature allows battery-operated pumps to be used thus providing the advantage of complete portability, particularly necessary in personal-type sampling.

The tubes are fabricated of glass tubing 7 cm long with a 6-mm O.D. and a 4-mm I.D. The absorbent consists of two sections of 20/40 mesh-sized activated charcoal. In the tube a 100 mg adsorbent section is separated by a 2 mm portion of urethane foam from a 50 mg backup, blank adsorbent section. A 3 mm portion of urethane foam is placed between the 50 mg section and the outlet end of the tube. A plug of glass wool is placed in front of the 100 mg section at the inlet end of the tube. Both ends of the tube are immediately flame-sealed after being packed to restrict contamination of the charcoal prior to sampling (Figure I.49b).

This modified tube has also been used for atmospheric sampling of mercury vapor exposures and has potential use in sampling for contaminants other than solvent vapors and mercury vapor. The tube was developed in a continuing effort to assist industrial hygienists and other interested persons in carrying out inspection responsibilities and will be of particular interest to those involved in inspection activities under the recent Occupational Safety and Health Act of 1970 (PL 91-596).*

An additional tentative specification has been added to denote charcoal quality based on reten-

*From Kupel, R. E. and White, L. D., *Am. Ind. Hyg. Assoc. J.*, 32, 456, 1971. With permission.

tion of the target material at a concentration of two times the TLV from a 10-l total sample volume. Under these conditions there should be less than 0.1 mg of the target material retained in the backup charcoal column.

These (organic, and mercury vapor) tubes are commercially available.[149,150] The following analytical procedure has been recommended:[151]

Atmospheric solvent vapors collected on activated charcoal are desorbed with carbon disulfide prior to separation and analysis using a gas chromatograph with a flame ionization detector.

1. A straight 4-mm I.D. glass tube packed with two sections of 20-40 mesh activated charcoal.[148] Charcoal source: Barneby-Chaney, Columbus, Ohio[150] (Figure 49.b).

2. Each section of activated charcoal is placed in a small glass stoppered microcentrifuge tube. Carbon disulfide (0.5 ml) is added to the charcoal and, after desorption is complete (30 min with agitation), a known aliquot is withdrawn with a microliter syringe and introduced into the gas chromatograph.

3. Calibration curves are established by injecting known amounts of each identified material into the gas chromatograph and measuring the areas under the peaks. A curve is established using concentration vs. peak area.

Sampling Precautions and Instructions:

1. The ends of the tubes should be broken so as to provide an opening at least 2 mm in diameter (i.e., 1/2 the I.D. of the tube). A smaller opening causes the pressure drop to exceed 1 in. of mercury.

2. The smaller section of charcoal is used as a backup and should therefore be positioned nearest the sampling pump. Sampling should be done with the charcoal tube in a vertical position.

3. These charcoal tubes have been tested at flow rates of up to 2.5 l/min and have demonstrated 100% trapping efficiency. The recommended sampling rate is 1 lpm.

4. In most cases, a 10-l sample of ambient air is adequate.

5. Treat one charcoal tube in the same manner as the sample tubes (break, seal, and transport) except for the taking of an air sample. This tube will be used as a blank.

6. After breaking and sampling, the polyethylene caps provided should be used for capping these charcoal tubes. Masking tape is the only acceptable substitute; rubber caps should never be used.

7. Return the tubes with a sample of the solvent(s) considered to be the source of contamination. If this is impossible (e.g., the contaminant is from an unknown source or is a gas at room temperature), saturate a charcoal tube by drawing a large volume of contaminant-laden air through it.

8. Liquid samples and charcoal tube samples (with broken ends) should never be transported or mailed in the same container.

The foregoing procedure was developed by Kupel, White, and co-workers at the Bureau of Occupational Safety Health in Cincinnati, Ohio.[151] A good bibliography of charcoal sampling prior to 1970 was included in the publication of their findings. Since a solvent vapor is seldom found by itself in the industrial environment, the analytical method should be capable of simultaneously separating and quantitatively measuring all compounds in the solvent mixture with a minimum of time and effort. Optimum GC operating conditions for separation and analysis of 14 selected solvent vapors of occupational health significance were developed. The 20 ft x 1/8 in. stainless steel column was packed with 10% FFAP stationary phase on 80-100 mesh acid-washed DMCS Chromosorb® W solid support, programmed from 100 to 140°C (10 min elapsed time per sample) and flushed with helium as carrier gas. Retention data are summarized in Table I.28.

The 14 selected solvent vapors were collected with 100% efficiency on activated charcoal, either alone or in any combination, including all 14 in a mixture, in the concentration range one half to 0.5 two times TLV for each solvent. The overall efficiency was equivalent to the efficiency of the desorption phase for each solvent vapor and was over 90% for each of the solvents except ethanol and pyridine. Polar compounds tended to be displaced from the first into the second charcoal section of the collecting tube by high concentrations of nonpolar organic solvents. For example, approximately 25% of acetonitrile collected was displaced when an equal concentration (100 ppm) of petroleum ether was present, regardless of sampling rate (0.5 to 2 lpm). Chloroform,

TABLE I.28

Retention Data and Boiling Points

Compound	Retention time (seconds)	Boiling point (°C)	Retention volume (ml)
Ethyl ether	89	35	125
Isooctane	97	99	136
Carbon disulfide	120	46	168
Carbon tetrachloride	162	77	227
2-Butanone	181	80	253
Ethanol	190	79	266
Benzene	213	80	298
Trichloroethylene	248	87	347
Chloroform	266	61	372
Perchloroethylene	293	121	410
Toluene	321	111	450
n-Butyl acetate	344	127	482
p-Dioxane	362	101	507
p-Xylene	445	138	620
m-Xylene	466	139	648
o-Xylene	522	144	725
Pyridine	560	115	775

From White, L. D., Taylor, D. G., Maner, P. A., and Kupel, R. E., *Am. Ind. Hyg. Assoc. J.*, 31, 225, 1970. With permission.

2-butanone, and ethanol were similarly affected. Sampling rate and volume did not alter the adsorption (collection) efficiency.

Further refinements in the charcoal sampling technique contributed by this group of investigators included trapping the GC peak in a charcoal capillary tube and transfer to a time-of-flight mass spectrometer via the hot filament tube for qualitative identification.[152] The limits of detection were also further improved (Table I.29). As described, the system is a completely integrated unit for collection, identification, and quantitation of solvent vapors from industrial atmospheres.

The relative effectiveness of activated charcoal for the removal of gases and vapors for clean room filter installations as classified by N. R. Rowe of the Barneby-Chaney Company should be of interest to air pollution investigators.[153]

1. High capacity group – acetic acid, asphalt fumes, some Freons®, acetone, some amines, engine exhausts, chlorinated hydrocarbons, some mercaptans, phosphine, CS_2, HCl, methyl ethyl ketone, HNO_3, smog, $COCl_2$, tobacco pyrolysis vapors, and oil vapor.

2. Low capacity group – some Freons, formic acid, SO_2, some amines, some mercaptans, arsine, stibine, CO, CO_2, NO, NH_3, and CH_2O.

TABLE I.29

Limits of Detection

Compound	Parts per million	Micrograms
Isoamyl acetate	10	2.4
Benzene	20	2.9
2-Butanone	20	2.6
n-Butyl acetate	20	4.2
Isobutyraldehyde	10	1.5
n-Butyraldehyde	25	3.3
Carbon tetrachloride	20	5.6
Chloroform	20	4.4
Crotonaldehyde	10	1.3
1,2-Dichloroethane	10	1.8
p-Dioxane	20	3.2
Ethanol	50	4.2
Ethyl acetate	25	4.0
Ethylbenzene	10	1.9
Ethyl ether	10	1.3
Heptaldehyde	25	5.2
Heptane	10	1.8
Hexane	20	3.1
Methylene chloride	20	3.1
Perchloroethylene	20	6.1
Propionaldehyde	25	2.7
Isopropyl alcohol	20	2.2
Toluene	25	4.2
Trichloroethylene	20	4.9
Isovaleraldehyde	10	1.6
n-Valeraldehyde	25	3.9

From Cooper, C. V., White, L. D., and Kupel, R. E., *Am. Ind. Hyg. Assoc. J.*, 32, 383, 1971. With permission.

Monitoring pumps that have been adapted to sampling periods up to 8 hr with the standard NIOSH charcoal tube are commercially available (Figure I.49.c).[153a]

d. Other Absorbents

The AIHA-ACGIH Joint Committee on Respirators in 1971 compiled a list of solid absorbents for the substances listed in the 1970 TLV table.[154] Although primarily intended as guidelines for use in the selection of respiratory system protective devices, many of the recommendations could be applied directly to air sampling. Therefore, the entire list is reproduced here with the permission of the copyright holder (Table I.30). The use of ordinary GC packing for collection of samples for insertion directly in the

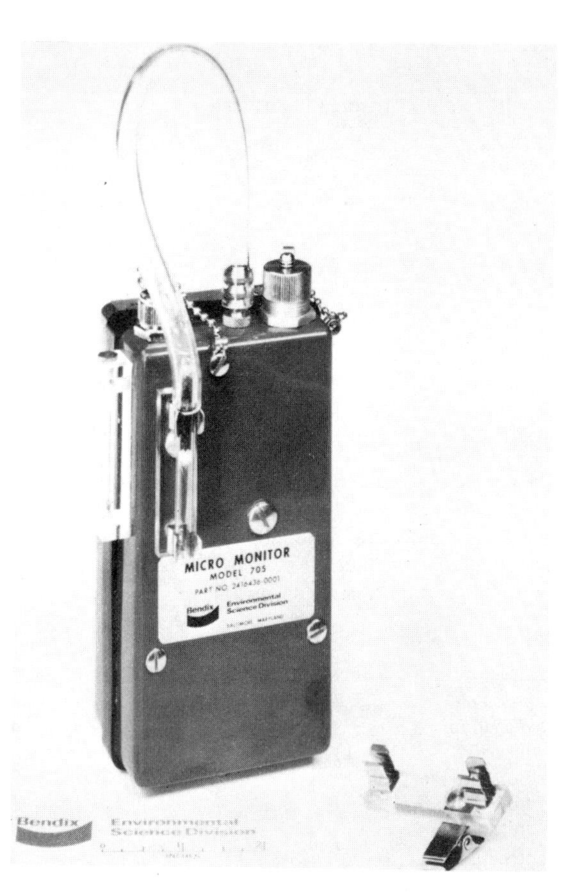

FIGURE I.49.c. Personnel monitor for charcoal absorption operation. (Courtesy of Bendix Environmental Science Division, Baltimore.)

thermal desorption port of the analyzing column has been reported, but information on this technique is scanty.[155]

A prototype contaminant sensor based on three sorbents – Poropak® Q, charcoal, and molecular sieve 5A – was developed by the Perkin-Elmer Corporation for detection of contaminants during space flight.[155a] The sorbents are contained in an accumulator cell for extraction of contaminants from air at room temperature. By heating the cell, the contaminants are desorbed into a mass spectrometer for identification and determination. A palladium-coated charcoal operated successfully as a CO detector. The lower range of detectable concentrations varied from 0.5 to 5 ppm. This system has sufficient advantages to justify adaptation to personnel monitoring.

3. Collection by Diffusion

An unusual gaseous contaminant sampling system was described by Palmes and Gunnison during the May, 1972 AIHA Conference.[173] The design incorporated the principle of molecular gas diffusion into a miniature personnel monitoring device which requires no power source for operation. The diffusion chamber was a flat-bottomed cylinder, 10 mm diameter x 60 mm length (5 ml volume), which was fitted with a capillary sealed through the flat top. Two ml of an appropriate collection reagent or solvent was used. Passive sampling with SO_2 was selected for measuring time-weighted average exposure over a concentration range of 0.1 to 10 ppm. Calculated and observed values agreed very well over a considerable range of orifice dimensions (length and diameter are controlling variables). Water vapor collection also demonstrated good precision and accuracy.

The quantity of gas transferred by diffusion from the environment through an orifice of known dimensions into a chamber maintained at zero concentration by a suitable collecting medium can be used as the basis for calculating average concentrations during the time the sampler remains in the environment.

$$J = D\ dx/dc = \text{diffusion flux in mol/cm}^2$$

Although solid adsorbents such as silica gel were not specifically mentioned, this approach to sample collection would appear to be an ideal solution to the power supply problem for positive displacement samplers.

4. The Portable Field Kit Laboratory

The "laboratory in a suitcase" has been a concept that has intrigued the investigator in the field who finds himself beyond the reach of even the basic analytical necessities such as distilled water, electrical power, balances, spectrophotometers, etc. A number of innovations have been proposed, but few have stood the test of time and even fewer have been applied successfully to personnel or biological monitoring. Those models that have survived have done so on their ability to deliver results intermediate in accuracy and precision between the relatively imprecise length of stain gas detector tubes and highly precise laboratory analysis, and on their convenience with on-the-spot results. Length of stain gas detector tube kits is not included in this category as almost without exception the colorimetric

TABLE I.30

Sorbents for Contaminants Listed in ACGIH 1970 Threshold Limit Values

Sorbent code:

1. Activated carbon
2. Soda lime
3. Combination of activated carbon and soda lime
4. Silica gel
5. Hopcalite (with drier)
6. Impregnated carbon, e.g., carbon treated with metallic salt to absorb ammonia
7. Special sorbent (manufacturers proprietary mix)
8. Filter, for highly toxic dusts
9. Filter, for toxic dusts

Note: Where two or more sorbent code numbers are listed after a substance, the sorbents are in decreasing order of preference. Where a plus (+) sign is used, both are required.
Very strong oxidizing substances in low concentrations may be sorbed by activated or impregnated carbon; however, carbon should not be used because of a possible explosion hazard with these combinations. For respiratory protection against very strong oxidizers, a self-contained breathing apparatus should be used.

List of Contaminants with Sorbent Code

Substance	Sorbent
Abate	9
Acetaldehyde	1
Acetic acid	1,2
Acetic anhydride	3,2
Acetone	1
Acetonitrile	3,1
Acetylene dichloride, see 1,2-Dichloroethylene	
Acetylene tetrabromide	1
Acrolein	1
Acrylamide – skin	1,6
Acrylonitrile – skin	3,1
Aldrin – skin	8+1
Allyl alcohol – skin	1
Allyl chloride	1,3
Allyl glycidyl ether (AGE)	1
Allyl propyl disulfide	3
2-Aminoethanol, see Ethanolamine	
2-Aminopyridine	1,7
Ammonia	7,6,4
Ammonium sulfamate (Ammate)	9
n-Amyl acetate	1
sec-Amyl acetate	1
Aniline – skin	1,7
Anisidine (*o,p*-isomers) – skin	9+1
Antimony & compounds (as Sb)	9
ANTU® (Alpha naphthyl thiourea)	9+1
Arsenic & compounds (as As)	6+9
Arsine	6,7
Azinphosmethyl – skin	9+1
Barium (soluble compounds)	9+6
Benzene (Benzol) – skin	1
Benzidine – skin	8+1
p-Benzoquinone, see Quinone	
Benzoyl peroxide	9+1
Benzyl chloride	1,3
Beryllium	8
Biphenyl, see Diphenyl	
Boron oxide	9
Boron tribromide	5+3
Boron trifluoride	5+3
Bromine	3
Bromine pentafluoride	2,7
Bromoform – skin	1
Butadiene (1,3-Butadiene)	1
Butanethiol, see Butyl mercaptan	
2-Butanone	1
2-Butoxy ethanol (Butyl Cellosolve®) – skin	1
Butyl acetate (*n*-Butyl acetate)	1
n-Butyl acetate	1
sec-Butyl acetate	1
tert-Butyl acetate	1
Butyl alcohol	1
sec-Butyl alcohol	1
tert-Butyl alcohol	1
Butylamine – skin	7,6,4
tert-Butyl chromate (as CrO_3) – skin	8
n-Butyl glycidyl ether (BGE)	1
Butyl mercaptan	1
p-tert-Butyltoluene	1
Cadmium (metal dust and soluble salts)	9,6
Cadmium oxide fume	8
Calcium arsenate	9
Calcium oxide	9
Camphor	9+1
Carbaryl (Sevin®)	9+1
Carbon black	9
Carbon dioxide	2
Carbon disulfide – skin	1
Carbon monoxide	5
Carbon tetrachloride – skin	1

From AIHA-ACGIH Joint Committee on Respirators, *Am. Ind. Hyg. Assoc. J.*, 32, 404, 1971. With permission.

TABLE I.30 (continued)

Sorbents for Contaminants Listed in ACGIH 1970 Threshold Limit Values

Substance	Sorbent
Chlordan – skin	9+1
Chlorinated camphene – skin	1
Chlorinated diphenyl oxide	1
Chlorine	3,6
Chlorine dioxide	2,7
Chlorine trifluoride	2,7
Chloroacetaldehyde	1
Chloroacetophenone (Phenacyl chloride)	1
Chlorobenzene (Monochlorobenzene)	1
O-Chlorobenzylidene malononitrile (OCBM)	3,1
Chlorobromomethane	1
2-Chloro-1,3-butadiene, see Chloroprene	
Chlorodiphenyl (42% chlorine) – skin	1,3
Chlorodiphenyl (54% chlorine) – skin	1,3
1-Chloro-2,3-epoxypropane, see Epichlorohydrin	
2-Chloroethanol, see Ethylene chlorohydrin	
Chloroethylene, see Vinyl chloride	
Chloroform (Trichloromethane)	1
1-Chloro-1-nitropropane	1
Chloropicrin	1
Chloroprene (2-Chloro-1,3-butadiene) – skin	1
Chromic acid and chromates (as CrO_3)	8+2
Chromium, soluble chromic or chromous salts (as Cr)	9
Metal & insoluble salts	9
Coal tar pitch volatiles (benzene soluble fraction – anthracene, BaP, phenanthrene, acridine, chrysene, pyridine)	9+1
Cobalt, metal fume & dust	8
Copper fume	8
Dusts and mists	9
Cotton dust (raw)	9
Crag herbicide	9+1
Cresol (all isomers) – skin	1
Crotonaldehyde	1
Cumene – skin	1
Cyanide (as CN) – skin	9+2
Cyanogen	3
Cyclohexane	1
Cyclohexanol	1
Cyclohexanone	1
Cyclohexene	1
Cyclopentadiene	1
2,4-D	9+1
DDT – skin	9+1
DDVP – skin	9+1
Decarborane – skin	9+7
Demeton – skin	8+1
Diacetone alcohol (4-Methyl-2-pentanone)	1
1,2-Diaminoethane (Diazomethane), see Ethylenediamine	
Diborane	7
1,2-Dibromoethane (Ethylene dibromide) – skin	1
Dibutyl phosphate	8+1
Dibutyl phthalate	1
Dichloroacetylene	3,7
o-Dichlorobenzene	1
p-Dichlorobenzene	1
Dichlorodifluoromethane	1
1,3-Dichloro-5-dimethylhydantoin	3
1,1-Dichloroethane	1
1,2-Dichloroethane	1
1-2-Dichloroethylene	1
Dichloroethyl ether – skin	1
Dichloromethane, see Methylene chloride	
Dichloromonofluoromethane	1
1,1-Dichloro-1-nitroethane	3
1,2-Dichloropropane, see Propylenedichloride	
Dichlorotetrafluoroethane	1
Dieldrin – skin	9+1
Diethylamine	7,6,4
Diethylaminoethanol – skin	7
Diethylenetriamine – skin	7, 6, 4
Diethylether, see Ethyl ether	
Difluorodibromomethane	1
Diglycidyl ether (DGE)	1
Dihydroxybenzene, see Hydroquinone	
Diisobutyl ketone	1
Diisopropylamine – skin	7,6,4
Dimethoxymethane, see Methylal	
Dimethyl acetamide – skin	1
Dimethylamine	7,6,4
Dimethylaminobenzene, see Xylidene	
Dimethylaniline (N-Dimethylaniline) – skin	7
Dimethylbenzene, see Xylene	1
Dimethyl-1,2-dibromo-2,2-dichloroethyl phosphate (Dibrom)	9+1
Dimethylformamide – skin	1
2,6-Dimethylheptanone, see Diisobutyl ketone	
1,1-Dimethylhydrazine – skin	7,6,4
Dimethyl phthalate	1
Dimethyl sulfate – skin	1,3
Dinitrobenzene (all isomers) – skin	9+3
Dinitro-*o*-cresol – skin	9+3
Dinitrotoluene – skin	9+3
Di-*sec*-octyl phthalate [Di(2-ethylhexyl) phthalate]	1+8,9
Dioxane (Diethylene dioxide) – skin	1
Diphenyl	1
Diphenylamine	9+1,7
Diphenylmethane diisocyanate, see Methylenebisphenyl isocyanate (MDI)	
Dipropylene glycol methylether – skin	1

TABLE I.30 (continued)

Sorbents for Contaminants Listed in ACGIH 1970 Threshold Limit Values

Substance	Sorbent
Endosulfan (Thiodan®)	8+1
Endrin – skin	8+1
Epichlorohydrin – skin	1
EPN – skin	9+1
1,2-Epoxypropane, see Propylene oxide	
2,3-Epoxy-1-propanol, see Glycidol	
Ethanethiol, see Ethylmercaptan	
Ethanolamine	7,6,4
2-Ethoxyethanol – skin	1
2-Ethoxyethyl acetate (Cellosolve® acetate) – skin	1
Ethyl acetate	1
Ethyl acrylate – skin	1
Ethyl alcohol (Ethanol)	1
Ethylamine	7,6,4
Ethyl *sec*-amyl ketone (5-Methyl-3-heptanone)	1
Ethyl benzene	1
Ethyl bromide	1
Ethyl butyl ketone (3-Heptanone)	1
Ethyl chloride	1
Ethyl ether	1
Ethyl formate	1
Ethyl mercaptan	1
Ethyl silicate	1
Ethylene chlorohydrin – skin	1
Ethylenediamine	7,6,4
Ethylene dibromide, see 1,2-Dibromoethane	
Ethylene dichloride, see 1,2-Dichloroethane	
Ethylene glycol dinitrate and/or Nitroglycerin – skin	1
Ethylene glycol monomethyl ether acetate, see Methyl Cellosolve® acetate	
Ethyleneimine – skin	7,6,4
Ethylene oxide	1
Ethylidene chloride, see 1,1-Dichloroethane	
N-Ethylmorpholine – skin	1
Ferbam	9+3
Ferrovanadium dust	9
Fluoride (as F)	9+3
Fluorine	3,2
Fluorotrichloromethane	1
Formaldehyde	1
Formic acid	1
Furfural – skin	1
Furfuryl alcohol	1
Gasoline	1
Glycidol (2,3-Epoxy-1-propanol)	1
Glycol monoethylether, see 2-Ethoxyethanol	
Guthion®, see Azinphos methyl	
Hafnium	9
Heptachlor – skin	9
Heptane (*n*-Heptane)	1
Hexachloroethane – skin	1
Hexachloronaphthalene – skin	9+1
Hexane (*n*-Hexane)	1

Substance	Sorbent
2-Hexanone	1
Hexone	1
sec-Hexyl acetate	1
Hydrazine – skin	7,6,4
Hydrogen bromide	2,3+9
Hydrogen chloride	2,3+9
Hydrogen cyanide – skin	2,7
Hydrogen fluoride	2,3,7
Hydrogen peroxide, 90%	2,7
Hydrogen selenide	6
Hydrogen sulfide	3,6
Hydroquinone	9+1
Indene	1
Indium and compounds (as In)	8
Iodine	9+1
Iron oxide fume	9
Iron salts, soluble (as Fe)	9
Isoamyl acetate	1
Isoamyl alcohol	1
Isobutyl acetate	1
Isobutyl alcohol	1
Isophorone	1
Isopropyl acetate	1
Isopropyl alcohol	1
Isopropylamine	7,6,4
Isopropyl ether	1
Isopropyl glycidyl ether (IGE)	1
Ketene	1
Lead	9
Lead arsenate	8
Lindane – skin	9
Lithium hydride	8
L.P.G. (Liquified petroleum gas)	1
Magnesium oxide fume	9
Malathion – skin	9+1
Maleic anhydride	9+3
Manganese	9
Mercury – skin	6,7
Mercury (organic compounds) – skin	8+6,7
Mesityl oxide	1
Methanethiol, see Methyl mercaptan	
Methoxychlor	9+1
2-Methoxyethanol, see Methyl Cellosolve®	
Methyl acetate	1
Methyl acetylene (propyne)	1
Methyl acetylene-propadiene mixture (MAPP)	1
Methyl acrylate – skin	1
Methylal (dimethoxymethane)	1
Methyl alcohol (methanol)	1
Methylamine	7,6,4
Methyl amyl alcohol, see Methyl isobutyl carbinol	
Methyl *n*-amyl ketone (2-Heptanone)	1
Methyl bromide – skin	1
Methyl butyl ketone, see 2-Hexanone	

TABLE I.30 (continued)

Sorbents for Contaminants Listed in ACGIH 1970 Threshold Limit Values

Substance	Sorbent	Substance	Sorbent
Methyl Cellosolve® – skin	1	Paraquat – skin	9+7
Methyl Cellosolve acetate – skin	1	Parathion – skin	8+1
Methyl chloride	1	Pentaborane	7,1
Methyl chloroform	1	Pentachloranaphthalene – skin	9+1
Methylcyclohexane	1	Pentachlorophenol – skin	9+1
Methylcyclohexanol	1	Pentane	1
o-Methylcyclohexanone – skin	1	2-Pentanone	1
Methyl ethyl ketone (MEK), see 2-Butanone		Perchloroethylene	1
		Perchloromethyl mercaptan	1
Methyl formate	1	Perchloryl fluoride	2,7
Methyl iodide – skin	1	Petroleum distillates (Naphtha)	1
Methyl isoamyl ketone	1	Phenol – skin	1
Methyl isobutyl carbinol – skin	1	*p*-Phenylenediamine – skin	7
Methyl isobutyl ketone, see Hexone		Phenyl ether (vapor)	1
Methyl isocyanate – skin	1	Phenyl ether-biphenyl mixture (vapor)	1
Methyl mercaptan	1	Phenylethylene, see Styrene	
Methyl methacrylate	1	Phenylglycidyl ether (PGE)	1
Methyl propyl ketone, see 2-Pentanone		Phenylhydrazine – skin	7,6,4
Methyl silicate	1	Phosdrin® (Mevinphos) – skin	8+1
Methyl styrene	1	Phosgene (Carbonyl chloride)	3,6
Methylenebisphenyl isocyanate (MDI)	8+1	Phosphine	6
Methylene chloride (Dichloromethane)	1	Phosphoric acid	9+1
Molybdenum, soluble compounds	9	Phosphorus (yellow)	8+2
Insoluble compounds	9	Phosphorus pentachloride	9+3
Monomethylaniline – skin	1,7	Phosphorus pentasulfide	9+3
Monomethylhydrazine – skin	7,6,4	Phosphorus trichloride	9+3
Morpholine – skin	1	Phthalic anhydride	3
Naphtha (coal tar)	1	Picric acid – skin	8+3
Naphthalene	1	Pival® (2-Pivalyl-1,3-indandione)	8
β-Naphthylamine	7,6,4	Platinum (soluble salts)	8
Nickel carbonyl	3+5	Polytetrafluoroethylene decomposition products	8+3
Nickel, metal and soluble compounds	9	Propane	1
Nicotine – skin	1	β-Propiolactone	1
Nitric acid	2,3	Propargyl alcohol – skin	1
Nitric oxide	3	*n*-Propyl acetate	1
p-Nitroaniline – skin	1	Propyl alcohol	1
Nitrobenzene – skin	1	*n*-Propyl nitrate	1,7
p-Nitrochlorobenzene – skin	1	Propylene dichloride	1
Nitroethane	1	Propylenimine – skin	1
Nitrogen dioxide	2,7	Propylene oxide	1
Nitrogen trifluoride	2,7	Propyne, see Methylacetylene	
Nitroglycerin – skin	1	Pyrethrum	9
Nitromethane	1	Pyridine	1
1-Nitropropane	1	Quinone	1
2-Nitropropane	1	RDX® – skin	8
N-Nitrosodimethylamine (Dimethyl-nitrosoamine) – skin	7,6,4	Rhodium, metal fume and dusts,	8
Nitrotoluene – skin	1	Soluble salts	8
Nitrotrichloromethane, see Chloropicrin		Ronnel	9+1
Octachloronaphthalene – skin	1	Rotenone (commercial)	9+1
Octane	1	Selenium compounds (as Se)	8
Oil mist (mineral)	9+1	Selenium hexafluoride	3
Osmium tetroxide	8	Silver, metal and soluble compounds	8
Oxalic acid	9+3	Sodium fluoroacetate (1080) – skin	8
Oxygen difluoride	3,6	Sodium hydroxide	9
Ozone	7	Stibine	6

TABLE I.30 (continued)

Sorbents for Contaminants Listed in ACGIH 1970 Threshold Limit Values

Substance	Sorbent
Stoddard solvent	1
Strychnine	8
Styrene monomer (Phenylethylene)	1
Sulfur dioxide	2
Sulfur hexafluoride	3
Sulfuric acid	2
Sulfur monochloride	9+3
Sulfur pentafluoride	3
Sulfuryl fluoride	7,3
Systox®, see Demeton	
2,4,5-T	9
Tantalum	9
TEDP – skin	8+1
Teflon® decomposition products	8+3
Tellurium	8
Tellurium hexafluoride	3
TEPP – skin	8
Terphenyls	1
1,1,1,2-Tetrachloro-2,2-difluoroethane	1
1,1,2,2-Tetrachloro-1,2-difluoroethane	1
1,1,2,2-Tetrachloroethane – skin	1
Tetrachloroethylene, see Perchloroethylene	
Tetrachloromethane, see Carbon tetrachloride	
Tetrachloronaphthalene – skin	9+1
Tetraethyllead (as Pb) – skin	8+1
Tetrahydrofuran	1
Tetramethyllead (TML) (as Pb) – skin	8+1
Tetramethyl succinonitrile – skin	1
Tetranitromethane	3
Tetryl (2,4,6-Trinitrophenyl-methylnitramine) – skin	9+3
Thallium (soluble compounds) – skin	8
Thiram	9+1
Tin (inorganic compounds, except SnH_4 and SnO_2)	9
Tin (organic compounds)	8+1
Titanium dioxide	9
Toluene (toluol)	1
Toluene-2,4-diisocyanate (TDI)	9+1
o-Toluidine – skin	1
Toxaphene, see Chlorinated camphene	
Tributyl phosphate	9+3
1,1,1-Trichloroethane, see Methyl chloroform	
1,1,2-Trichloroethane – skin	1
Trichloroethylene	1
Trichloromethane, see Chloroform	
Trichloronaphthalene – skin	9+1
1,2,3-Trichloropropane	1
1,1,2-Trichloro-1,2,2-trifluoroethane	1
Tri-*o*-cresyl phosphate	8+3
Triethylamine	7,6,4
Trifluoromonobromomethane	1
Trimethyl benzene	1
2,4,6-Trinitrophenol, see Picric acid	
2,4,6-Trinitrophenylmethylnitramine, see Tetryl	
Trinitrotoluene – skin	9+1
Triphenyl phosphate	9+3
Tungsten & compounds, water soluble	9
Insoluble	9
Turpentine	1
Uranium (natural) soluble & insoluble compounds (as U)	8
Vanadium, V_2O_5 dust	8
V_2O_5 fume	8
Vinylbenzene, see Styrene	
Vinyl chloride	1
Vinyl cyanide, see Acrylonitrile	
Vinyl toluene	1
Warfarin	8
Xylene (Xylol)	1
Xylidine – skin	1
Yttrium	9
Zinc chloride fume	9
Zinc oxide fume	9
Zirconium compounds (as Zr)	9

indicator is self-contained and requires no chemical manipulations (see Section I.A.2.d, "Direct Reading Length of Stain Gas Detecting Tubes"). The field kits available currently:

a. Aromatic Isocyanates in Air – Toluene Diisocyanate (TDI) and Methylene-bis-(4-phenylisocyanate) (MDI)

A kit based on the method developed by Marcali as modified by Grim and Linch for the detection of airborne TDI and MDI in the concentration range 0.01 to 0.08 ppm has produced consistently reliable results in the field and is commercially available.[157] The diisocyanate absorbed in the acid reagent (aqueous HCl and acetic acid) as the air sample is drawn through the sampler is hydrolyzed quantitatively to toluene diamine. The addition of sodium nitrite reagent diazotizes the aromatic amine which is then coupled to "diamine" (1-*α*-naphthylethylene-diamine) to produce a blue-red azo dye. The color density is translated to ppm of either TDI or MDI by visual comparison with hermetically sealed, nonfading liquid color standards retained in a

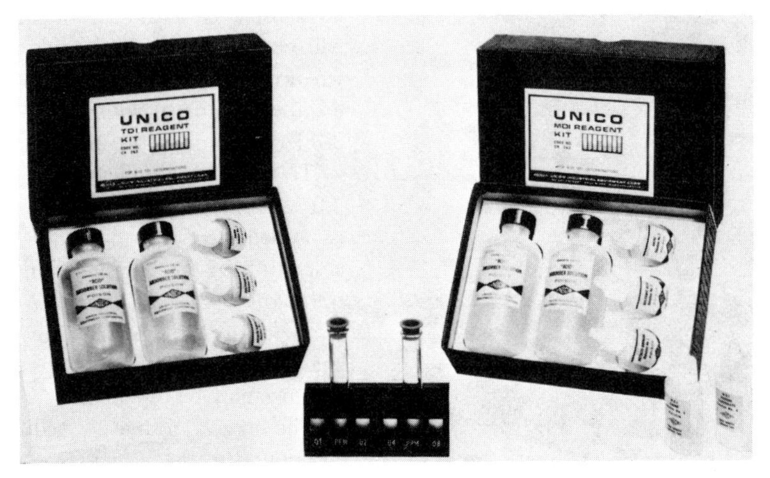

FIGURE I.50. TDI/MDI reagents and color standards. Developed by the E.I. Du Pont de Nemours Company, the Uncio® TDI/MDI reagents and color standards kit provides the sensitivity required for the rapid colorimetric determination of atmospheric concentrations of toluene diisocyanate in the threshold limit range. Results within 0.01 to 0.08-ppm limits can be obtained in less than 20 min. (Courtesy of Unico Environmental Instruments, Inc., Fall River, Mass.)

comparator block. Although intended primarily for use with the midget impinger, by volumetric adjustments the assembly can be used directly with the microimpinger recommended for personnel monitoring. Personnel monitoring kits complete with battery-operated samplers, micro-impingers, reagents, and color standards also are available (Figure I.50).[157]

A similar TDI in air kit based on the orange-brown color formed by the reaction of the toluene diamine that is produced by quantitative hydrolysis of the diisocyanate in the acidic absorbing reagent with glutaconic aldehyde generated *in situ* has been reported.[158] The dye is concentrated by absorption on ion-exchange beads and compared visually with colorimetric standards. The instantaneous readout of the color, which forms as rapidly as TDI is absorbed, which permits detection of short duration excursions of TDI concentrations above the TLV, is a distinct advantage. Application to MDI was not mentioned. However, the investigator loses the option of laboratory confirmation of the color density by spectrophotometric analysis. A commercial kit has been offered to the trade.[159] Application to personnel monitoring would be quite simple and would provide immediate warning of excessive concentrations to the wearer and his associates. Sensitivity as related to sampling time was reported to be

0.005 ppm in 10 min at 2.8 lpm sampling rate
0.16 ppm in 1¼ min at 2.8 lpm sampling rate.

A test paper method also has been described, but little is known relative to its performance in the field.[160] A 5-l sample was drawn at 1 lpm through a 1-cm diameter paper treated with sodium nitrite and 2-hydroxy-11H-benzo(*a*) carbazole-3-carboxy-*p*-anisidide (Brenthol® GB) to produce a red-gray stain at 0.1 ppm.

b. Tetraalkyl Lead Compounds in Air – Tetraethyl Lead (TEL) and Tetramethyl Lead (TML)

The dithizone procedure on which the commercially available Telmatic 150® TEL/TML lead in air analyzer was based involves an immiscible solvent (chloroform) extraction to isolate the lead as the red lead dithizone complex, which requires a somewhat more elaborate suitcase laboratory[163] (Figures I.51 and I.52). The Telmatic sampling unit was designed for "walk-around" surveys to evaluate the efficiency of lead gasoline storage tank cleaning operations. The "laboratory" is complete from deionized water to permanent glass color standards. The legends for Figures I.51 through I.53 tell the story rather completely.

A measured sample of air, 2 ft^3, is drawn at a constant rate of 0.1 ft^3/min through a scrubber containing an iodine in methanol solution. As the

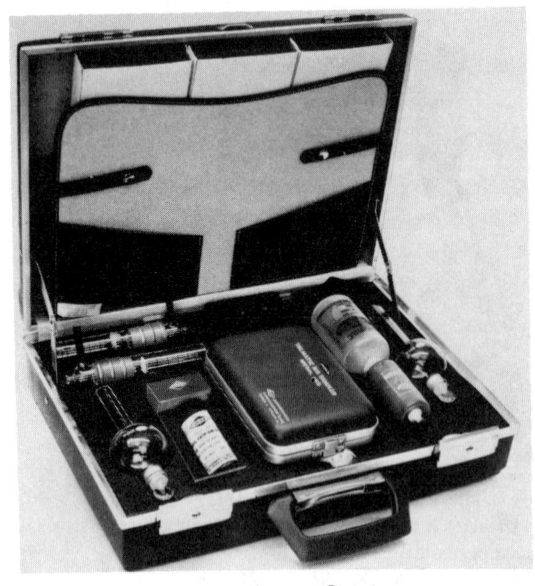

FIGURE I.51. The Telmatic 150® lead in air analyzer. The outer case is open to show compact storage and accessibility of equipment. (Courtesy of Unico Environmental Instruments, Inc., Fall River, Mass.)

air sample is scrubbed, organic lead compounds will combine with the iodine to form lead iodide when the reaction is complete. The amount of lead in the scrubber solution is determined by a dithizone colorimetric analysis.

To insure a complete reaction to lead iodide, a 2.0 *N* iodine solution in aqueous potassium iodide is added to the methanol solution following collection of the sample. This mixture must then be warmed to at least 26.67°C and held at that temperature for 5 min before proceeding with the analysis. Unless these steps are followed, low readings may be obtained.

The excess iodine is reduced to a colorless compound in the comparator tube by the addition of sulfite solution "A." The resulting aqueous solution is then shaken with a chloroform solution of dithizone which reacts with the lead compounds to form colored complexes. The concentration of lead is determined by comparing the color density with permanent glass color standards in a Unico comparator which is calibrated to read directly in $\mu g/ft^3$ when a 2-ft^3 sample is collected. (The color discs are arranged so that determinations can be made quickly in the field within ± 1

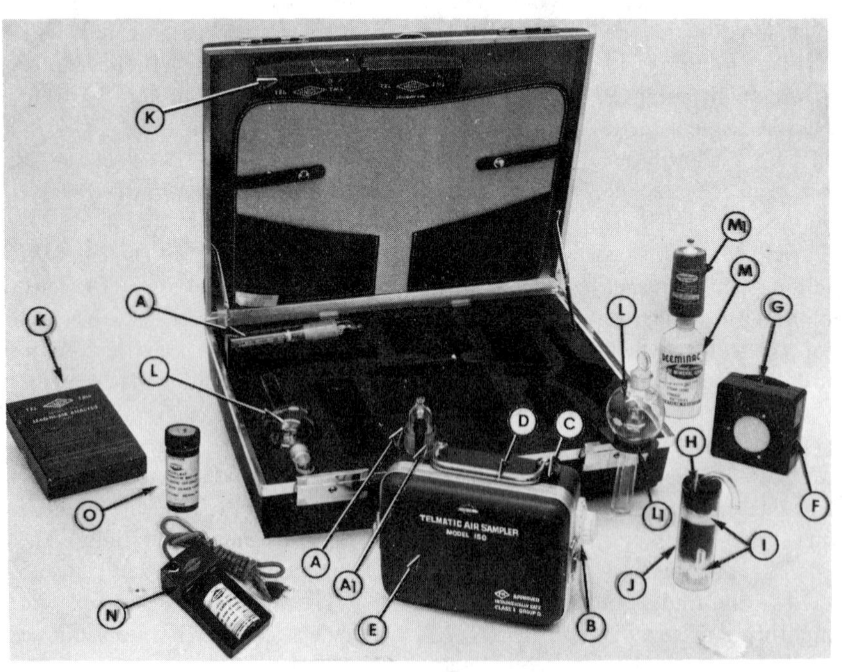

A. Standard midget impinger-scrubber
A_1. Opening in air sampler case to permit scrubber entry
B. Timer
C. Potentiometer "on-off" and adjust
D. Carrying handle
E. Telmatic air sampler
F. Unico comparator
G. Permanent glass color standard
H. Charcoal trap
I. Glass wool (above and below charcoal)
J. Activated carbon charcoal
K. Reagents
L. Comparator tube
L_1. Comparator tube holder
M. Deeminac water purifier
M_1. Deeminac water purifier element
N. 115-V AC battery charger
O. 8.4-V 500-mah battery (rechargeable nickel cadmium)
P. 115-V AC charging cord

FIGURE I.52. The Telmatic 150® complete kit components. (Courtesy of Unico Environmental Instruments, Inc Fall River, Mass.)

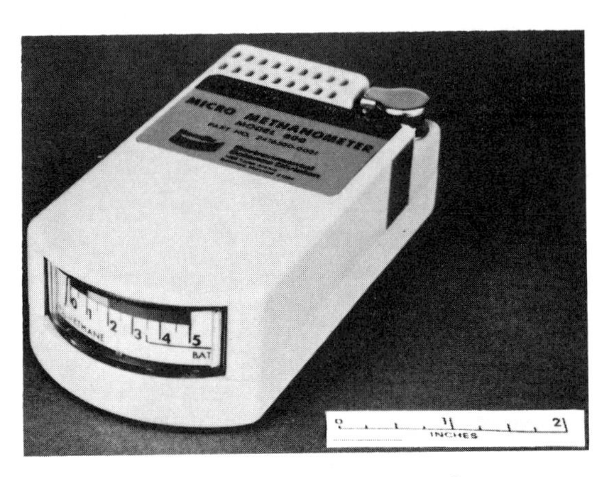

FIGURE I.53. The Micro Methanometer®, Model 800. (Courtesy of Bendix Environmental Science Division, Baltimore.)

$\mu g/ft^3$ in the range between 0 and 10.)

By substituting the personnel monitor assembly employed for the evaluation of tetraalkyl lead exposure in the manufacturing environment and iodine monochloride reagent for the alcoholic iodine, direct conversion from "walk-around" to personnel monitoring could be attained with a minimum of volumetric changes.[96] No equipment modifications would be necessary.

Conversion to personnel monitoring for mercury and its compounds, including alkyl mercury, also would require only a minimum of adjustments to the reagents and volumetric requirements.[164] The only limitation to the flexibility of this basic field kit is the ingenuity of the investigator, as many other "wet" colorimetric analysis procedures can be adapted with but few alterations in the glassware, and calibration of additional color standards. For example, the Reilly procedure for MDI in air would fit into this system with only recalibration of the color standards needed.[162]

c. The Unsaturated Perfluoroalkenes – Perfluoroisobutylene (PFIB) and Hexafluoropropene (HFP)

The yellow color produced when PFIB or HFP reacts with a mixture of pyridine and piperidine in methanol provided the basis for a laboratory kit, but this development was not commercially exploited.[165] A 28-l sample collected at a 2.8-lpm rate was sufficient to detect 0.1 ppm PFIB with certainty. Permanent liquid visual color standards were made up from aqueous $K_2Cr_2O_7$. The original specifications were based on the midget impinger for collection of the sample, but only minor volumetric adjustments would be necessary for conversion to personnel monitoring with the microimpinger.[121]

d. Inorganic Lead

A technique for the detection of lead-containing particulate retained on filter paper by reaction with tetrahydroxyquinone to produce a pink to red color[28] has been incorporated into a self-contained air sampler and is available commercially.[29] For details see Section I.A.l.e.7, "Lead."

e. Gas Chromatographs

Battery-operated self-contained instruments for the field have become available recently. One packaged unit which weighs less than 10 kg provides a 1.8 m x 3.2 mm stainless steel column packed with 3% GC grade SE-30 on 80-100 mesh Chromosorb W H.P., lecture bottle of carrier gas, a choice of electron capture, H_2 flame ionization or thermal conductivity detector, and oven for heating the column.[166] An ambient temperature model fitted with thermistor-type thermal conductivity detector, 1.6 m x 3.2 mm column packed with molecualr sieve 5A, and 44-hr carrier gas supply provides an alternate selection.[167]

f. Water Analysis Kits

This area of application has been exploited to a much greater extent than air analysis. One manufacturer in particular has developed a number of ingenious kits, a suitable electronic colorimeter, and a continuous color scale visual comparator. Complete units for acidity, alkalinity (pH), halogens, halides, CO_2, Ca, Ba, Al, Cr, Cu, cyanide, detergents, hydrazine, H_2S, Fe, Mn, Ni, NH_3, $-NO_3$, $-NO_2$, O_2, phenol, phosphate, Se, Si, Ag, $-SO_4$, $-SO_3$, tannin, turbidity, and Zn are available.[168] Collection of any one of these common contaminants in an appropriate reagent in a microimpinger combined with visual color comparison readout would provide a very convenient "package" for in-the-field monitoring.

g. Electronic Hazardous Gas Detectors

The development of a pocket-size, handheld, battery-operated methane detector should be mentioned as a personnel monitor to detect the presence of explosive gases in the ambient work environment. The Model 800 Micro Methanometer® measures

36 x 60 x 117 mm, weighs 255 g, and operates on rechargeable batteries (Figure I.53). The all solid-state circuitry provides a direct readout as percent (0 to 5%) methane on an illuminated dial.[174]

C. REFERENCES FOR CHAPTER I

1. **Stokinger, H. E.,** Threshold Limit Values of Airborne Contaminants and Physical Agents with Intended Changes Adopted by ACGIH for 1971, Committee on Threshold Limits, Cincinnati, O., 1971.
2. Kitagawa® SO_2 Test Paper, National Environmental Instruments, Inc., P.O. Box 590, Pilgrim Station, Warwick, R.I., 1972.
3. Metronics Colortec (T. M.) H_2S Detector, Metronics Associates, Inc., 3201 Porter Drive, Palo Alto, Cal., 1971.
4. **Morgenstern, A. S., Ash, R. M., and Lynch, J. R.,** The evaluation of detector tube systems. I. Carbon monoxide, *Am. Ind. Hyg. Assoc. J.,* 31, 630, 1970.
5. Joint ACGIH-AIHA Committee on Direct Reading Gas Detecting Systems, Direct reading gas detecting tube systems, *Am. Ind. Hyg. Assoc. J.,* 32, 488, 1971.
6. The Results of Survey of the SO_2 Concentration Distribution in Yokohama City by Kitagawa Test Paper, Yokohama Municipal School Pharmacists Assoc., 1968.
7. Hydrogen Sulfide Detector, Del Mar Products, 6306 Schevers, Houston, Tex.
8. Hydrogen Sulfide Detector, Davis Emergency Equipment Co., Inc., Newark, N. J.
9. Hydrogen Sulfide – Methods for the Detection of Toxic Substances in Air, Booklet No. 1, H.M.S.O., London, 1958.
10. Hydrogen Cyanide Vapour – Methods for the Detection of Toxic Gases in Industry, Leaflet No. 2, H.M.S.O., London, 1938.
11. Sulfur Dioxide – Methods for the Detection of Toxic Gases in Industry, Leaflet No. 3, H.M.S.O., London, 1939.
12. Phosgene – Methods for the Detection of Toxic Substances in Air Booklet No. 8, H.M.S.O., London, 1939.
13. Arsine – Methods for the Detection of Toxic Gases in Industry, Leaflet No. 9, H.M.S.O., London, 1939.
14. **Linch, A. L., Lord, S. S., Kubitz, K. A., and DeBrunner, M. R.,** Phosgene in air – Development of improved detection procedures, *Am. Ind. Hyg. Assoc. J.,* 26, 465, 1965.
15. **Moureu, H., Chovin, P., and Triffert, L.,** Sur la transformation photochemique de la chloropicrin en phosgene, *Compt. Rend.,* 228, 1954, 1949.
16. **Liddel, H. F.,** A detector paper for phosgene, *Analyst,* 82, 375, 1957.
17. **Lamouroux, A.,** A new specific color reaction of phosgene, *Mem. Poudres,* 38, 383, 1956.
18. **Dixon, B. E. and Hands, G. C.,** A field method for the determination of phosgene, *Analyst,* 84, 463, 1959.
19. **Leithe, W.,** *The Analysis of Air Pollutants,* Ann Arbor Science, Ann Arbor, Mich., 1970.
20. **Jacobs, M. B.,** *The Analytical Toxicology of Industrial Inorganic Poisons,* Interscience, New York, 1967.
21. Arsine Detector Kit, Catalog No. DA-81100, Mine Safety Appliances Co., 201 North Braddock Ave., Pittsburgh, 1962.
22. *Reagent Chemicals – ACS Specifications,* 4th ed., American Chemical Society, Washington, D.C., 1968.
23. Hands, G. C. and Bartlet, A. F. F., A field method for the determination of sulfur dioxide in air, *Analyst,* 85, 147,
23. **Hands, G. C. and Bartlet, A. F. F.,** A field method for the determination of sulfur dioxide in air, *Analyst,* 85, 147, 1960.
24. British Ministry of Labor, Sulfur Dioxide, Booklet No. 5, H.M.S.O., London.
25. Tintometer Ltd., Salisbury, Wiltshire, England.
26. Intersociety Committee for a Manual of Methods for Air Sampling and Analysis, Tentative method of analysis for sulfur dioxide content of the atmosphere – Colorimetric 42401-01-69T, *Health Lab. Science,* 7, 4, 1970.
27. **Hemeon, W. C. and Haines, G. F.,** Automatic sampling and determination of micro-quantities of mercury vapor, *Am. Ind. Hyg. Assoc. J.,* 22, 75, 1961.
28. **Quino, E. A.,** Field method for the determination of inorganic lead fumes in air, *Am. Ind. Hyg. Assoc. J.,* 20, 134, 1959.
29. Unico Lead-In-Air Detector, National Environmental Instruments, Inc. (formerly Union Industrial Equipment Corp.), P.O. Box 590, Pilgrim Station, Warwick, R.I., 1960.
30. **Yeager, D. W., Cholak, J., and Henderson, E. W.,** Determination of lead in biological and related material by atomic absorption spectrophotometry, *Environ. Sci. Technol.,* 5, 1020, 1971.
31. **Matson, W. R., Allen, H. E., and Mancy, K. H.,** Trace metal characterization in aquatic environments by anodic stripping voltammetry, *J. Water Pollut. Control Fed.,* 42, 573, 1970.
32. The Multiple Anodic Stripping Analyzer, MASA 2014, Environmental Sciences Associates, 49 Hampshire Street, Cambridge, Mass., 1971.

33. **Keenan, R. G.,** Determination of Lead in Air and in Biological Materials: The "USPHS" Double Extraction, Mixed Color Dithizone Method, ACGIH, Cincinnati, O., 1963; also *Am. Ind. Hyg. Assoc. J.*, 24, 481, 1963.
34. **Leroux, J. and Mahmud, M.,** Flexibility of X-ray emission spectrography as adapted to microanalysis of air pollutants, *J. Air Pollut. Control Assoc.*, 20, 402, 1970.
35. **Judd, S. H. and Tebbens, B. S.,** Air sampling and analysis of lead on paper, *Am. Ind. Hyg. Assoc. J.*, 22, 86, 1961.
36. AISI Samplers, Bulletins 2322A and 2331, Research Appliance Co., Craighead Rd., Allison Park, Pa., 1971.
37. Tape Samplers, 2800 Series, Models 80TS-1, H_2S, and HF, National Environmental Instruments, Inc., P.O. Box 590, Pilgrim Station, Warwick, R.I., 1971.
37a. **Katz, M., Ed.,** Tentative Method of Analysis for Particulate Fluorides in the Atmosphere (Separation and Collection with a Double Paper Tape Sampler), *Intersociety Committee on Methods for Air Sampling and Analysis, Am. Public Health Assoc.*, Washington, D.C., 1972.
38. **Stone, I.,** Detection of fluorides using the zirconium lake of alizarin, *J. Chem. Educ.*, 8, 347, 1931; see also Mavrodineanu, R., U.S. Patent 2,823,984.
39. **Leithe, W.,** *The Analysis of Air Pollutants,* Ann Arbor Science, Ann Arbor, Mich., 1970, 210.
40. **Bennet, E. L., Gould, C. W., and Swift, E. H.,** Systematic qualitative tests for certain acid elements in organic compounds, *Anal. Chem.*, 19, 1035, 1947.
41. Hydrogen Fluoride-In-Air Detector, Catalog No. DH-81400, Mine Safety Appliances Co., 201 North Braddock Ave., Pittsburgh, 1962.
42. Unsymmetrical Hydrazine Detector Kit, Catalog No. DH-82050, Mine Safety Appliances Co., 201 North Braddock Ave., Pittsburgh, 1962.
43. **Plantz, C. A., McConnaughey, P. W., and Jenca, C. J.,** Colorimetric personal dosimeter for hydrazine fuel handlers, *Am. Ind. Hyg. Assoc. J.*, 29, 162, 1968.
44. Boranes-In-Air Detector, Catalog No. DA-82100, Mine Safety Appliances Co., 201 North Braddock Ave., Pittsburgh, 1962.
45. **Ruch, W. E.,** *Chemical Detection of Gaseous Pollutants,* Ann Arbor Science, Ann Arbor, Mich., 1966.
46. Chromic Acid Mist Detector, Catalog No. DA-79011, Mine Safety Appliances Co., 201 North Braddock Ave., Pittsburgh, 1962.
47. Photovolt Photometers, Models 200-M, 502-M, 520-M, and 610, Bulletin 810, Photovolt Corp., 1115 Broadway, New York, 1968.
48. Paper Tape Densitometer, Catalog No. 1401, National Environmental Instruments, Inc., P.O. Box 590, Pilgrim Station, Warwick, R.I., 1971.
49. RAC Spot Evaluators, Bulletins 2296, 2332A, and 2348 (models numbered same as bulletins), Research Appliance Co., Craighead Rd., Allison Park, Pa., 1971.
50. **Silverman, L. and Gardner, G. R.,** Potassium pallado sulfide method for carbon monoxide detection, *Am. Ind. Hyg. Assoc. J.*, 26, 97, 1965.
51. Deadstop Carbon Monoxide Detector, Devco Engineering, Inc., 36 Pier Lane West, Fairfield, N.J., and Deco Systems, Inc., P.O. Box 55562, Houston, Tex., 1968.
51a. **McFee, D. R., Lavine, R. E. and Sullivan, R. J.,** Carbon monoxide, a prevalent hazard indicated by detector tabs, *Am. Ind. Hyg. Assoc. J.*, 31, 749, 1970.
51b. Air Quality Criteria for Carbon Monoxide, Publication No. AP-62, Natl. Air Pollut. Control Admin., Environmental Health Service, Public Health Service, Department of Health, Education and Welfare, Washington, D.C., March 1970.
52. Portable Personnel Carbon Monoxide Alarm, Unico® Model 555, National Environmental Instruments, Inc. (formerly Union Industrial Equipment Corp.), P.O. Box 590, Pilgrim Station, Warwick, R.I., 1964.
53. **Linch, A. L. and Pfaff, H.,** Carbon monoxide – Evaluation of exposure potential by personnel monitor surveys, *Am. Ind. Hyg. Assoc. J.*, 32, 745, 1971.
54. **Falgout, D. A. and Harding, C. I.,** Determination of H_2S exposure by dynamic sampling with metallic silver filters, *J. Air Pollut. Control Assoc.*, 18, 15, 1968.
55. **Sanderson, H. P., Thomas, R., and Katz, M.,** Limitations of the lead acetate impregnated paper tape method for hydrogen sulfide, *J. Air Pollut. Control Assoc.*, 16, 329, 1966.
56. **Pfeil, R. W.,** Crayon for Detection of G Agents (Chemical Warfare Gases), U.S. Patent 2,929,791, 1960.
57. **Witten, B. and Prostak, A.,** Sensitive detector crayons for phosgene, hydrogen cyanide, cyanogen chloride and lewisite, *Anal. Chem.*, 29, 885, 1957.
58. Aromil Detector Crayons, Aromil Chemical Company, 5646 Belle Ave., Baltimore, 1957.
59. Gaseous Oxygen Analyzer, BMI Series OA-200, Model 322R with audible alarm, Bio-Marine Industries, Inc., 303 West Lancaster Ave., Devon, Pa., 1971.
60. Oxygen Indicator, Model 244R, Mine Safety Appliances Co., 201 North Braddock Ave., Pittsburgh, 1972.
61. Thin film devices as oxygen sensors, *Am. Ind. Hyg. Assoc. J.*, 32, 672, 1971; NASA-CR-1182 (N68-35149) Clearinghouse for Federal Scientific and Technical Information, Springfield, Va.
62. **Saltzman, B. E.,** Direct Reading Colorimetric Indicators, in *Air Sampling Instruments for Evaluation of Atmospheric Contaminants* 4th ed., ACGIH, Cincinnati, O., 1972.

115. **Lippmann, M., Ed.,** *Air Sampling Instruments for Evaluation of Atmospheric Contaminants,* 4th ed., ACGIH, Cincinnati, O., 1972.
116. **Roberts, L. R. and McKee, H. C.,** Evaluation of absorption sampling devices, *J. Air Pollut. Control Assoc.,* 9, 51, 1959.
117. *Recommended Practices for Sampling Atmospheres for Analysis of Gases and Vapors,* ASTM designation D, Am. Soc. Testing Materials, Philadelphia, 1960, 1605.
118. **Elkins, H. B.,** *The Chemistry of Industrial Toxicology,* 2nd ed., John Wiley and Sons, New York, 1959.
119. **Calvert, S. and Workman, W.,** The efficiency of small gas absorbers, *Am. Ind. Hyg. Assoc. J.,* 22, 318, 1961.
120. **Monkman, J. L., Ed.,** Analytical Guides, AIHA, Akron, O., 1965.
121. **Linch, A. L. and Corn, M.,** The standard midget impinger – design improvement and miniaturization, *Am. Ind. Hyg. Assoc. J.,* 26, 601, 1965.
122. **Neale, E. and Perry, B. J.,** The sampling and analysis of toxic vapors in the field, *Analyst,* 84, 226, 1959.
123. **Yaffee, C. D., Beyers, D. H., and Horsey, A. D.,** *Encyclopedia of Instrumentation for Industrial Hygiene,* University of Michigan, Ann Arbor, 1956, 211.
124. **West, P. W. and Gaeke, G. C.,** Fixation of sulfur dioxide as disulfitomercurate (II) and subsequent colorimetric estimation, *Anal. Chem.,* 21, 361, 1960.
125. **Linch, A. L.,** The spill-proof microimpinger, *Am. Ind. Hyg. Assoc. J.,* 28, 497, 1967.
126. **Linch, A. L., O'Connor, G. B., Barnes, J. R., Killian, A. S., and Neeld, W. E.,** Methylene-bis-ortho-chloro-aniline (MOCA®): Evaluation of hazards and exposure control, *Am. Ind. Hyg. Assoc. J.,* 32, 802, 1971.
127. **Katz, M.,** Ed., Tentative method of analysis for nitrogen dioxide content of the atmosphere (Griess-Saltzman reaction) (42602-01-68T), *Health Lab. Sci.,* 6, 106, 1972; see also Katz, M., *Methods of Air Sampling and Analysis,* Intersociety Committee, Am. Public Health Assoc., Washington, D.C., 1972.
128. **McCaldin, R. D. and Hendrickson, E. R.,** Use of a gas chamber for testing air samplers, *Am. Ind. Hyg. Assoc. J.,* 20, 509, 1959.
129. Micro-glass-spheres ("Uni-Spheres"), Class 4, No. 203, Microbeads Division, Cataphote Corp., Jackson, Miss., 1969.
129a. **Scheflan, L. and Jacobs, M. B.,** *The Handbook of Solvents,* Van Nostrand Reinhold, New York, 1953.
130. **Peterson, J. E., Hoyle, H. B., and Schneider, E. J.,** The analysis of air for halogenated hydrocarbon contaminants by means of absorption on silica gel, *Am. Ind. Hyg. Assoc. Quart.,* 17, 429, 1956.
131. **Moffet, D. A., Doherty, T. F., and Monkman, J. L.,** Collection and determination of micro amounts of benzene and toluene in air, *Am. Ind. Hyg. Assoc. Quart.,* 17, 186, 1956.
132. **Erley, D. S.,** Infrared analysis of air contaminants trapped on silica gel, *Am. Ind. Hyg. Assoc. J.,* 23, 388, 1962.
133. **Urone, P. and Smith, J. E.,** Analysis of chlorinated hydrocarbons with the gas chromatograph, *Am. Ind. Hyg. Assoc. J.,* 22, 36, 1961.
134. **Altshuller, A. P., Bellar, T. A., and Clemons, C. A.,** Concentration of hydrocarbons on silica gel prior to gas chromatographic analysis, *Am. Ind. Hyg. Assoc. J.,* 23, 164, 1962.
135. **Pernell, C.,** The collection and analysis of halogenated hydrocarbon vapors employing silica gel as an absorbing agent, *J. Ind. Hyg. Toxicol.,* 26, 331, 1944.
136. **Fahy, J. P.,** Determination of chlorinated hydrocarbons in air, *J. Ind. Hyg. Toxicol.,* 30, 207, 1948.
137. **Jacobs, M. B.,** *The Analytical Chemistry of Industrial Poisons, Hazards and Solvents,* 2nd ed., Interscience, New York, 1949.
138. **Campbell, E. E. and Ide, H. M.,** Air sampling and analysis with micro columns of silica gel, *Am. Ind. Hyg. Assoc. J.,* 27, 323, 1966.
139. Individual Protective and Detection Equipment, Technical Manual TM3-290, Department of the Army; Technical Order TO39C-10C-1, Department of the Air Force, 1953, 56.
140. **Hay, E. B.,** Exposure to aromatic hydrocarbons in a coke oven by-product plant, *Am. Ind. Hyg. Assoc. J.,* 25, 386, 1964.
141. **Sherwood, R. J.,** The monitoring of benzene exposure by air sampling, *Am. Ind. Hyg. Assoc. J.,* 32, 840, 1971.
142. **Feldstein, M., Bolestrieri, S., and Leoaggi, D. A.,** The use of silica gel in source testing, *Am. Ind. Hyg. Assoc. J.,* 28, 381, 1967.
143. **Van Mourik, J. H. C.,** Experiences with silica gel as adsorbent, *Am. Ind. Hyg. Assoc. J.,* 26, 498, 1965.
144. Charcoal tubes for OSHA compliance of certain chemicals in air, Environmental Compliance Corp., Venetia, Pa., 1972.
145. **Whitman, N. E. and Johnston, A. E.,** Sampling and analysis of aromatic hydrocarbon vapors in air: A gas-liquid chromatographic method, *Am. Ind. Hyg. Assoc. J.,* 25, 464, 1964.
145a. **Durham, W. F. and Wolfe, H. R.,** Measurement of the exposure of workers to pesticides, *Bull. W.H.O.,* 26, 75, 1962.
146. **Franst, C. L. and Hermann, E. R.,** Charcoal sampling tubes for organic vapor analysis by gas chromatography, *Am. Ind. Hyg. Assoc. J.,* 27, 68, 1966.
146a. **Reid, F. H. and Halpin, W. R.,** Determination of halogenated and aromatic hydrocarbons in air by charcoal tube and gas chromatography, *Am. Ind. Hyg. Assoc. J.,* 29, 390, 1968.
146b. **Roper, P.,** Specifications for Charcoal Sampling Tubes, private communication, January 17, 1973.
147. **Moffitt, A. E. and Kupel, R. E.,** A rapid method employing impregnated charcoal and atomic absorption spectrophotometry for the determination of mercury, *Am. Ind. Hyg. Assoc. J.,* 32, 614, 1971.

148. **Kupel, R. E. and White, L. D.**, Report on a modified charcoal tube, *Am. Ind. Hyg. Assoc. J.*, 32, 456, 1971.

149. MSA Charcoal Sampling Tubes, Catalog Nos. 459,004 (organic vapors) and 459,003 (mercury vapor), Mine Safety Appliances Co., 400 Penn Center Blvd., Pittsburgh, Pa. (1972).

150. Charcoal Sampling Tubes, Barneby-Chaney, Cassady at 8th, Columbus, Ohio.

151. **White, L. D., Taylor, D. G., Maner, P. A., and Kupel, R. E.**, A convenient optimized method for the analysis of selected solvent vapors in the industrial atmosphere, *Am. Ind. Hyg. Assoc. J.*, 31, 225, 1970.

152. **Cooper, C. V., White, L. D., and Kupel, R. E.**, Qualitative detection limits for specific compounds utilizing gas chromatographic fractions, activated charcoal and a mass spectrometer, *Am. Ind. Hyg. Assoc. J.*, 32, 383, 1971.

153. **Rowe, N. R.**, Removal of contaminant gases by adsorption, *Ind. Hyg. News Rep.*, No. 6, 3, June 1963; Second Annu. Meeting Am. Assoc. Contamination Control, May 1–3, 1963.

153a. Micro Monitor, Model 700, Bendix Environmental Science Division, 1400 Taylor Ave., Baltimore.

154. AIHA-ACGIH Joint Committee on Respirators, sorbents for contaminants listed in ACGIH 1970 threshold limit values, *Am. Ind. Hyg. Assoc. J.*, 32, 404, 1971.

155. **Cropper, F. R. and Kaminsky, S.**, Determination of toxic organic compounds in admixtures in the atmosphere by gas chromatography, *Anal. Chem.*, 35, 735, 1963.

155a. Perkin-Elmer Corp., Laboratory Contaminant Sensor – Final Report, Contract NAS-1-7266, NASA CR-66606, March 1968, CFSTI: N68-23461.

156. **Grim, K. E. and Linch, A. L.**, Recent isocyanate-in-air analysis studies, *Am. Ind. Hyg. Assoc. J.*, 25, 285, 1964.

157. TDI/MDI Reagents and Color Standards, Unico, Catalog No. 5965, National Environmental Instruments, Inc., P.O. Box 590, Pilgrim Station, Warwick, R. I.

158. **Belisle, J.**, A portable field kit for the sampling and analysis of toluene diisocyanate in air, *Am. Ind. Hyg. Assoc. J.*, 30, 41, 1969.

159. The TDI Analyzer, John A. Pendergrass Co., 2643 South Riviera Dr., White Bear Lake, Minn.

160. **Reilley, D. A.**, Test paper method for the determination of toluene diisocyanate vapour in air, *Analyst*, 93, 178, 1968.

161. **Meddle, D. W. and Wood, R.**, A method for the determination of aromatic isocyanates in air in the presence of primary aromatic amines, *Analyst*, 95, 402, 1970.

162. **Reilly, D. A.**, A field method for determining 4,4′ di-isocyanatodiphenylamine (MDI) in air, *Analyst*, 92, 513, 1967.

163. "Telmatic 150" TEL/TML Lead-in-Air Analyzer, National Environmental Instruments, Inc., P.O. Box 590, Pilgrim Station, Warwick, R. I.

164. **Linch, A. L., Stalzer, R. F., and Lefferts, D. T.**, Methyl and ethyl mercury compounds – recovery from air and analysis, *Am. Ind. Hyg. Assoc. J.*, 29, 79, 1968.

165. **Marcali, K. and Linch, A. L.**, Perfluoroisobutylene and hexafluoropropene determination in air, *Am. Ind. Hyg. Assoc. J.*, 27, 360, 1966.

166. Portable Model 510 Gas Chromatograph, Analytical Instruments Development Co., 250 S. Franklin, West Chester, Pa. 1971.

167. Portable Model 7500 Gas Chromatograph, Carle Instruments, Inc., 1141 E. Ash Ave., Fullerton, Cal., 1972.

168. Water and Waste Water Procedures, 2nd revised ed., Hach Chemical Co., P.O. Box 907, Ames, Iowa, 1969.

169. **Townsend, C. R., Giarrusso, G. A., and Silverman, H. P.**, Thin Film Personal Dosimeters For Detecting Toxic Propellants, Research and Development Division, Magna Corp., Redondo Beach, Cal., Contract AF-33(615)-1751, Project 6302, Task 630203, AMRL-TR-66-231, February 1967, CFSTI, DDC: Ad 652849.

170. **Taguchi, N.**, Gas Detecting Device, British Patent 1,259, 566, January 5, 1972.

171. **Riehl, W. A. and Hager, K. F.**, Rapid detection of aniline vapors in air, *Anal. Chem.*, 27, 1768, 1955.

172. **Merz, O.**, Orientierende Messungen mit Gaspruefroehrchen bei Luft – und Ofentrocknung (Informative measuring with gas test tubes in air and oven drying), *Staub. Reinhaltung Luft*, 31(10), 399, 1971.

173. **Palmes, E. D. and Gunnison, A. F.**, Personal monitoring device for gaseous contaminants, *Am. Ind. Hyg. Assoc. J.*, 34, 78, 1973.

174. Micro Methanometer®, Model 800, Bendix Environmental Science Division, 1400 Taylor Ave., Baltimore.

D. PREFERRED LITERATURE FOR INDUSTRIAL HYGIENE SURVEYS

1. *American Industrial Hygiene Association Journal.*
2. *Archives of Environmental Health* (AMA).
3. *Journal of Occupational Medicine.*
4. *The Analyst* (British).
5. *Analytical Chemistry* (ACS).
6. **Stokinger, H. E.**, Ed., *Documentation of Threshold Limit Values*, ACGIH, Cincinnati, O., 1971.
7. Hygienic Guides, AIHA, Akron, O.
8. Analytical Guides, AIHA, Akron, O.

will be a loss of liquid from splashing or from entrainment.

2. Willson Impinger: This type samples at a rate of 5 lpm at a pressure drop of about 42 inches of water. The nozzle velocity is about 100 meters per second (same as the large impinger). The higher nozzle velocity gives a greater collection efficiency for fine particulate matter. The greater height of this impinger lessens the chance of loss through splashing liquid.[7] Collection is in 10 ml of liquid.

All glass midget impingers are available in both modifications with Teflon® baffles[6] and with impingement plates attached to the jet. These must be used when rubber dissolved from the stopper will interfere with the chemical analysis, or when an ultraviolet or infrared absorption method is used. The high cost of all-glass impingers prevents the exclusive use of this type.

Since the midget impinger is more convenient to use, it has largely replaced the bulky large impinger for field work. Moreover, the development of more sensitive chemical methods has lessened the need to sample the larger volumes of air that are collected with the large impinger. This is especially true when sensitive analytical techniques, such as ultraviolet absorption, are used. A further modification of the impinger, a micro impinger, which requires only 2–3 ml of liquid, has been described.[8] *

A flexible O-ring joint in place of the fragile ring seal to support the impinger tube in the head of the all-glass standard, tapered, jointed midget impinger not only alleviated the breakage problem but also provided lateral adjustment of the orifice to impinging surface distance and interchangeable orifices by sealing short lengths of precision bore tubing to the tip of the impinger tube. These relatively inexpensive impinger tubes can be interchanged easily, and permit selection or matching of orifices with pressure drops (ΔP) – flowrate tolerances, which make individual calibrations unnecessary. (For a detailed description of the improved midget impinger and microimpinger construction, see Chapter I, Section B.1, "Impinger Systems – Liquid Reagents," and Figures I.40 and I.41.)

2. Tolerances

The tolerances (Figure I.38) between impingers should be as small as possible in order to duplicate the rate of air flow and efficiency of collection between impingers. The inside diameter of the impinging nozzles should be close to 1.0 ± 0.025 mm for the midget impinger and 2.3 mm for the large impinger; the distance from the tip of the nozzle to the bottom of the flask (or to the attached plate) should be 5.0 ± 0.5 mm.[5,6] Each impinger purchased should be checked because wide variations are found between impingers received from various commercial suppliers.

The measurement of air flow through each impinger should be determined at a definite pressure drop across the impinger by means of a water manometer and a wet or dry gas meter. Since impinger nozzles are constant size limiting orifices, this method will provide an indirect measurement of the diameter of the orifice. The air flowrate tolerance for a set of impingers should be less than ± 5 per cent if they are to be used interchangeably. Units selected within these tolerances are available commercially.[7] Variations as high as 50 per cent have been found, particularly in all-glass impingers.

Impingers which fall outside the ± 5 per cent tolerance can be adjusted by restricting the orifice opening in a flame or by carefully enlarging it with a very fine abrasive. With reasonable care, a set of impingers can be selected or adjusted to deliver air flows within a tolerance of ± 2 per cent. If tips of impingers are chipped during use, the air flow may be affected. Chipped impingers should be rechecked and adjusted if necessary.

The 5 mm distance between the tip and the impinging surface is critical for standard air flow required for the rated collection efficiency. With large impingers, an error of ± 2 mm does not influence the efficiency,[9] but with the midget impinger the 5 mm distance should be maintained at less than ± 0.5 mm.[6] This distance can be adjusted with rubber stopper impingers, but all-glass impingers failing to meet this tolerance should be returned to the supplier. A calibration ring should be engraved on the flask to delineate this distance.[10]

3. Airflow Measurement

Measurement of air flow through an impinger for sampling purposes is easy and reliable as the impinger itself is a limiting orifice. Any of the following devices, calibrated in advance by comparison with a wet or dry gas meter or some other equally reliable primary standard, can be used:

a. A manometer or vacuum gauge to measure pressure drop across the impinger(s).

b. A limiting orifice between the impinger(s) and the pump (this requires a high pressure drop).

c. A rotameter between the impinger(s) and the pump. It may be necessary to dry the air entering the rotameter to prevent condensation of moisture therein.

d. A rotameter at the inlet of the impinger where it measures air at atmospheric pressure – not recommended for atmospheres containing dusts, mists, fumes or corrosive gases. For corrosive atmospheres an all-glass rotameter can be used. Preferably this technique should be reserved for the laboratory.*

4. Efficiency of Collection for Dust and Fume

Impingers are highly efficient for the collection of particulate matter above 1 micron in size, but the efficiency drops off rapidly for smaller particles.[9,11] For

*From Monkman, J. L., Ed., *Analytical Guides,* AIHA, Akron, O., 1965. With permission.

the latter it is preferable to use a filter paper collector or an electrostatic precipitator. Impinger efficiency is high for mists such as sulfuric acid mists.[12]

The efficiency can be increased slightly for particulates by (a) maintaining a jet velocity of 100 meters per second, (b) using a midget impinger flask which can sample at 5 lpm instead of 2.8, or (c) using a reagent which will react with and dissolve the particulate, e.g., dilute nitric acid for collecting lead fume. Two impingers in series containing a reagent may be used for collecting fume if it is recognized that the collection efficiency may be only 70 to 80 per cent.*

5. Evaluation of the Microimpinger

The following material is presented by permission of the copyright holder.[8]

Theoretical calculation of the paths followed by particles carried along with air passing through a slit and impinging on a plate indicated that the theoretical efficiency of impingement is related to the particle inertial parameter, Ψ, as given below.[13,14]

$$\Psi = (C \rho_p d_p^2 V)/(18 \mu D_c)$$

where

- C = Cunningham correction factor, dimensionless.
- ρ_p = density of aerosol particle, gm/cm^3.
- d_p = effective diameter of aerosol particle, cm.
- V = velocity of aerosol jet, cm/sec.
- μ = viscosity of air, poise.
- D_c = diameter of round jet or width of rectangular jet, cm.

The inertial parameter has a physical meaning: It is the ratio of the stopping distance – that is, the distance a particle will penetrate into still air when given an initial velocity, V – to the diameter or width of the aerosol jet. It can also be demonstrated to be the ratio of the force necessary to stop a particle initially traveling at velocity V in the distance $D/2$, to the fluid resistance at a relative particle velocity, V. The association between Ψ and collection of particles by impaction has been demonstrated experimentally.[13-15]

Therefore, to obtain particle collection efficiencies with the microimpinger which are comparable to the collection efficiencies of the midget impinger, the inertial parameters, Ψ, characterizing performance of the instruments for the particle sizes of interest, should be equivalent. It is also necessary that the spacing of the orifice (h) from the instrument base should be such that 2 times the ratio of h divided by the orifice diameter (D) is between 1.0 and about 4.0 for values of Ψ between 0.01 and 1.0.[13,15] For a jet operating under a given set of conditions, there is a minimum particle size below which impaction does not occur, and there is a maximum particle size above which all particles are impacted.

The important features of dynamic design for the micro and midget impingers are shown in Table II.1 along with performance data on three different particulate clouds. The fluid Reynolds number at the jet (Re_j) was calculated as follows:

$$Re_j = D V \rho / \mu$$

where

- D = orifice diameter, cm.
- V = velocity of air at the jet, cm/sec.
- ρ = density of air, gm/cm^3.
- μ = viscosity of air, poise.

Jet air velocities of the three capillaries at different volumetric rates of flow are shown in Figure I.42. Performance characteristics of selected capillaries at a single jet velocity, 60 m/sec, are shown in Figure II.1, which is typical of a family of curves which would be drawn to characterize capillary performance at a given jet velocity.

Table II.2 also contains comparisons of results of performance of two different capillary microimpingers and the standard all-glass midget impinger when sampling three different types of particulate clouds. The silica dust was reported by the Bureau of Mines to contain over 99% free silica (SiO_2) and to pass through a 325-mesh sieve (which has square openings 43 microns on a side). The bituminous coal dust was obtained from the Bureau of Mines experimental mine in the Pittsburgh coal bed and was also screened through a 325-mesh sieve. The New York State talc dust was received from the Stanford Research Institute Particle Bank.

Clouds were established by air blast atomization of the bulk material in a stainless-steel chamber 4 feet in diameter by 4 feet high. After 5 minutes of tranquil settling of particles in the cloud, sampling was begun with calibrated orifices and the rates of air flow specified in Table II.1. Samples were evaluated with the aid of a microscope equipped with a 16-mm achromatic objective (10X) and a 30X eyepiece. The bright-line Spencer hemocytometer cell (0.1 mm deep) was used for counting.[16] Counting was begun after 1-minute allowance for particles to settle to the cell base. Particles having projected area diameters between 1 and 5 microns were counted. The counting standard deviations were estimated from $N^{1/2}$, where N is the total particle count for a given sample.[17] Particle counts were converted to particle concentrations in the air of the chamber expressed as millions of particles per cubic foot of air (mppcf) at 760 mm and 20°C.

It was not the purpose of these studies to demonstrate rigorously the equivalence of the all-glass microimpingers and the all-glass midget impinger, as, for instance, Brown and Schrenk did for the Greenburg-Smith impinger and the all-glass midget impinger.[17] The data in Table II.1 indicated that a sampling rate of 0.02 cfm the inertial parameter, Ψ, was much higher for the microimpingers than it was for the midget impinger. Silica dust counts with the microimpingers were roughly twice those obtained by sampling with the midget impinger. At a sampling rate of 0.01 cfm the dust counts obtained with

*From Lippmann, M., *Air Sampling Instruments for Evaluation of Atmospheric Contaminants,* 4th ed., ACGIH, Cincinnati, O., 1972. With permission.

TABLE II.2

Asbestos Concentration by Operation, 1969

Work practice	Environmental conditions	Average asbestos fiber levels: Personal samples Fibers/ml	Fibers/ml	Area samples distance from source (ft)	
Asbestos cement[1]	High-ceilinged room. Louver venting.	2.4		.45	2
Asbestos cement[2]	Low-ceilinged room. Poor ventilation.	2.6		–	–
Asbestos cement[3]	Access tunnel.	6.1		–	–
Asbestos cement[4]	Power house. Low ceiling. Poor ventilation.	3.9		2.5	3–5
Cutting calcium silicate block, and pipe[1]	Table and hand saws in power house. Open.	1.2		–	–
Cutting calcium silicate block, and pipe[2]	Table and hand saws in industrial building. Good ventilation.	4.1		–	–
Cutting calcium silicate block, and pipe	Apartment house boiler room. No ventilation. Work 3 to 18 in. from breathing zone.	11.5		–	–
Cutting calcium silicate block and pipe	Limited ventilation.	9.4	1.6	3–4	
Spraying insulation	Turbines in power plant. Very high ceiling. Good ventilation.	47.7 28.0	19.5	3 6	

Fibers/ml >5 μ in length

[1] Conditions usually variable: Cement is mixed dry and applied wet; rapid changes occur in local ventilation; composition of material may vary; number of men on job may vary.
[2] Average of counts (excluding spray insulation): ⩾5 fibers/ml = 64.5%; 5 to 12 fibers/ml = 25.5%; >12 fibers/ml = 10.0%.
[3] Information prepared by Reitze, Nicholson, and Holaday.

the microimpinger capillaries shown were consistently lower, by 10 to 25%, than the counts made of samples obtained with the midget impinger operating at 0.1 cfm. The comparative performance data in Table II.1 clearly demonstrated that the microimpingers, as predicted by impaction theory, can be operated so that their particle collection characteristics are comparable to those of the midget impinger. The 0.01-cfm sampling rate appeared to be slightly below that required for perfect matching of performance to that of the midget impinger, and further tests are required to focus on the ideal sampling rate for the microimpinger.

A threefold reduction in the dimensions of this "O" ring model midget impinger gave a microimpinger which performed efficiently at one-fifth the flow rate and with only one-fifth of the sample and reagent volumes required for the larger counterpart. Results for aniline vapor, sulfur dioxide, and dust collection confirmed performance equivalent or superior to that of the standard midget impinger. Replacement of the conical ground orifice with a short section of precision-bore capillary permitted fabrication precision consistently within tolerances which permitted interchangeability without calibration to ensure matched performance. As many as 5 units could be operated in parallel with no increase in aspirating capacity over that required for a single midget impinger, or need for individual flow control meters. Combined with a recently available battery-operated miniature pump, the microimpinger provides a compact personnel monitoring instrument which delivers standard midget impinger results.*

Dust collection efficiency of the microimpinger in the range of 1 to 5 microns can be matched with the performance of the standard midget impinger by reducing the sampling rate to 0.10 to 0.012 cfm. At higher rates, some increase in efficiency in this range was obtained, but the increased fine particle collection, less than 1 micron in range, approximately

*See Chapter III of this book.

doubled the total collection. However, evaluation of the dust hazard in industry by the assessment of airborne particle counts is not an exact science, and the correlation of dust counts with disease risk is at best approximate. In view of these limitations, operation of the microimpinger at a flow rate of 0.01 cfm and acceptance of the resultant particle counts as indicative of levels of dustiness as specified for the midget impinger did not appear to us to be a severe infraction of good judgment on the part of the industrial hygienist.*

The microimpinger is readily adaptable to personnel monitoring, as shown elsewhere.** When collection of particulate is desired, the filter mounted on the impinger inlet would of course be removed. (For the "spill-proof" microimpinger modification, see Figures 1.44 and 1.45.)

6. Collection Medium

The selection of "trapping" liquid for the midget impinger can have a significant effect on the collection efficiency, especially at the lower end of the particle size range. Jacobson and co-workers at the Bureau of Mines in Pittsburgh found that collection efficiency for coal dust below 1 μ in diameter was significantly greater in isopropyl alcohol than in water or ethyl alcohol.[19] The following is the authors' abstract:

Coulter Counter techniques were used to determine the collection efficiency of the midget impinger for coal dust samples with isopropanol, ethanol, and water as the collecting fluids. The collection efficiency of the impinger with the alcohols and water was approximately 95% and 68%, respectively, for particles 1 micrometer and greater in diameter, and 77% and 64%, respectively, when the particle size range was expanded to include particles down to 0.68 micrometer in diameter. The midget impinger samples were also analyzed by standard microscopic techniques. Equivalent counts were not obtained for the two alcohols, indicating the effect fluid viscosity has on standard microscopic techniques. Samples counted in a controlled-temperature chamber, using the microscopic technique, showed a change in count of 2.4% and 1.2% per degree centigrade for isopropanol and ethanol, respectively.

7. Counting

The technique for counting the collected particulate is well described by Paul M. Giever in *Air Pollution,* and is reproduced here.

Techniques of sampling and counting were originally developed for evaluating mineral dusts related to pneumoconiosis. The light field counting procedure was standardized and published in U.S. Public Health Service Bulletin No. 217 in 1935. Discrepancies between the light field counting method and other techniques have been well documented over the years but because of the correlation of light field counts with incidence of human pneumoconiosis in this country, this method has continued to be used for evaluation of these health hazards.

Where the sample is to be used for estimation of exposure to pneumoconiosis-producing dust, the standard light field counting technique should be used. Samples should be collected with the Greenburg-Smith midget impinger or microimpinger[8] using dust-free distilled water, isopropyl or ethyl alcohol, or a mixture of alcohol and water. The use of alcohol is preferable for collection of dusts that may be soluble or flocculate in water, and where oil mists may interfere with the counting of mineral dusts.

Optical Microscope and Accessories

The standard biological microscope or its equivalent should be used for dust counting. An adjustable draw tube is preferable to the fixed tube type, as calibration can be more easily achieved. The microscope should be equipped with an Abbe condenser and iris diaphragm. The objectives, 16 mm, 4.4 mm, and 1.8 mm, used with a 10X ocular usually provide the necessary range of magnification for evaluating particles in the 0.5–25 μ size range.

The counting area is usually defined by a "Whipple disk" or similar type of ocular grid, or by rulings on the counting cell. The portion of the grid to be evaluated must be calibrated with a stage micrometer, a glass plate upon which accurately spaced lines have been ruled, so that the exact area is known. Adjusting the tube length of the microscope will aid in the precise calibration of an area to be counted.

Illumination for the light field counting technique consists of a 15-watt substage lamp placed under the condensor or a 75-watt microscope lamp positioned 10 inches in front of a plain mirror. A daylight filter is required for this type of counting. Kohler illumination is considered to be one of the most satisfactory illuminating techniques.[25]

Counting Cells

The Committee on Standard Methods of the American Conference of Governmental Industrial Hygienists recommends that the standard counting cell have a liquid depth of 1.0 mm.

The Sedgwick-Rafter cell consists of a glass plate slightly larger than an ordinary microscope slide on which a rectangular area has been set off by cementing to the base plate strips precisely 1.0 mm high. A cover glass completes the cell. The Sedgwick-Rafter cell is durable, but the raised sides make the cell difficult to clean.

The Dunn cell is composed of three separate parts: a rectangular glass plate, exactly 1 mm thick containing a circular hole for holding the sample liquid. This sets on a thick glass base plate, with a cover glass completing the

* From Linch, A. L. and Corn, M., *Am. Ind. Hyg. Assoc. J.,* 26, 601, 1965. With permission.

** Linch, A. L., *Biological Monitoring for Industrial Chemical Exposure Control,* CRC Press, Cleveland, in press, 1973.

Most of the filters made of fibrous material are unsuitable for collecting airborne particulate matter for count and/or size determinations, as the particles generally penetrate deeply into the body of the filtering medium, and thus cannot be readily evaluated microscopically.

Samples are generally collected on filters of a type that do not permit deep penetration, and that can be rendered suitable for the analytical technique to be employed. Molecular or membrane filters, from which either count or size analysis can be made, are widely used for this purpose. They are capable of uniformly high efficiencies, greater than 99% for all sizes of particles and are useful for the collection of submicron particles, especially when those particles are present in relatively low concentrations. The particles are deposited on the surface of the membrane in the same state in which they exist when suspended in air or gas.[22]

The primary collection mechanism consists of screening by membrane pores of controlled size. Membrane filters ranging in pore size from 0.005 up to 10 μ may be used. The collection efficiency of membrane filters is absolute for particles larger than the pore size of the membrane. Many particles smaller than the absolute size of the surface pores are also collected, due to electrostatic effects and impaction on the walls of pore passages formed by the interlocking of cell layers.[23]

The following criteria may be useful in selecting the filter medium:

1. The collection efficiency of the filter should be 90% or greater for the smallest size range of interest.
2. The efficiency of the filter should be uniform.
3. The resistance to air flow should be within acceptable ranges for the sampling procedure to be used.

Careful consideration must also be given to the filter holder, as numerous commercial models have been found to leak air through threads and connections in such quantities as to render them useless for obtaining accurate samples.[21] *

The main requirements for physical analysis of collected particulate are (a) that the filter be held tightly at the edges to prevent by-pass of the airstream through the seal at the filter edge, and (b) that all jointed sections be tight to prevent air from leaking inward. These two faults are common and it is advisable to check each filter holder assembly by applying a vacuum with a manometer coupled to the filter inlet under static conditions and record leakage as mm/Hg/min.[21] Some multisection filter holders such as the Millipore® Type AA require taping when conducting airflow rate calibration (Figures II.2–II.4).[22] The following description of the Millipore Type AA was furnished by the manufacturer:

Aerosol Analysis Monitors

Molded of Tenite plastic. A filter-retaining ring between the top and bottom halves permits removing the top for "open" aerosol sampling, using an aerosol adapter containing a flow-limiting orifice for connection to the vacuum source. Average background counts are indicated on the shipping carton.

MAWG 037 AO Type AA white, gridded filter, for particle counting (50/ctn).
MABG 037 AO Type AA black, gridded filter, for particle counting (50/ctn).
MAWP 037 AO Type AA white, plain filter, for gravimetric analysis (50/ctn).
XX62 000 04 Aerosol adapter

Function

Clean (or sterile) transparent disposable plastic filter holders, preassembled with filters in place, for vacuum-filtration use in particulate or microbiological sampling applications.

A 37 mm Millipore filter disc is sealed tightly between the top and bottom halves of the monitor (Figure II.2), with a cellulose pad beneath the filter for support and to distribute fluid flow over the entire filter surface (Figure II.3). Effective filtration area is approximately 9.0 cm². All monitors are assembled in an ultra-clean environment to minimize background counts.**

Other similar filters are available from other sources, also.[23-24a]

b. Calibration

Airflow rate calibration should be carried out by attaching a flowrate meter, preferably a calibrated rotameter or spirometer which introduces a minimum pressure drop (ΔP) across the filter, to the inlet connection, and a vacuum source (see Chapter III, Section F, "The Personnel Monitor") to the outlet connection of the filter holder assembly. A water manometer connected to a "T" tube between the filter outlet and the vacuum source should be included to furnish ΔP data.

c. Measurement of Volume of Air Samples

The pressure drop across the filter will increase as the sample load on the filter increases; therefore, the air flow

*From Giever, P. M., in *Air Pollution, Vol. 2, Analysis, Monitoring and Surveying,* 2nd ed., Stern, A. C., Ed., Academic Press, New York, 1968, chap. 21. With permission.

**From Aerosol Analysis Monitors, Catalog No. MAWG 037-AO, type AA, Catalog No. MC/1:2A360, Millipore Corp., Bedford, Mass., 1972. With permission.

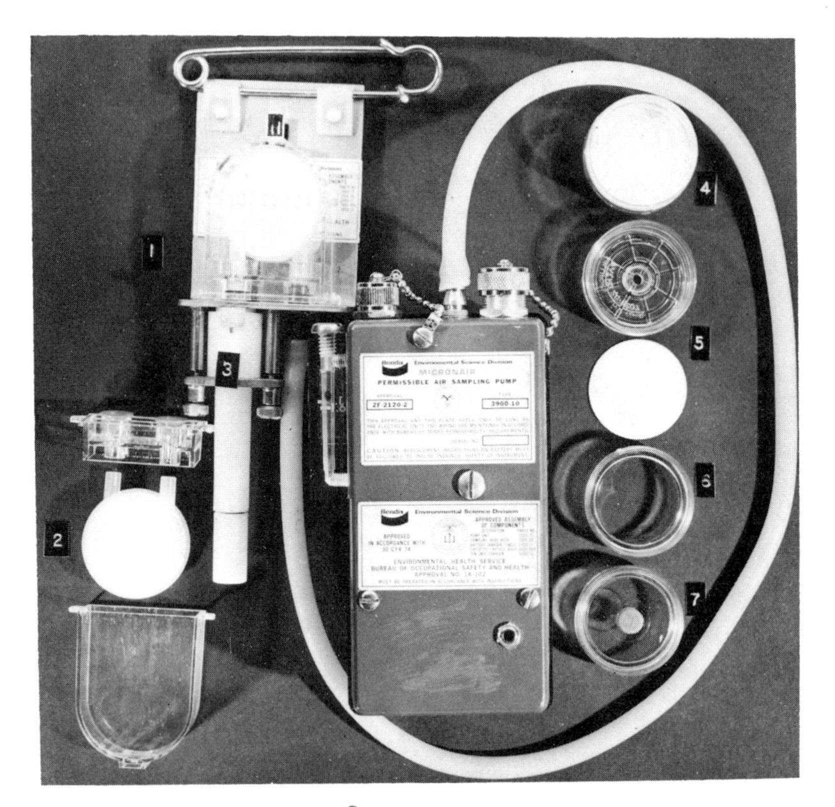

FIGURE II.2. The Micronair® dust sampling pump component parts. 1. Respirable dust collection assembly. 2. Filter cassette (preweighed). 3. Cyclone separator. 4. Open-face filter with cover removed, ready for sampling. 5. Filter membrane and base plate. 6. Retaining ring. 7. Cover with plug in place. (Courtesy of National Environmental Instruments, Inc., P.O. Box 590, Pilgrim Station, Warwick, R.I.)

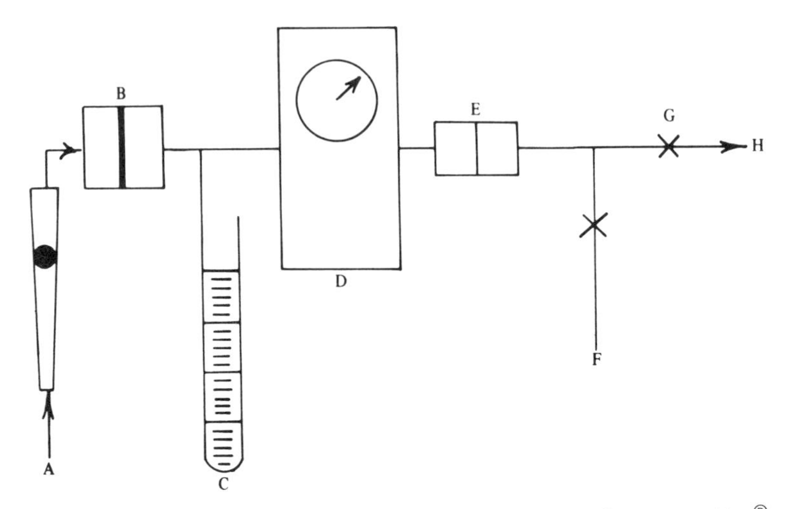

FIGURE II.3. Dust filter flowrate calibration system. A. Rotameter (Ace® Glass, Model 2-15-3). B. Filter – Millipore® (0.8-μ pore) Catalog No. MAWG 037-AO – aerosol type. C. Water manometer. D. Dry test meter. E. Limiting orifice (glass, 0.508 mm I.D. x 10 mm). F. By-pass bleed-micrometer valve. G. Throttling valve. H. Vacuum supply (water jet).

f. An Example – Asbestos Fiber Monitoring

The following statements extracted from the *Federal Register* set the stage for the following monitoring program recommended by NIOSH.[30] The adopted standard requires periodic monitoring at intervals no longer than 6 months and prescribes the use of the membrane filter method, which is an acceptable method for determination of asbestos fibers.

As of July 1, 1976, TWA concentrations of asbestos fibers longer than 5 micrometers will not be allowed to exceed two fibers/cc., with a ceiling value of 10 fibers/cc. The current TWA concentrations of five fibers, and ceiling concentrations of 10 fibers/cc., will be permitted until July 1, 1976, during what will be a transitional period deemed necessary to allow employers to make the needed changes for coming into compliance with the more stringent standard.

Methods of compliance. It has been pointed out by many persons, that protection against asbestos fibers is best obtained by controlling the generation of fibers first, and secondly, by controlling the dispersion of released fibers into the ambient air of the workplaces. Therefore, the standard requires feasible technological controls and appropriate work practices as the primary means of compliance. Rotation of employees as a way of meeting the TWA concentration requirement is allowed only in stated exceptional circumstances, because, as a general rule, it would be difficult to implement. Personal protective equipment, such as respirators, cannot be relied upon because, among other reasons, they may be so uncomfortable as to be burdensome, except for short periods of time. Therefore, it is expected that respirators and shift rotation will be used during the period necessary to install engineering controls and to train employees in sound work practices, but, after technological compliance has been achieved, their use must be limited to special work situations and emergencies. Where both are practicable, shift rotation is required.

1. Medical Examinations

The adopted standard requires medical examinations both at the beginning and the termination of employments exposed to concentrations of asbestos fibers, and also requires annual medical examinations of every employee exposed to airborne concentrations of asbestos. It has been pointed out that in certain industries, such as construction, an employee may work for several employers during the same year. Accordingly, the standard does not require either preemployment, or termination, or periodic examination of any employee who has been examined in accordance with the standard within the past year.

2. Air Sampling Methods

In the study of asbestosis conducted by Dreessen et al.[31] midget impinger count data were used as an estimate of dust exposure. All of the dust particles seen, both grains and fibers, were counted since too few fibers were seen to give an accurate measurement. The resulting count concentration was a measure of overall dust levels rather than a specific measurement of the asbestos concentration. This method was satisfactory at that time since exposures were massive and the control measures installed to reduce overall dust levels also reduced the asbestos dust levels.

As dust levels were reduced, it became necessary to measure the biologically appropriate attribute of the dust cloud. At equal levels of overall dustiness, the concentration of asbestos could vary considerably from textile manufacture (75–85%) to insulation (5–15%). Furthermore, if the limit were lowered below the 5 mppcf used previously and dust counts taken by the impinger technique, it would be necessary to consider the effect of background dust, which could be as high as 1 mppcf.

A number of methods for measurement of asbestos dust concentrations have been used in the NIOSH epidemiological study of the asbestos product industry.[32-35] Based on these data, the preferred index of asbestos exposure is the concentration of fibers longer than 5 μm counted on membrane filters at 430X with phase contrast illumination.[36,37] This index is utilized in the method adopted as the standard field sampling method by the Public Health Service.

Fibers longer than 5 μm in length are counted in preference to counting all fibers seen in order to minimize observer/microscope resolving power variability. Furthermore, the British define a "fibre" as a particle, "of length between 5 μm and 100 μm and having a length-to-breadth ratio of at least 3:1, observed by transmitted light by means of a microscope at a magnification of approximately 500X."[38]

Although the British have refrained from standardizing on a single method of measurement, recent measurements have been performed by a method essentially identical to the fiber-count method described in detail below, and the British hygiene standards for use with their asbestos regulations are stated in these terms.[38]

3. Principles of Sampling

A dust sampling procedure msut be designed so that samples of actual dust concentrations are collected accurately and consistently. The results of the analysis of these samples will reflect, realistically, the concentrations of dust at the place and time of sampling.

In order to collect a sample representative of airborne dust, which is likely to enter the subject's respiratory system, it is necessary to position a collection apparatus near the nose and mouth of the subject or in his "breathing zone."

The concentration of dust in the air to which a worker is exposed will vary, depending upon the nature of the operation and upon the type of work performed by the operator and the position of the operator relative to the source of the dust. The amount of dust inhaled by a worker can vary daily, seasonally, and with the weather. In order to obtain representative samples of workers' exposures, it is necessary to collect samples under varying conditions of weather, on different days, and at different times during a shift.

The percentage of working time spent on different tasks will affect the concentration of dust the worker inhales since the different tasks usually result in exposure to different concentrations. The percentage can be determined from work schedules and by observation of work routines.

The daily average weighted exposure can be determined by using the following formula:

$$\frac{\text{(Hours x conc. task A) + (Hours x conc. task B) + etc.}}{\text{8 hours (or actual hours worked)}}$$

The concentration of any air contaminant resulting from an industrial operation also varies with time. Therefore, a longer sampling time will better approximate the actual average.

With the following recommended sampling procedure, it is possible to collect samples at the workers' breathing zones for periods from 4 to 8 hours, thus permitting the evaluation of average exposures for a half or full 8-hour shift – a desirable and recommended procedure. Furthermore, dust exposures of a more normal work pattern result from the use of personal samplers. In evaluating daily exposures, samples should be collected as near as possible to workers' breathing zones.

4. Collecting the Sample

The method recommended in this report for taking samples and counting fibers is based on a modification of the membrane filter method described by Edwards and Lynch.[36]

The sample should be collected on a 37-millimeter Millipore type AA* filter mounted in an open-face filter holder. The holder should be fastened to the worker's lapel and air drawn through the filter by means of a battery-powered personal sampler pump similar to those approved by NIOSH under the provisions of 30 CFR 74. The filters are contained in plastic filter holders and are supported on pads which also aid in controlling the distribution of air through the filter. To yield a more uniform sample deposit, the filter-holder face-caps should be removed. Sampling flow rates from 1.0 liter per minute (lpm) up to the maximum flow rate of the personal sampler pump (usually not over 2.5 lpm) and sampling time from 15 minutes to eight hours are acceptable provided the following restraints are considered:

a. In order to obtain an accurate estimate of the number of fibers the statistical error resulting from the random distribution of the fibers must be kept to an acceptably low level. Since fiber counts follow a Poisson distribution, a count of 100 fibers in a sample would have a standard deviation of $\sqrt{100}$ or 10 fibers or ± 10%. Thus the 95% confidence limits would be approximately 2 standard deviations or ± 20%. Since the 37 mm filter has an effective collecting area of 855 mm^2 and the projected field area of the Porton reticle is 0.005 mm^2, each field represents 1/171,000 of the sample. Based on this ratio the following number of fields must be counted to measure the various limits in various sampling times:

Sampling time minutes	Flow rate lpm	Number of fields for 100 fibers 0.2 fibers/ml	2.0 fibers/ml	10 fibers/ml
10	2	4350	435	91
15	2	2860	286	58
30	2	1430	143	29
90	1	1000	100	20
90	2	500	50	10
240	1	260	26	7
240	2	180	18	4
480	1	180	18	4

b. Do not count a field containing over 20 fibers because in addition to the fibers being counted, there are also present a number of grains, which interfere with the accuracy of the count.

Based on these restraints, i.e., number of fields to be counted and maximum number of fibers per field, acceptable sampling parameters for the various limits are underlined in the above table.

The following conclusions may be drawn from this analysis:

1. The short-term limit should be for a period of at least 15 minutes and preferably 30 minutes.

2. The 2.0 fiber/cm^3 limit may be evaluated over periods of from 90 to 480 minutes.

As many fields as required to yield at least 100 fibers should be counted. In general the minimum number of fields should be 20 and the maximum 100.

5. Mounting the Sample

The mounting medium used in this method is prepared by dissolving 0.05 g of membrane filter per ml of 1:1 solution of dimethyl phthalate and diethyl oxalate. The index of refraction of the medium thus prepared is ND = 1.47.

To prepare a sample for microscopic examination, a

*Mention of commercial products does not constitute endorsement by the Public Health Service or U.S. Department of Health, Education and Welfare.

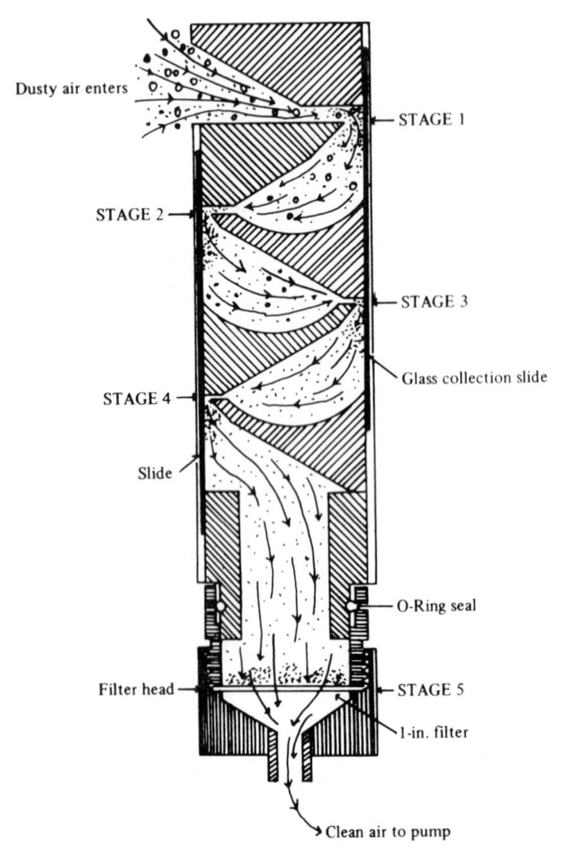

FIGURE II.6. The cascade impactor size selective dust collector. (Courtesy of National Environmental Instruments, Inc., P.O. Box 590, Pilgrim Station, Warwick, R.I.)

Recommendations

1. Scope – To be effective, this program must encompass all facets of the work performed in the shops in the field, and all sites should be included.

2. Breathing zone personnel monitoring should be conducted in the insulation shops and in the installation areas at each site location. One employee selected at random should be assigned to wear the sampler one 8-hr day sometime during each month on a rotation basis.

3. Trained personnel under the supervision of a certified industrial hygienist or equivalent technical personnel should be responsible for the control, operation, adjustment, and maintenance of the air sampling equipment, recording of required data associated with the sample collection, forwarding the dust samples to a certified laboratory for counting, and compiling records of analysis reports.

4. When selecting personnel who will wear the monitor, consideration should be given to sampling in the area of greatest potential dust exposure on the day selected for monitoring.

5. The air sampling pump should be capable of drawing the air sample through the filter at a 2-lpm rate for not less than 8 hr. Either the Unico® Model C115 or the Unico Micronair® pump is suitable (see Chapter III).[40] The filter should be an 0.8-μ pore size membrane filter that is compatible with the recommended counting system (Millipore Catalog No. MAWG 037-AO, Type AA complete with plastic retainer assembly).[22]

6. The filter sealed in its plastic retainer, upon completion of sample collection, should be forwarded to a laboratory certified for asbestos fiber counting by phase contrast microscopy and total dust by gravimetric analysis.

7. Reports that indicate either asbestos concentrations greater than or approaching 5 fibers/ml 75 μ in length or total dust greater than or approaching 10 mg/m^3 require that corrective action be taken to reduce dust loading in the work place. This may require additional monitoring to delineate the sources of dust generation. After improvements have been installed, additional 8-hr personnel samples should be taken to evaluate the effectiveness of the improvements.

8. Interpretation of the monitoring results and adequacy of the program should be the responsibility of a certified industrial hygienist or an equivalent technical person.

Instructions

Prepare filter:

A. Remove plug from filter deck end and attach the open tubulature to the gum rubber sampling hose. *Slight* moistening of the plastic tubulature assists with hose attachment. Excessive moistening will cause hose to immediately slip off.

B. Remove the top (uppermost) section of the filter assembly without disturbing the seal of the center ring against the filter paper on the filter support deck. Do not remove the plug from this top section at any time before attachment of the sampling hose in preparation for sample collection.

C. After the sampling has been completed, replace the top section with plug in place and seal securely. Tape the plugs in place by passing tape around the unit axially several times under slight tension. Any self-adhesive tape will serve.

D. Identify each plastic filter case by applying a label containing the site serial number and date sample was taken. Label to be applied on filter case shelf for protection in handling.

E. Wrap the filter unit in paper toweling or other bulky paper packing and place in the mailing container.

F. Do not attempt, under any circumstances, to remove or replace the filter membrane or use a filter which has been exposed by either plug becoming dislodged or by separation of either assembly member before use. The background blank supplied with each lot of filters is valid only for sealed sampling units.

Prepare sampler (Unico C115):[40]

A. Insert a fully charged battery into the bracket located on the inside of the front panel.

1. Make sure the positive (+) pole is in contact with the positive contact of the bracket. Although the pump will operate on reversed polarity, excessive strain and wear will soon destroy the pump assembly.
2. If the pump fails to start when the switch is closed in the "on" position, check the battery contacts for open circuit due to distortion or bending of the bracket.
3. Failure to start also may be due to a discharged battery:

    a. Neglecting to recharge after last use.
    b. Accidentally discharged – switch positioned left in the "on" position.
    c. Dead battery – not rechargeable, return for repair.
    d. Fault in the charger – see charger instructions.
    e. Check by replacing the inoperative battery with a freshly charged unit.

4. If a hum is heard, but air is not drawn through the rotameter, the motor-pump linkage has failed – return for repair.

B. Install the clothing clip just under the filter unit.

C. Attach the open end of the sampling hose to the vacuum tubulature. Again, moistening the bore of the hose will facilitate assembly.

Install the assembly:

A. Pass the wearer's trouser belt through the loops provided in the rear of the air sampler.

1. The selection of location is up to the discretion of the wearer, unless the job assignment will determine the location. In any event, locate in a position that will least interfere with the employee's work motions, and that will involve the least potential damage to the sampler unit.

B. Attach the filter unit to the wearer's clothing as close as practical to his breathing zone by means of the fabric clip. If the air sampler is worn in back, the sampling hose should be brought up over either the right or left shoulder.

1. Adjust hose length, position, and attachment to the clothing to eliminate loops which may catch on protruding objects in the work environment. Extra loops to further secure the hose can be fabricated quickly from self-adhesive tape and attached to the clothing with safety pins.

Sampling:

A. Advance the flow control valve on the right upper corner of the sampler until the rotameter indicates 1,200 cm^3/min (1,200 cm^3/min = 1.2 lpm = 0.0424 ft^3/min = 2.54 ft^3/hr) or the calibration value equivalent to 1,200 cc/min (use calibration furnished).

1. The rotameter is calibrated to read the bottom of the float.

respiratory tract which differ markedly in structure, size, and function, and which have different mechanisms for particle elimination. Thus, a complete determination of dose from an inhaled toxicant depends on the regional deposition and the retention times at the deposition sites and along the elimination pathways, in addition to the properties of the particles.

c. Anatomical and Physiological Factors in Respiratory Tract Particle Deposition and Clearance

The succeeding paragraphs present a brief summary of the factors controlling particle deposition and clearance. More complete descriptions of the anatomy of the respiratory tract and of some of the factors controlling particle deposition and clearance are presented by Hatch and Gross.[51]

d. Nasal Passages

Air enters through the nares or nostrils, passes through a web of nasal hairs, and flows posteriorly toward the nasopharynx while passing through a series of narrow passages winding around and through shelflike projections called turbinates. The air is warmed and moistened in its passage and partially depleted of particles. Some particles are removed by impaction on the nasal hairs and at bends in the air path, and others by sedimentation. Except for the anterior nares, the surfaces are covered by a mucous membrane composed of ciliated and goblet cells. The mucus produced by the goblet cells is propelled toward the pharynx by the beating of the cilia, carrying deposited particles along with it. Particles deposited on the anterior unciliated portion of the nares and at least some of the particles deposited on the nasal hairs are usually not carried posteriorly to be swallowed, but rather are removed mechanically by nose wiping, blowing, sneezing, etc.

e. Oral Passages, Pharynx, Larynx

In mouth breathing, some particles are deposited, primarily by impaction, in the oral cavity and at the back of the throat. These particles are rapidly eliminated to the esophagus by swallowing.

f. Tracheo-bronchial Tree

The conductive airways have the appearance of an inverted tree, with the trachea analogous to the trunk and the subdividing bronchi to the limbs. In Weibel's[53] simplified model, there are 16 generations of bifurcating airways. . . the diameter decreases from generation to generation, but because of the increasing number of tubes the total cross section for flow increases and the air velocity decreases toward the ends of the tree. In the larger airways, particles too large to follow the bends in the air path are deposited by impaction. At the low velocities in the smaller airways, particles deposit by sedimentation, and if small enough by diffusion.

Ciliated and mucus secreting cells are found at all levels of the tracheo-bronchial tree. Inert non-soluble particles deposited in this region are thus carried within hours toward the larynx on the moving mucus sheath which is propelled proximally by the beating of the cilia. Beyond the larynx, the particles enter the esophagus and pass through the G.I. tract.

g. Alveolar Region

The region beyond the terminal bronchioles is the region in which gas exchange takes place. The epithelium is non-ciliated, and, therefore, insoluble particles deposited in this region by sedimentation and diffusion are removed at a very slow rate, with clearance half times on the order of a month or more. The mechanisms for particle clearance from this region are only partly understood and their relative importance is a matter of some debate. Some particles are engulfed by phagocytic cells which are transported onto the ciliary "escalator" of the bronchial tree in an undefined manner. Others penetrate the alveolar wall and enter the lymphatic system. Still others dissolve slowly in situ. "Insoluble" dusts all have some finite solubility, which is greatly enhanced by the large surface to volume ratio characteristic of particles small enough to penetrate to the alveolar region of the lung. Morrow[54] has demonstrated that the clearance half times of many "insoluble" dusts in the lung are proportional to their solubilities.

h. Regional Deposition and Clearance Dynamics

For the purpose of estimating toxic dose from inhaled particles, the respiratory tract can be divided into a number of functional regions, which differ grossly from one another in retention time at the deposition site, the elimination pathway, or both. These are

1. Alveolar region (for both nose and mouth breathing)
2. Tracheo-bronchial tree (for both nose and mouth breathing)

3A. Oral cavity, pharnyx and larynx (for mouth breathing)

3B. Nasopharynx, pharynx and larynx (for nose breathing)

4. Ciliated nasal passages (for nose breathing)
5. Anterior unciliated nares (for nose breathing)

The fractional deposition in each of these regions is dependent on the aerodynamic particle size* and the subject's airway dimensions and respiratory characteristics (flowrate, breathing frequency, tidal volume, etc.).

Ideally, air sampling data should provide data on the deposition to be expected in each functional region, or at least in regions (1, 2 and 3–5 inclusive). Unfortunately, the sparse experimental data available on regional human deposition are inconsistent. These data have been summarized in a recent paper[50] prepared for background

*Aerodynamic size – the diameter of a unit density sphere having the same settling velocity as the particle in question of whatever shape and density.

and documentation for the "Interim Guide on 'Respirable' Mass Sampling"[52] of the AIHA Aerosol Technology Committee.

i. Measurement of Mass Concentrations within Size Graded Aerosol Fractions

Since the dose from inhaled toxicants is dependent on the regional deposition, which is dependent on particle size, the best dose estimates for a material whose toxicity is proportional to absorbed mass can be derived from a knowledge of the mass concentrations within various size ranges. Such information can be obtained in several ways: (1) by separating the aerosol into size fractions corresponding to anticipated regional deposition during the process of collection; (2) by making a size distribution analysis of the airborne aerosol, e.g., with a conifuge, cascade impactor, light scattering aerosol spectrometer, etc.; and (3) by making a size distribution analysis of a collected sample.

The most reliable information can be obtained using methods in which the aerosol is fractioned on the basis of aerodynamic diameters in much the same manner as it is fractionated within the respiratory tract. Thus, differences in particle shape and density are automatically compensated for.

Light scattering instruments which sort the pulses resulting from the scattered light from individual particles can provide information on the distribution of airborne particle diameters. In converting this information to a size-mass distribution, an average particle density must be assumed. Furthermore, the accuracy of the diameter distribution is dependent on the particle shape, index of refraction and surface roughness. For example, Whitby and Vomela report that for India Ink particles, which absorb light and have a rough surface, the indicated size was 1/2 to 1/5 of the true size for the three different instrument designs tested.

Further opportunities for error arise when the size distribution analysis is performed on collected samples. It is almost impossible to examine the sample in the original state of dispersion. Thus, particles which were unitary in the air may be analyzed as aggregates and vice versa. Furthermore, particles analyzed by microscopy will be graded by a linear dimension or by projected area diameter, and these are normally larger than the true average diameter.

j. Standards and Criteria for Respirable Dust Samples

British Medical Research Council (BMRC) – In 1952, the British Medical Research Council adopted a definition of "respirable dust" applicable to pneumoconiosis producing dusts. It defined respirable dust as that reaching the alveoli. The BMRC selected the horizontal elutriator as a practical size selector, defined respirable dust as that passing an ideal horizontal elutriator, and selected the elutriator cut-off to provide the best match to experimental lung deposition data. The same standard was adopted by the Johannesburgh International Conference on Pneumoconiosis in 1959.

In order to implement these recommendations, it was specified that:

1. For purposes of estimating airborne dust in its relation to pneumoconiosis, samples for compositional analysis, or for assessment of concentration by a bulk measurement such as that of mass of surface area, should represent only the 'respirable' fraction of the cloud.
2. The 'respirable' sample should be separated from the cloud while the particles are airborne and in their original state of dispersion.
3. The 'respirable fraction' is to be defined in terms of the free falling speed of the particles, by the equation $C/C_o = 1\text{-}f/f_c$, where C and C_o are the concentrations of particles of falling speed f in the 'respirable' fraction and in the whole cloud, respectively, and f_c is a constant equal to twice the falling speed in air of a sphere of unit density 5μ in diameter.

A sampling device which meets these requirements would have a sampling efficiency vs. size curve suggested by Davies.[55] It is illustrated in Figure II.5 and defined as follows:

% Deposition –	10	20	30	40	50	60	70	80	90	100
Diam. (μ) –	2.2	3.2	3.9	4.5	5.0	5.5	5.9	6.3	6.9	7.1

(for spheres of unit density)

U.S. Atomic Energy Commission (AEC) – A second standard, established in January 1961 at a meeting sponsored by the AEC Office of Health and Safety,[51] defined "Respirable Dust" as that portion of the inhaled dust which penetrates to the non-ciliated portions of the lung. This application of the concepts of respirable dust and concomitant selective sampling was intended only for "insoluble" particles which exhibit prolonged retention in the lung. It was not intended to include dusts which have an appreciable solubility in body fluids and those which are primarily chemical intoxicants. Within these restrictions, "respirable dust" was defined as follows (Figure II.7):

Particle Size vs. Respirability

Size* (μ)	10	5	3.5	2.5	2
% Respirable –	0	25	50	75	100

*Sizes referred to are equivalent to an aerodynamic diameter having the properties of a unit density sphere.

ators meeting the BMRC criteria and through parallel 10 mm nylon cyclones operating at 1.3, 1.65, 1.95 and 2.64 lpm. They found that comparable estimates of the "respirable" mass concentration were obtained when the cyclone is operated at 1.4 lpm. After reviewing all of these data, the AIHA Aerosol Technology Committee has recommended a flowrate of 1.7 lpm for this cyclone in their Interim Guide.[52]

Recently, Roesler, Stevenson and Nader[74] have reported on the size distribution of sulfate aerosols in the ambient air of Cincinnati and Chicago. Samples were collected with both Anderson cascade impactors at 1 cfm and two-stage samplers consisting of an Aerotex 2 cyclone and a glass fibre filter. Respirable sulfate aerosol, as determined from the samples recovered from stages 4-7 of the Anderson impactor and/or the filter following the Aerotec 2 cyclone, represented 80-90% of the total sulfate aerosol. On the same basis, 50% of the total aerosol mass was respirable.

Most of the two-stage sample data reported in the literature were collected at industrial operations and show larger fractions for the first stage collection.

In 1961, Lippmann and Harris[66] presented data relating respirable dust to total dust in uranium and beryllium refineries obtained using HASL cyclones at the initially specified flowrates. For beryllium most of the filter fractions fell within the range 15 to 50% while for uranium they were usually less than 15%.

Hyatt[74b] has presented data on samples taken with HASL cyclones in uranium mills. In most cases, the "respirable" fraction ranged between 1 and 10%. For samples collected in the ore crushing department, the concentration of uranium in "respirable" dust was much higher than in the parent material.

Schulte[74c] has reported the results of a comparison of sample pairs near room ventilation outlets in which one sample was a standard filter paper and the other a two-stage cyclone-filter combination. For the highest monthly average concentration samples collected in plutonium fabrication areas, the fraction on the second stage varied from about 10% to 50% of the total. In a uranium processing area only about 6% of the uranium reached the second state.

Fischoff[74d] has compared single stage and two-stage air sample concentration data with urine excretion data for uranium refinery workers. As expected, the urine uranium excretion correlated more closely to the "respirable" concentration of uranium than to the gross airborne concentration.

Donaldson, et al.[74a] presented data on samples collected with 8.5 cfm HASL cyclone and the 40 cfm Aerotec 2 in beryllium refining and foundry operations which was consistent with that of Lippmann and Harris.[66] Data on respirable dust concentrations in an ore processing plant and a metal fabrication plant were also presented.

Ayer[45] has reported that for airborne dust sampled in the Vermont granite sheds, the average percent of free silica in electrostatic precipitator samples was 18%, as compared to 7% in dust passing either an elutriator or a cyclone pre-collector. For a limited series of samples collected in foundries, the percent of free silica was 29% in gross air samples and 7.5% for dust passing a horizontal elutriator.

It should be noted that most of the two-stage sample data cited in the above summarized literature were obtained from cyclone-filter samplers operated at their originally recommended flowrates. Since later, more reliable calibration data indicate that these flowrates were too high, the reported "respirable" dust concentrations were, in most cases, too low.*

1. Pulsation Free Sampling

A very recent investigation reported by Lamonica and Treaftis from the Pittsburgh Field Health Group indicated that more dust is collected when pulsation free flow is used.[78a] The study of two types of approved coal mine dust personal samplers (Unico® piston type, Figure II.2, and MSA® diaphragm type, Figure III.6) disclosed that irregularities of flow due to pump-induced pulsation had a significant effect on the concentration of respirable dust collected in accordance with the Federal Coal Mine Health and Safety Act of 1969. Velocity fluctuations were reduced by means of a pulsation damper ("snubbers") previously described by Lamonica and Treaftis (Figure II.10).[79a]

Damper efficiency of approximately 85% was needed to reduce the pulsation ("ripple") to an acceptable level. Nonpulsating (orifice controlled rotary pump) flow was used to determine the flowrate necessary for the cyclone to operate according to the penetration curve recommended by the AEC (Figure II.7).[66] The relationship between the amplitude of the pulsation (maximum velocity/minimum velocity) and the mass of dust collected (pulsation mass/nonpulsation mass) is graphically shown in Figure II.8. The reduction in pulsation amplitude is demonstrated in Figure II.9.

Lamonica and Treaftis' conclusions are quoted directly as follows:

> Pressure and velocity measurements made in the line adjacent to the cyclone and in the cyclone entrance show that extreme velocity fluctuations occur due to the pumps used in the MSA and Unico personal samplers. The velocity varies from about 1,470 cm/sec to 270 cm/sec for the MSA pump. This represents flows of 3.75 and 0.68 liters per minute, while the average flow as indicated by a wet test meter is 2.0 liters per minute. The Unico

*From Lippmann, M., *Air Sampling Instruments for Evaluation of Atmospheric Contaminants,* 4th ed., ACGIH, Cincinnati, O., 1972. With permission.

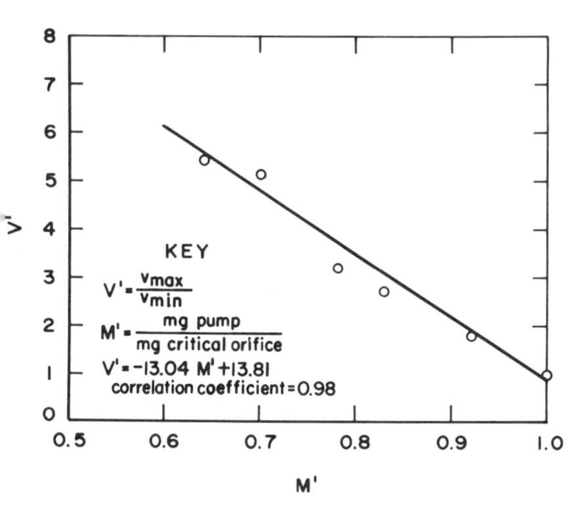

FIGURE II.8. Relationship between the ratio of masses of respirable dust collected with pulsating and non-pulsating flow to the ratio of maximum to minimum airflow velocities. (From Lamonica, J. A. and Treaftis, H. N., The Effect of Pulsation Damping on Respirable Dust Collected By Coal Mine Dust Personal Samplers, R. I. 7636, Bureau of Mines, Department of the Interior, 1972.)

pump velocity varies from about 1,050 cm/sec to 390 cm/sec, which corresponds to 2.68 liters and 1.0 liters per minute. These velocity fluctuations cause a decrease in the number of particles collected in the range from 2.0 to 6.0 μm. The effect is more pronounced in the MSA pump because of the greater amplitude of velocity fluctuations. (V′ is greater.) The mass measurements and the distributions determined by use of the Coulter counter show that the filters used with pumps equipped with satisfactory snubbers collect essentially the same number, size, and mass of particles as the filters using critical orifices to regulate the flow. The anemometer shows that this occurs when the snubbers are about 85 percent efficient or better, or when V′ has been reduced to 1.2 or less.

Therefore, in order for the cyclones used with the personal sampling pumps to follow any criteria set forth as to penetration curve characteristics and flow rate, they should be equipped with snubbers of 85 percent efficiency or greater or else of sufficient efficiency to lower V′ to about 1.2 or less, if this criterion is more appropriate (Figure II.8).*

An exploded view of the snubber is presented in Figure II.10 and the critical variables – effects of volume of the snubber and diaphragm elasticity on the efficiency of the pulsation damping – in Figure II.11. The conclusions relative to specifica-

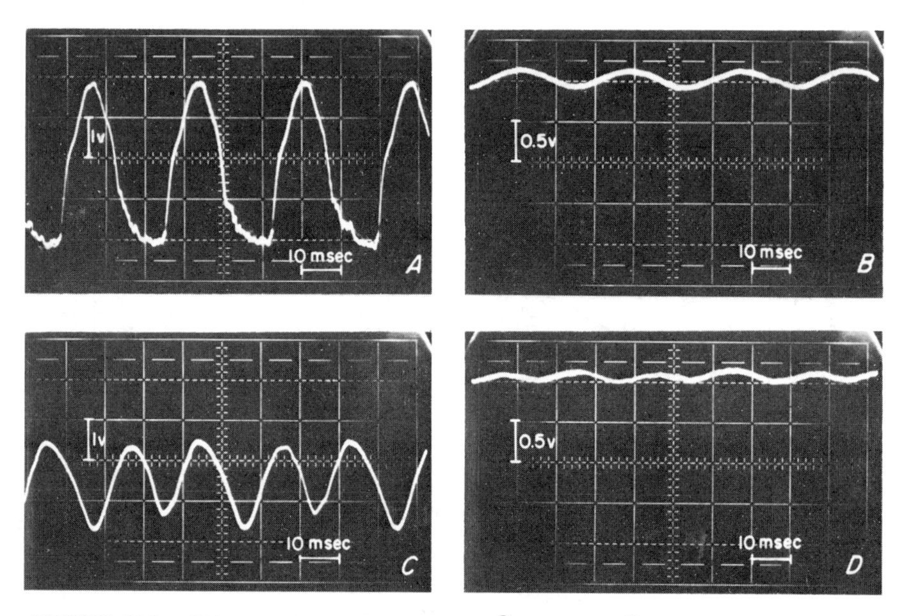

FIGURE II.9. Velocity fluctuations in M-S-A® and Unico® flowrate without and with snubbers. A. M-S-A pump without snubber. B. M-S-A pump with snubber. C. Unico pump without snubber. D. Unico pump with snubber. (From Lamonica, J. A. and Treaftis, H. N., The Effect of Pulsation Damping on Respirable Dust Collected By Coal Mine Dust Personal Samplers, R. I. 7636, Bureau of Mines, Department of the Interior, 1972.)

*From Lamonica, J. A. and Treaftis, H. N., The Effect of Pulsation Damping on Respirable Dust Collected By Coal Mine Dust Personal Samplers, R.I. 7636, Bureau of Mines, Department of the Interior, 1972.

tions are quoted directly from the report by Lamonica and Treaftis:[79a]

1. Use as large a volume as space permits.
2. Have as large a diaphragm area as the volume permits. Since shape is not important, a rather flat snubber would permit a large diaphragm area.
3. Use a thin flexible elastomer diaphragm, 0.008 inch or thinner.
4. Use the smallest center orifice diameter possible without affecting the flow, about 0.070 inch.
5. Use the combination of smallest inlet and outlet orifices that can be used without affecting the flow, about 0.13 inch for both the inlet and outlet orifices.

The data presented were gathered using one of the currently available personal sampler pumps. However, since the method for determining the snubbers' efficiency is dependent upon the amplitude of the pressure fluctuations, comparison tests were conducted using other available pumps. The amplitude of the pulsation for the various type pumps was reduced to the same level, even when the initial amplitude differed by a factor of 2.*

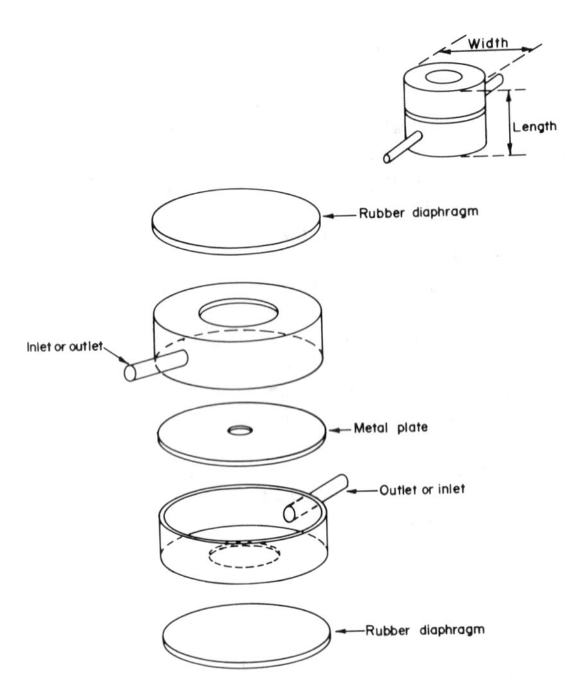

FIGURE II.10. Exploded view of snubber. (From Lamonica, J. A. and Treaftis, H. N., Investigation of Pulsation Dampers for Personal Respirable Dust Samplers, R. I. 7545, Bureau of Mines, Department of the Interior, 1971.)

m. Domestic Suppliers

Domestic producers currently are offering satisfactory two-piece, size-selective, 10-mm cyclones (10 mm O.D. x 102 mm long) fabricated from

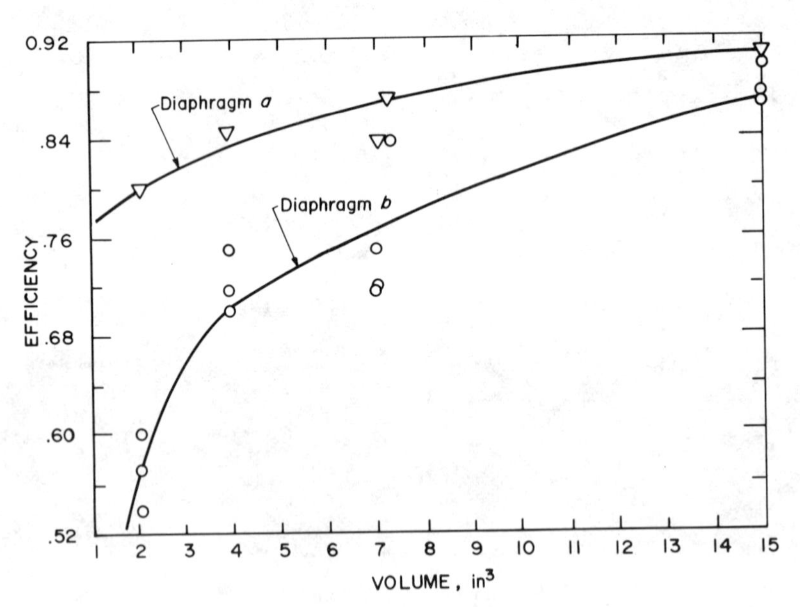

FIGURE II.11. Effect of diaphragm elasticity. (From Lamonica, J. A. and Treaftis, H. N., Investigation of Pulsation Dampers for Personal Respirable Dust Samplers, R. I. 7545, Bureau of Mines, Department of the Interior, 1971.)

*From Lamonica, J. A. and Treaftis, H. N., Investigation of Pulsation Dampers for Personal Respirable Dust Samplers, R.I. 7545, Bureau of Mines, Department of the Interior, 1971.

molded nylon.[40,52,75,76] In Figure II.2.3 the Unico cyclone is shown assembled with a tangential flow filter. This assembly meets specifications promulgated by the Federal Coal Mine Safety Act of 1969.[40]

Air enters the capsule tangentially and parallel to the face of the filter medium thus avoiding the possible rupture and loss of collected sample which can result with other, direct impingement samplers.

Tangential entry also means more uniform distribution of collected dust and a more gradual build-up of pressure across the filter. The load on the pump is thus reduced giving longer overall life and more battery reserve power on a day-to-day basis.

The lightweight capsule, together with its filter membrane, can be weighed by the user before and after sampling for an accurate determination of the particulate collected. Because the capsule need not be opened to weigh the collected solids, the risk of loss due to improper handling is virtually eliminated. The capsule design guarantees that the filter membrane will never be exposed to handling – before, during, or after sampling.

The uniform distribution provided by the tangential entry assures more accurate data from microscopic examination of the collected sample. The corrosion resistant properties of the capsule and membrane make it suitable for the sampling of a wide range of aerosols.*

The continuing uncertainty about the collection characteristics of such simple devices is unfortunate, since if the correct cut is not made during the process of collection, it cannot be made with much assurance later. A rough approximation could be made on the basis that an equivalent mass would have been calculated on the filter at another flowrate, as previously discussed. In this respect, multi-stage collectors such as the cascade impactor have an advantage over the two-stage collector, even though there are conflicting data in the literature on the collection efficiencies of the various collection stages. In this case, even if the stage constants used and the resulting size-mass distribution are incorrect, the basic stage collection date are valid, and a corrected size-mass plot can be made at a later time utilizing more reliable stage calibrations. Also, as discussed previously, a major advantage of multi-stage sampling is that overall size-mass distributions can be used to estimate deposition at all levels of the respirable tract, not just the non-ciliated level. Also, the mass median diameter (MMD), as determined from multi-stage sampling data, can serve as an indicator of the % "respirable" dust as proposed by the ICRP Task Group.

The most widely used type of multi-stage sampler is the cascade impactor which is available commercially in a variety of designs. One major limitation to their application is that only limited sample masses can be collected without re-entrainment. Further limitations, which are shared by other multi-stage instruments such as the cascade centripeter, are wall losses between collection stages and the increased number of analyses per sample which are required.**

n. Cascade Impactor Analysis

Cascade impactor analysis has been recognized as an inherently simple and straightforward method since first described by Dr. K. R. May, of Porton, England, in 1945. The advantages of this method include:

1. Particle size analyses can be performed with a series of mass determinations, rather than with the individual sizing of several hundred particles. A mass analysis can be a colorimetric or other chemical determination, a radiometric count, or any other procedure suitable to the material being studied.
2. The size distribution of one component of an aerosol can be determined in situations where the presence of background particulates would prevent accurate sizing by microscopic or other methods.
3. The size distribution of dusts which are altered by the process of collection, or which decompose or decay following collection, can be determined. A cascade impactor grades an aerosol on the basis of its airborne particle size, and as long as some fraction or derivative of the original material remains on the slides the size analysis will be the same as if the material were unchanged from its air-borne state. Thus, particles which swell or shrivel, or droplets which spread, or coalesce, or partially evaporate can be analyzed. The particles are graded by their effective aerodynamic diameter in both the cascade impactor and the respiratory system, since similar processes of inertial impaction take place in both cases.***

A small size cascade impactor developed by Lippmann and Harris[66] can be obtained commercially and is sufficiently compact (28 mm x 29 mm x 127 mm) for personnel monitoring (Figure II.6). This design incorporates the following features:

Calibration: Based on microscopic sizing of the particles collected on each stage.

Measures size of most airborne materials: Calibrated for particles of any density from 0.8 to 20 g/cc.

Efficiency over wide range of flowrates:

*From Unico Model C115 and Micronair Sampler, National Environmental Instruments, Inc., P.O. Box 590, Pilgrim Station, Warwick, R.I. With permission.

**From Lippmann, M., *Air Sampling Instruments for Evaluation of Atmospheric Contaminants,* 4th ed., ACGIH, Cincinnati, O., 1972. With permission.

***National Environmental Instruments, Inc., P.O. Box 590, Pilgrim Station, Warwick, R.I. With permission.

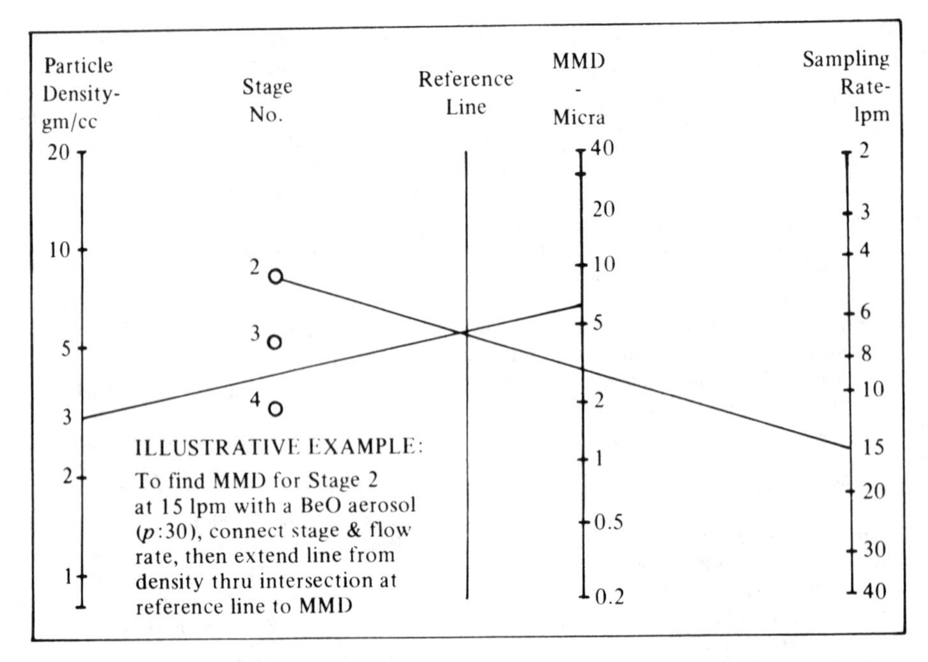

FIGURE II.12. Cascade impactor calibration nomograph. (From Micronair Gravimetric Dust Sampler, Catalog No. 3900-10; with Sampling Head (Catalog No. 3900-90), National Environmental Instruments, Inc., Box 590, Pilgrim Station, Warwick, R.I. With permission.)

Calibrated for any sampling rate between 2 and 40 lpm.

Measures wide range of sizes: With choice of appropriate flowrate, mass median sizes from 20 down to ½ μ can be measured.

Movable collection slides: Collects large samples without overloaded slides and re-entrainment.

Ruggedness and durability: Machined from solid blocks of corrosion resistant aluminum.

Preparation: Spread adhesive on two standard 1- x 3-in. glass microscope slides.

Analytical procedures: Any suitable method of mass determination may be used – chemical analysis, radiation measurement, etc.

Using the calibration nomogram (Figure II.12):

As an illustrative example, the Stage 2 Mass Median Diameter (MMD) for a Beryllium oxide (BeO) aerosol sampled @ 15 lpm would be determined as follows:

1. Draw a straight line connecting the Stage 2 circle with the flow rate (15 lpm).
2. Draw a second straight line connecting the intersection of the first line and the reference line with the density of BeO (3.0), and extend this line beyond the reference line until it intersects the MMD line.
3. Read the MMD at the intersection – For this illustrative example, it is 6.0 micra.
4. Repeat the procedure for Stages 3 and 4.

The stage MMD's obtained from the calibration nomogram are plotted against distribution data obtained from the mass analyses of the material collected on the various stages. These data, when plotted on logarithmic-probability paper, give a size-mass distribution curve describing the dust being sampled. Each stage MMD is plotted against the % of the sample smaller than that size, i.e., half of the material on that stage plus all of the material passing that stage. No meaningful calibration values can be assigned to stages 1 and 5 since the upper and lower size limits of the aerosol may be infinite.*

Interpreting the distribution curve (Figure II.13):

The illustrative distribution curve was obtained with a U_3O_8 aerosol. The size at the 50% intersection is the MMD of the over-all aerosol, and represents the size at which half of the mass is in particles larger than that size and half in particles smaller than that size. The standard deviation (σ) is equal to the size at the 84.16% intercept divided by the size at the 50% intercept. If the distribution of particle diameters is desired instead of the size-mass distribution, it can be calculated. The following

*National Environmental Instruments, Inc., P.O. Box 590, Pilgrim Station, Warwick, R.I. With permission.

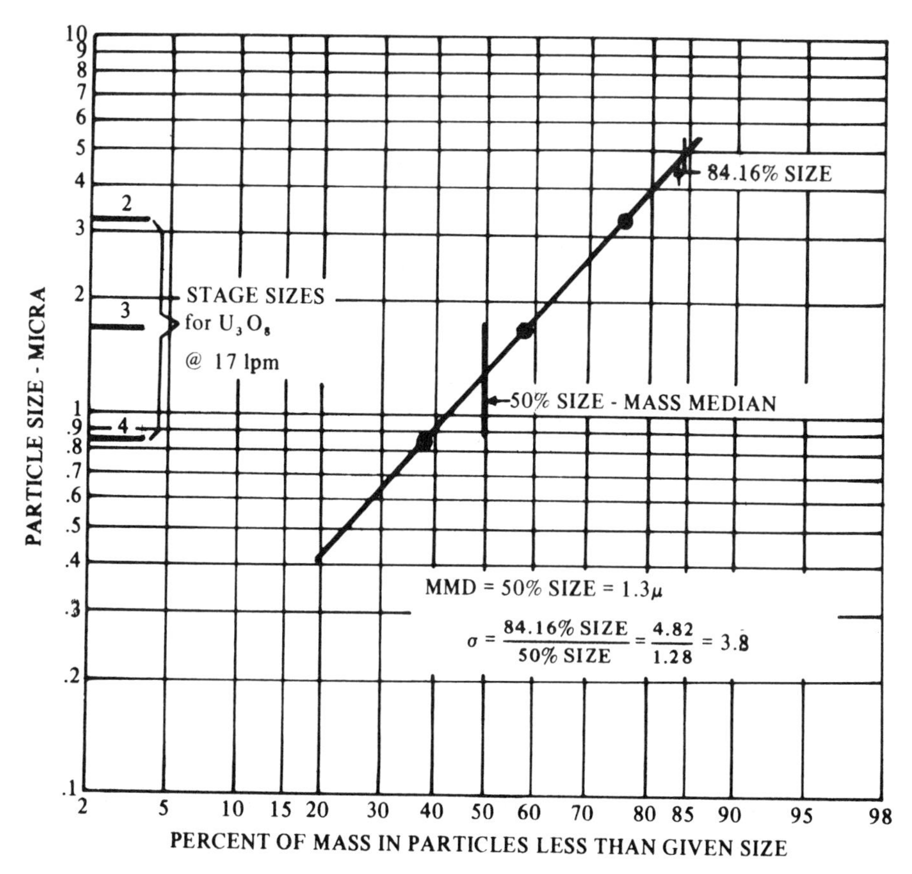

FIGURE II.13. Aerosol size distribution curve. (From Micronair Gravimetric Dust Sampler, Catalog No. 3900-10; with Sampling Head (Catalog No. 3900-90), National Environmental Instruments, Inc., Box 590, Pilgrim Station, Warwick, R.I. With permission.)

equation relates the median particle diameter (D) to the MMD:

$$\log D = \log MMD - 6.9 \log^2 \sigma^*$$

Probably the most sophisticated multi-stage cascade impactor offered to the industrial hygienist is the Andersen multi-jet Mini-Sampler®,[77,78] shown in Figures II.14 and II.15. This dust collector operates on the principle of aerodynamic particle sizing and accommodates all three dimensions of particle size, shape, and density. In brief, the following parameters are provided:

Collects and automatically sizes particles aerodynamically within a worker's breathing zone.

Small, lightweight anodized aluminum, 25 mm diameter x 25 mm high.

Can be worn by a worker while on the job.

Accommodates four 25.4 mm diameter jet and collection plates.

Calibrated at flowrate of 1.4 lpm. Design permits other flowrates.

Sizes particles in the respirable range from 5.5 down to 0.3 μ.

Collects particles in four fractions of varying respirability: 0, 20, 50, and 100%.

Provides a fast, simple method for evaluation of airborne particulate matter without tedious counting and sizing.

Actually simulates the collection of particles in the respiratory tract.

Circle of jets alternates with circle of deposits whose diameters vary from plate to plate.

A 25-mm membrane filter disc (see Figure II.15) can be inserted for collection of particles below 0.3-μ size.

*National Environmental Instruments, Inc., P.O. Box 590, Pilgrim Station, Warwick, R.I. With permission.

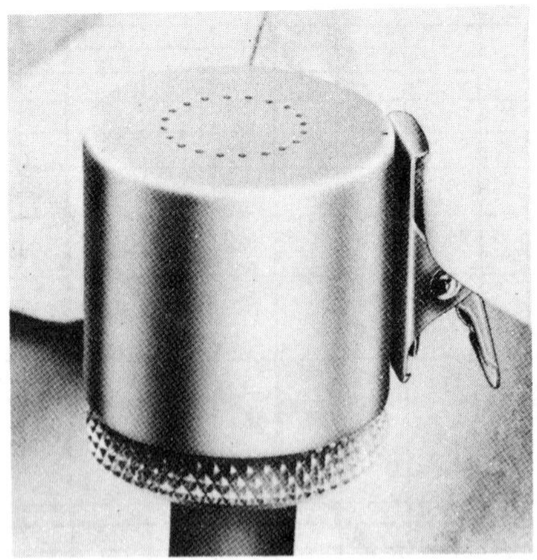

- Collects and automatically sizes particles aerodynamically within a worker's environment.
- Small, lightweight anodized aluminum, 1-1/8" diameter x 1" high.
- Can be worn by a worker while on the job.
- Accommodates four 1" diameter jet and collection plates.
- Calibrated at flow rate of 1.4 liters per minute. Design permits other flow rates.
- Sizes particles in the respirable range from 5.5 microns down to 0.3 micron.
- Collects particles in four fractions of varying respirability: 0% – 20% – 50% – 100%.
- Provides fast, simple method for evaluation of airborne particulate matter without tedious counting and sizing
- Actually simulates the collection of particles in the respiratory tract.

FIGURE II.14. The Andersen Mini-Sampler® multi-stage, multi-jet particle size analyzer. (Courtesy of Andersen Air Samplers, P.O. Box 20769, Atlanta, Ga.)

EXPLODED VIEW

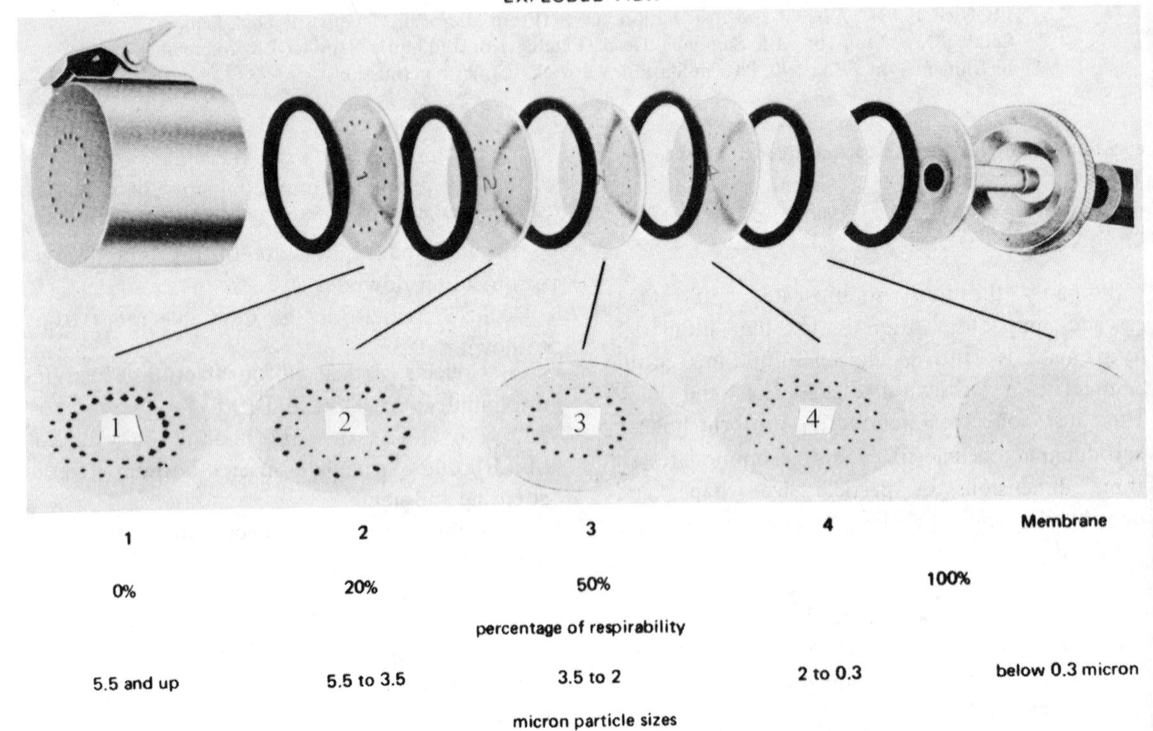

FIGURE II.15. The Andersen Mini-Sampler® particle size analyzer – exploded view. (Courtesy of Andersen Air Samplers, P.O. Box 20769, Atlanta, Ga.)

Aluminum plates can be easily weighed to determine the amount of material in each of the respirable categories.

o. Oil Mist Collection

Filters, both with and without the cyclone stage, and cascade impactors have been applied to the collection of oil mists as well as to solid particulate collection.[79] The glass fiber filter, which has delivered satisfactory reproducibility, resistivity, and efficiency, was found to be too fragile for use in the miniature assemblies suitable for personnel monitoring. The cellulose acetate membrane filter proved to be a suitable substitute in the 10-mm cyclone two-stage sampler. The most convenient method for determining particle size frequency distribution for oil mists was provided by the cascade impactor (Lippman-Harris model[66]) with preweighed membrane filter sheets on the stages. Sampling periods of at least 30 min and a microbalance were generally required. Microscopic methods for size frequency distribution were more tedious and the collection of samples more difficult.[79]

3. Gravimetric Analysis

Direct weighing of the total dust collected in membrane filters is rapidly increasing in popularity. Due to the hydroscopic nature of cellulose fibers, the collection of particulate on filter paper cannot be recommended unless the filter medium can be equilibrated in a constant humidity chamber (desiccator) before and after use. The membrane-type filter media are practically nonhydroscopic and are currently available in preweighed (tarred) form ready for field use.[24a,80]

a. Coal Dust

The Federal Coal Mine Health and Safety Act of 1969 (30 U.S.C. 811, Public Law 91-173) specified gravimetric personnel monitoring to demonstrate compliance with the respirable threshold limit of 2 mg/m^3. Part 1, "Mandatory Health Standards – Surface Work Areas of Underground Coal Mines and Surface Coal Mines," appeared later.[82] The Bureau of Mines designed a personnel monitor which met the requirements of Part 74 of the Coal Mine Safety and Health Act.[81,84] The MSA unit[80] is essentially a replica of this device, whereas the Unico model includes innovations (tangential sample flow across the filter and pulsation dampening) which greatly improve the performance of the sampler.[83] This equipment is well suited for gravimetric sampling in general and has been accepted for compliance with the nuisance particulate TLV (10 mg/m^3).[4] A partial list of nuisance particulates for which gravimetric sampling is acceptable is presented in Table II.3.

TABLE II.3[4]

Some Nuisance Particulates*

Alundum® (Al_2O_3)	Kaolin
Calcium carbonate	Limestone
Cellulose (paper fiber)	Magnesite
Portland cement	Marble
Corundum (Al_2O_3)	Pentaerythritol
Emery	Plaster of Paris
Glass, fibrous† or dust	Rouge
Glycerin mist	Silicon carbide
Graphite (synthetic)	Starch
Gypsum	Sucrose
Vegetable oil mists (except castor, cashew nut, or similar irritant oils)	Tin oxide
	Titanium dioxide

* When toxic impurities are not present, e.g., quartz < 1%
† < 5 to 7 μm in diameter

b. Cotton Dust

The OSHA regulations for the control of cotton dust in the work place require 4- and 6-hr continuous personnel monitoring with gravimetric filter equipment:[85]

8-hr time-weighted average: 1 mg/m^3.

Analytical method – Net weight of gross airborne dust, using balance with sensitivity of 0.01 mg.

Sampling equipment: Personal samplers – Cotton is to be determined primarily with the personal sampler attached to the workman's clothing with the face of the filter pointing down. The filter should not be covered. Our experience is that at least a 4- and preferably a 6-hour sample (at 1.5 lpm) is required in a textile plant opening or carding area to obtain reliable results. A preweighted filter (0.01 mg) should be used and reweighed to the same accuracy.

There is no ceiling or any limit on excursions.

Because of the nature of both the sampling method and the development of byssinosis, it is neither desirable nor possible to get grab samples.

Blanks – With each batch of samples collected, two additional filters should be subjected to exactly the same handling except that no air is drawn through them. These should be used as blanks.*

c. General Requirements

As indicated in the foregoing instructions, all that is required in the way of laboratory equipment other than the sampler itself is an analytical balance sensitive to 0.01 mg; for screening purposes, to determine whether dustiness is less than one half or more than twice the TLV, a sensitivity of 0.1 mg is sufficient, especially if the filter is to be analyzed for a discrete substance such as lead. However, prudence dictates checking the tare weight to confirm accuracy of the manufacturer's weight for accurate analysis or when legal implications may be involved. For a detailed discussion of respirable fraction and size selective sampling see Section II.B.2, "Cascade Collectors – Respirable Fraction," and for detailed instructions for carrying out a dust evaluation program see Section II.B.1.f, "An Example – Asbestos Fiber Monitoring." In general, when a toxic component is present in gross concentrations, sampling techniques should be limited to situations where the entire aerosol is absorbed, e.g., for some highly soluble aerosols, or where the particle size distribution is relatively constant and there is a known fixed ratio between the gross concentration and the concentration in the size range of interest.

d. Difficulties Encountered

As would be expected when the application of a new technique on a large scale is attempted, certain serious difficulties have been encountered. Morse summarized the problems faced by the coal mining industry at the outset of the adoption of gravimetric sampling for assessment of the coal dust hazard.[87] Since these problems would be common to most gravimetric dust sampling programs, they are outlined as follows:

1. Training problem
2. Flowrate disagreement
3. Unattended operation
4. Preweight filter control
5. Filter selection
6. Rock dust contribution
7. Moisture effects
8. Cyclone orientation
9. Weighing problems with a small mass
10. Quartz analysis
11. Lack of correlation with pulmonary response

Training problem – Only competent personnel should be assigned to the collecting of samples, weighing of filters, and maintaining the equipment in good operating condition. Large numbers of untrained people will doom a project to failure before it starts. Obtaining dust measurements of quality is not as simple as appearances would indicate. This problem not only affects the industrial work force, but is equally applicable to the governmental enforcement officers.

Flowrate disagreement – Where more than one type of collection instrument is available, or where flowrate is critical to the success of the program, agreement on operational parameters must be reached before sampling is initiated. The controversy generated by the two standard coal dust samplers (the British Medical Research Elutriator (MRE) vs. the American two-stage cyclone personal monitor) illustrates this point.

Unattended operation – Currently available samplers require flowrate checks several times (hourly) during the work shift. Supervisory personnel limitations or remote work locations may reduce this necessary control to very nearly zero. Some individual may utilize this opportunity to tamper with the personnel monitor he is wearing or a fellow worker's monitor.

Preweighed filter control – Differences of over 65% in the weights of filters from the same box have been found (17.62 to 10.64 g). The quality control exercised by the manufacturer must be validated by the user until confidence can be established. Otherwise, each filter must be weighed by the user both before and after sampling.

Filter selection – There may be considerable variation in pressure drop (ΔP) across membrane filters of the same nominal pore sizes under 5 μ. This variability must be compensated by flowrate adjustment to a constant fixed rate, or appreciable differences in the amount of respirable dust collected will be encountered. Impurities present in the filter matrix may introduce errors in subsequent analysis (e.g., silicates in filters used for silica determination).

*From Cotton Dust, OSHA Sampling Sheet #5, Occup. Safety and Health Admin. Standards, 29 CFR-1910.93, (b) Table G-1 of August 13, 1971.

Moisture effects – Although the filter specifications require nonhygroscopic elements, the collected dust may be hygroscopic. To quote:

> The specifications for dust sampling instruments state that the filter shall be nonhygroscopic and membrane filters have been placed in this category. We have loaded 37 mm filters with coal dust, arranged them in series with a fritted bubbler in water and a chamber for removal of water droplets. We drew air through the chain at 2.0 liters/min and weighed the filters after 30, 60, 90 and 120-minute intervals, returning the filter to the chain after the first three intervals. After 120 minutes, the first filter gained 16.52 mg of moisture with the subsequent filters in the chain showing lesser but appreciable moisture build-up. By placing the filters in a desiccator for 25 to 30 minutes, the filter weights will generally return to the dry weight.
>
> We have similarly added moisture to nondusted VM-1 filters. Even if one assumes a filter is nonhygroscopic, this does not mean that coal dust and rock dust are not hygroscopic.*

Cyclone orientation – Although the cyclone is insensitive to position orientation as far as operation is concerned, when worn by repairmen the device does not always hang in a vertical position. Under these conditions some nonrespirable dust collected in the cyclone trap may be transferred to the filter.

Weighing problems with a small mass – When a small mass (2 mg) of dust must be determined by difference on a filter mass of many magnitudes greater weight (5,000 mg) the problem of accuracy becomes critical.

Lack of correlation with pulmonary response – It should be fully recognized that occupational health problems will not be adequately resolved unless the physician and industrial hygienist work together as a team in undertaking epidemiological studies of the dose-response relationship to establish valid standards. No change in the standard for a dust concentration from one established epidemiologically can be validated alone by performance correlations among different sampling instruments. The establishment of a valid standard based on mass concentration (mg/m^3) in place of particle count can be done only by an in-depth epidemiologic study of the work environment together with the findings derived from the medical records.[87]

4. Chemical Analysis

The following general information from the AIHA *Analytical Guides*, relative to selection of filter media for the collection of aerosols for chemical analysis, is presented by permission of the copyright owner.

a. Filter Media

Aerosol sampling by filtration is one of the more common techniques for chemical studies in industrial health. Most aerosol samples may be stored without deterioration and are relatively simple to handle both in the field and in the laboratory. There are some excellent discussions on the characteristics of filter media. Lippmann[1] has prepared a comprehensive review of various filter media with their retentive characteristics, as well as a discussion of filtration theory and of the criteria to be considered in selecting filters. Pate and Tabor[86] have presented a critical review of glass fiber filters for the collection and analysis of atmospheric particulate matter. No attempt has been made to duplicate this subject. The main emphasis in this discussion is on the separation of the contaminant from the filter medium and the preparation of it for subsequent analysis.

No single type of filter is suited to all possible conditions of collection and analysis. The selection of a suitable filter depends on the physical and chemical nature of the aerosol (particle size, vapor pressure, reactivity, etc.), the sample volume (flow rate, size of filter holder, sampling period) and the chemical procedure used to prepare the sample for analysis which may introduce contaminants from the filter.

All filter media will contain contaminants of one sort or another which may interfere with the subsequent analysis; the amounts of contaminants may be variable. Contaminants found in glass fiber filters in appreciable quantities include chlorides, zinc, iron, chromium and others. The amounts of contaminants will vary depending on the source of the filter, and also from lot to lot from the same source.[86] Organic binders, if used in glass fiber filters, will interfere with the determination of organic materials; the binder must be burned off before use. Cellulose filters are generally low in ash, but there can be interference in some fluorescence analyses – particularly if the filter is leached with acid. A control analysis should be performed on the particular lot of filter media before use to determine interference blanks.

b. Isolation by Filter Destruction

Completely nonvolatile metals can be isolated from organic-base filters by wet or dry ashing, each technique having its place in the scheme of metal analysis. The advantages of wet ashing over dry or combustion methods are (a) the conversion of the more volatile metals to nonvolatile salts; (b) control of oxidation conditions. Closed oxidation systems with pure oxygen are successful with relatively small samples. Plastic fiber filters can be

*From Morse, K. M., Problems in the gravimetric measurement of respirable coal mine dust, *J. Occup. Med.*, 12, 400, 1970. With permission.

destroyed but require longer ashing times than paper or small membrane filters. Granular beds are difficult to ash and their use is limited.

c. Isolation by Leaching

Since many contaminants are readily soluble in specific reagents or organic solvents, they can be recovered quantitatively from filter paper or other types of filters by treating the filter with an appropriate solvent or reagent. The usual procedure consists of macerating the filter in the chosen solvent, separating the filtrate from the filter material by filtering or centrifuging the mixture, and repeating this process several times. Glass and asbestos fiber filters especially require this type of treatment. Care should be exercised, however, to prevent the contaminant from being converted to a form that will be tenaciously adsorbed on the fibers of the filters. The leaching technique can be used on plastic fiber filters if a solvent is selected which does not dissolve the plastic filter.

d. Isolation by Dissolution of the Fiber

Plastic and membrane filters adapt readily to dissolution when the contaminant also is soluble in the solvent system chosen. The filter may be macerated in a solvent. A solvent that is inert to the filter will dissolve the contaminant; then the filter may be dissolved and extracted with an immiscible solvent. Glass fiber filters, which have been fired to remove their organic binder before sampling, can be dissolved in hydrofluoric acid and the solution analyzed for certain substances.

e. Direct Determination of Contaminant

Particle-sizing, the direct radiometric measurement of gross activity, and the use of color-indicating filters are examples of direct methods. For particle-sizing by light microscopy, only flat-surfaced filters are useful. Plastic filters can be dissolved in a solvent. If the particles are truly inert to the solvent, they can be concentrated by centrifugation and other techniques, and subsequently measured by light microscopy. Color-indicating filters may be made of any base, but most techniques use some form of paper. The paper may be impregnated with sensitive reagents which develop a color when the contaminant is present, or the color may be produced by adding the reagents after the sample is collected.

Radiometric techniques are applicable to all filter media. However, alpha-active aerosols must be collected on an impenetrable filter medium to minimize the reduction in alpha activity resulting from penetration of the particles into the filter media. The alternative is to apply a large correction factor but the uncertainty of this factor precludes an absolute determination. Self-absorption in the filter is usually not an important factor when the aerosol particles are beta or gamma emitters; therefore, little or no correction is required.

The ease of treatment of several types of filter media is indicated in Table II.4. In addition to the filters listed in the table some filter samples are collected on wool pads, cotton matting, and other substances commonly used in respirator filters.*

f. Methods

1. Silica Dust

The inhalation of "active" silica (cristobalite, quartz, fused silica, and tridymite) presents no acute health hazard, but does produce a high chronic health risk over a period of years.

Inhalation of extreme concentrations of particles of one micron or less may lead to a diffuse, fulminating lung fibrosis in a few months. Development of the usual chronic type of silicosis takes many months or years. Effects of repeated inhalations of silica dusts are cumulative and progressive. The different crystalline forms produce varying responses, in animals, with tridymite and cristobalite more severe than quartz. Disease is characterized by an initial generalized linear or perivascular increase in pulmonary density, progressing to small nodules diffusely scattered throughout the lung fields. As exposure continues, these nodules attain sufficient size to interfere with normal pulmonary function. Dyspnea and emphysema are characteristic of the more advanced stages. Silicosis seldom, if ever, produces death. However, common complications are tuberculosis, chronic bronchitis, and other bacterial infections.**

Silica-bearing dusts may be collected by impinger or filters, and determined by petrographic (microscopic), X-ray, or chemical techniques. The electron microscope also is useful for evaluating properties of the dust. Filters employed for silica collection must be essentially silica free (less than 10% of the amount collected). The sampling and analysis procedure for crystalline silica, adopted by OSHA for high-volume, fixed-station monitoring, consists of:[41]

Standard: 29 CFR 1910. 93(b), Table G-3 of December 7, 1971

Form

Quartz (respirable): $\frac{10\ mg/m^3}{\%SiO_2+2}$

Cristobalite: One half the value for quartz

Tridymite: One half the value for quartz

Analytical Method

Gravimetric personal samples for worker exposure should be tared and reweighed by DOL or Cahn balance. The percent of free silica is determined by NIOSH on the fixed location sample using a colorimetric wet chemical method, minimum detectable amount. 0.01 mg/filter.

*From Monkman, J. L., Ed., *Analytical Guides,* AIHA, Akron, O. With permission.

**From Hygienic Guide Series for Silica, AIHA, Akron, O., 1957. With permission.

TABLE II.4

Treatment of Various Filter Media

Filter media	Ease of ashing		Direct isolation of contaminant			Direct determination of radioactive aerosols		
	Wet	Dry	Organic solvent extraction	Water or reagent extraction	Dissolution of media	Alpha	Beta	Gamma
Cellulose fiber filter (paper)	G	G	S	S	O	P	G	G
Glass fiber filter (with and without binder)	O	O	S	S	O	P	G	G
Asbestos fiber filter (cellulose binder)	O	O	S	S	O	O	S	G
Plastic fiber filters								
Polyvinyl chloride	P	P	P	S	G	P	S	G
Polystyrene	P	P	P	S	G	P	S	G
Membrane filters (cellulose esters)	G	G	P	S	G	E	E	E

E = Excellent
G = Good
S = Satisfactory
P = Poor
O = Impractical

From Monkman, J. L., Ed., *Analytical Guides,* AIHA, Akron, O., 1965. With permission.

Sampling Equipment

For worker exposure:

Personal sampling pump plus preweighed MSA type FWS-B or Gelman type VM-1 filter, using 2-piece filter cassette and 10-mm nylon cyclone. Two filters should be subjected to the same handling as the samples except that no air is drawn through them; these should be used as blanks. Sampling rate: 1.8 l/m.

For determining % quartz:

Collect at fixed locations near workers, using NIOSH or preweighed MSA type EWS-B filters, 3-piece cassettes plus Gast. 1531 pump with 9-l/m critical orifice and 1/2-in. steel cyclone. The vacuum release valve should be adjusted to 10 to 15-in. Hg suction. Two blanks of the same filter type should be submitted with the sample. Sampling rate: 9 l/m.

Sample Size

Minimum: 4 hours. Maximum: 8 hours.

The colorimetric chemical method developed by Talvitie appears to be the method of choice.[42] The abstract of the recommended refinements in the original method[43] follows:

Refinements contributing to the precision and accuracy of the phosphoric acid method can be made by the use of commercially available apparatus comprising a rotator and a voltage-controlled heater. Additional refinements include optimizing the conditions of analysis by digestion of orthoclase with phosphoric acid for varying lengths of time, use of correction curves to compensate for loss of quartz, and identification of undissolved minerals by means of a polarizing microscope. An eight-year study of the precision of the phosphoric acid method gave a mean standard deviation of 0.34 in the analysis of typical samples containing from 3 to 50 per cent quartz. The recovery of quartz from the less-than-5-micron fraction of air-borne dust is about 95 per cent when the digestion time is reduced to 8 minutes. The method can be extended to samples of air-borne dust as small as 2 milligrams by spectrophotometric determination of the quartz as molybdosilicic acid.[44] *

By digestion 2 mg of airborne dust may be analyzed for 8 min in 25 ml of 85% phosphoric acid, diluted with water at 60 to 70°C, and filtered through a 47-mm membrane filter (0.45 μ pore

*From Crystalline Silica – OSHA Sampling Data Sheet #3, Occup. Safety and Health Admin., Department of Labor, December 21, 1972.

size). The filter containing the quartz residue is ashed in a platinum crucible and the quartz dissolved by moistening with 1:1 H_2SO_4 and 10 ml 48% HF, and heating gently for 30 to 60 min. The soluble silica then is quantitated by conversion to the yellow molybdosilicic acid.[44] Good agreement with the gravimetric procedure was found, although precision was only fair. One evaluation of the sampling methods, prepared by Ayer, led to the following conclusions:[45]

> Dust measurements should give results related closely to the health hazard, be simple, reproducible, and inexpensive. Measurement of total airborne mass concentration or, preferably, respirable mass concentration is simpler, cheaper and gives more reproducible results than the impinger. Comparative sampling indicates that a respirable dust limit of 10/(% Respirable Quartz + 2) mg/m³ would be equivalent to the present impinger TLV. A total dust limit proposed is: 30/(% Quartz + 2) mg/m³. Either mass limit permits personal sampling techniques to be used. Quartz analyses are made on the respirable fraction for the respirable limit, and on total dust for the total dust limit. Until safety of present quartz limits is demonstrated, the limits should be used conservatively.

During its 1968 annual meeting the American Conference of Governmental Industrial Hygienists (ACGIH) accepted a report of its Committee on Threshold Limit Values[52,57] giving a quartz "TLV for respirable dust in mg/m³." The method of collecting this dust was specified as:

> "Both concentration and per cent quartz for the application of this limit are to be determined from the fraction passing a size-selector with the following characteristics:"

Aerodynamic Diameter (μ) (Unit density sphere)	% Passing Selector
⩾2	90
2.5	75
3.5	50
5.0	25
10	0

This limit was published[59] as part of *Safety and Health Standards for Federal Supply Contracts* of the U.S. Department of Labor. Also, it was reported in December 1968 that the Consumer Protection and Environmental Health Service, Public Health Service, had recommended a respirable mass criterion for dust collected in a size-selective instrument developed by the British Mining Research Establishment.*

Physical methods of analysis also have been developed. X-ray diffraction has been rather widely used.[46,47] However, sensitivity is not entirely adequate for the quantitation of the small amounts of silica usually found in personnel monitor samples.[42] Filters fabricated from silver have been recommended for X-ray diffraction determinations directly on the filter bed. X-ray fluorescence analysis probably could be applied directly to the sample.[48,48a,49] Petrographic analysis by polarized light can yield relative data, and is most useful as a method to determine completeness of extraction by wet chemical extraction methods.[42]

2. Lead Determination

An acceptable dithizone procedure, in which the filter medium can be substituted for blood directly in the procedure, is described elsewhere.[93,94] A variation of this method is described in Reference 89. The method adopted by OSHA follows:[88]

Standard: 29 CFR 1910.93(b), Table G-2 of August 13, 1971 & ANSI Z37.11-1969

8-hour time-weighted average: 0.2 mg/m³.

Analytical Method:

Atomic absorption.

Detection Limit: 0.0003 mg/filter (0.3 μg).

Sampling Equipment:

Personal sampling pump plus unweighted 37mm Millipore type HA filter. Either 2- or 3-piece filter cassette may be used. Sampling rate: 1.5 l/m.

Sample Size:

A minimum sampling period of 60 minutes is recommended, and longer periods, up to full shift, are preferable. Care must be taken to prevent material from falling off heavily loaded filters during shipment. This may be done by placing another filter on top of the one with the sample, exercising care not to dislodge any of the sample in the process.

Blanks:

With each batch of samples, submit two filters which are subjected to exactly the same handling as for the samples except that no air is drawn through them. Label these as blanks.

Shipping:

The cassettes in which the samples are collected should be shipped in a suitable container, designed to prevent damage in transit.**

A field kit for inorganic lead particulate in air

*From Ayer, H. E., The proposed ACGIH mass limits for quartz. Review and evaluation, *Am. Ind. Hyg. Assoc. J.*, 30, 117, 1969. With permission.

**From Lead and its Inorganic Compounds (except lead arsenate) – OSHA Sampling Data Sheet #4, Occup. Safety and Health Admin., Department of Labor, December 21, 1972.

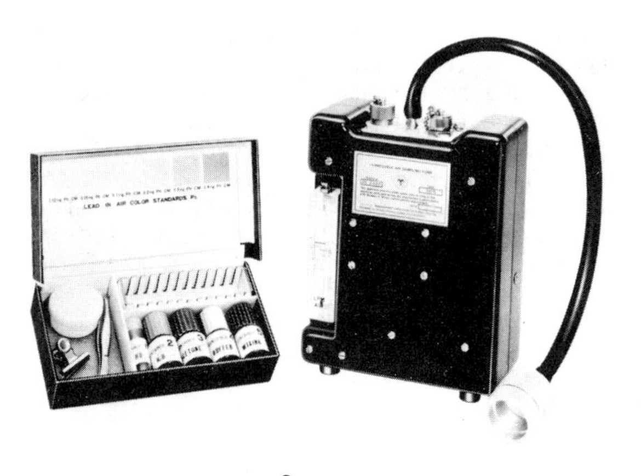

FIGURE II.16. Unico® lead dust in air field kit. (Courtesy of National Environmental Instruments, Inc., P.O. Box 590, Pilgrim Station, Warwick, R.I.)

based on the work reported by Quino[95] has been developed and is available commercially.[96] The red stain produced with tetrahydroxyquinone on filter paper is compared visually with prepared color standards. The detailed instructions are outlined in Chapter I, Section A.1.e.7, "Lead." The components of the field kit are shown in Figure II.16.

3. *Analysis of Particulate from the Ambient Atmosphere*

The following procedures for the analysis of particulates collected from the ambient atmosphere have been recommended by the Intersociety Committee on Methods of Air Sampling and Analysis[89] and can be easily adapted to personnel monitoring by adjustment of sample size. (The following material is used by permission of the copyright holder.)

a. Tentative Method of Analysis for Antimony Content of the Atmosphere 12102-01-69T

Health Lab. Sci. 7(1), 92-95 (1970)

Pentavalent antimony in the presence of a large excess of chloride ion reacts with Rhodamine B to form a colored complex, which may be extracted with organic solvents such as benzene, toluene, xylene, or isopropyl ether. Trivalent antimony will not react with Rhodamine B and must be oxidized to the pentavalent state. The antimony sample is collected by filtration of a measured volume of air through membrane cellulose, or glass fiber filters. Phosphoric acid is used to minimize interference by iron. The absorbance of the pink extract is measured in a spectrophotometer at 565 nm or in a colorimeter, using a green filter, and the amount of antimony present is determined by reference to a calibration curve prepared from known amounts of antimony.

With a 1 cm light path cell, the method covers a range from 0 to 10 μg antimony/10 ml solvent.

Since the sensitivity is 1.0 μg antimony/determination (0.05 absorbance units), a concentration of 0.05 μg antimony/m^3 can be measured if a 20 m^3 air sample is taken for analysis.

Of the commonly encountered ions, only iron is likely to interfere. Iron interference can be eliminated by extracting with isopropyl ether instead of benzene.

Three samples containing 3.0, 5.0, and 8.0 μg antimony/ml, analyzed in triplicate by 10 collaborating laboratories, in addition to the referee, showed standard deviations in the range of 0.113 to 0.245.

b. Tentative Method of Analysis for Arsenic Content of Atmopsheric Particulate Matter 12103-01-68T

Health Lab. Sci. 6(2), 57-60 (1969)

This method is suitable for the analysis of particulate samples collected on either membrane or glass fiber filters.

The arsenic compounds are dissolved from the sample of particulate matter with HCl followed by reduction of arsenic to the trivalent state with KI and $SnCl_2$. The trivalent arsenic is further reduced to arsine, AsH_3, by zinc in acid solution in a Gutzeit generator. The evolved arsine is then passed through an H_2S scrubber which consists of glass wool impregnated with lead acetate and then into an absorber containing silver diethyldithiocarbamate dissolved in pyridine. In this solution, the arsine reacts with silver diethyldithiocarbamate, forming a soluble red complex which is suitable for photometric measurement with a maximum absorbance at 535 nm.

Since the level of arsenic in ambient air usually lies between 0.1 and 1.0 μg/m^3, an aliquot representing at least 10 m^3 air is required if the sample absorbance is to fall on the calibration curve. The amount of filter used must be limited or the filter itself may physically and/or chemically interfere in the analysis. No difficulties will be encountered if the filter area used does not exceed 2 sq in. Glass fiber filters gave better recovery of arsenic than Whatman No. 40 or 541 paper. Average recovery of the arsenic from the paper was 86±4% and from the glass fiber 103±5%.

If suitable samples (10 m^3 air) are available, concentrations as low as 0.1 μg As/m^3 can be measured by this procedure. The maximum measurable concentration with a comparable sample is 1.5 μg As/m^3. Higher concentrations can be measured if smaller samples are used.

High concentrations of nickel, copper, chromium, and colbalt interfere in the generation of arsine. The presence of interferences can be determined by the internal standard technique in which known amounts of arsenic are added to a sample and the recovery is determined. The filters used to collect atmospheric particulates may contain measurable quantities of arsenic. Portions of unused filters should be analyzed and corrections made if necessary. This can be done separately or as part of the reagent blank procedure.

Spiked samples containing 0.10, 1, 5, and 10 μg arsenic were analyzed with an accuracy of ±0.04 μg, based upon 7 replicate determinations at each concentration.

c. Tentative Method of Analysis for Selenium Content of Atmospheric Particulate Matter 12154-01-69T

Health Lab. Sci. 7(1), 96-101 (1970)

Selenium may exist in the atmosphere most probably as the element, the oxide, or the selenite. Particulate samples are collected by filtration of measured volumes of air through cellulose, membrane, or glass fiber filters, followed by oxidation of the sample with concentrated nitric acid. The Se (IV) reacts with 2,3-diaminonaphthalene (DAN) in acid solution to form the red-colored and strongly fluorescent 4,5-benzopiazselenol. The latter is extracted with toluene. The fluorescence intensity of the solution sample is measured at an excitation wavelength of 390 nm and a fluorescent wavelength of 590 nm, using a sample containing 1.0 μg selenium as the reference standard.

In the fluorimetric procedure, a linear calibration curve is observed over the range of 0 to 1 μg selenium in 10 ml solvent (1 cm light path cell). In the spectrophotometric procedure, the calibration curves follow Beer's law over the range of 0 to 20 μg selenium/5 ml solvent (1 cm light path cell) at a wavelength of 390 nm, using a reagent blank.

The sensitivity of the spectrophotometric procedure is 0.0032 μg/sq cm for log $I_0/I = 0.001$, but the fluorimetric procedure with DAN is over 10X as sensitive as the spectrophotometric procedure. The concentration of selenium in the atmosphere over an urban-industrial area may be expected to average 0.2 μg/200 m^3 air.

In the presence of masking agents, the primary interference which is observed at a 2000-fold excess of foreign ion to selenium is due to substances like hypochlorite, which oxidize the reagent, or reducing agents like Sn (II), which reduce selenium to the elemental state. Ions such as Al, Zn, Cu, Ca, Cd, Mn, Ni, Mg, Ba, and Sr are separated from Se (IV) by ion exchange of the sample solution through Dowex 50 W-X8 resin. Tellurium in large excess does not interfere. Nitrite ion causes interference. In preparing the sample for analysis, the minimum amount of nitric acid is used to dissolve the sample and the oxides of nitrogen are removed by boiling. Nitrate ion does not appreciably interfere with the spectrophotometric method and not at all with the fluorimetric method.

Both the spectrophotometric and the fluorimetric procedures show a relative standard deviation of 9.9% for the analysis of selenium samples containing a large excess of copper.

d. Tentative Method of Analysis for Beryllium Content of Atmospheric Particulate Matter 12105-01-70T

Principle of the method – The beryllium content of ambient air has been found to range from undetectable to as high as 0.01 μg/m^3. The analytical method described herein involves measuring the fluorescence of an aqueous beryllium solution with ultra violet light in the presence of morin. It is well suited to the analysis of air particulate samples in this range.

Range and sensitivity – The recommended concentration range for analysis is 0.01 μg–1.0 μg of beryllium in 10 ml of solution. Greater sensitivity can be achieved depending on instrumentation. Detection of as little as 0.0004 μg Be has been reported. Theoretically, a minimum concentration of 0.00002 μg/m^3 could be detected if a 20 m^3 sample of air particulates is taken for analysis; but from a more realistic point of view, the minimum detectable concentration is probably one order of magnitude greater. Higher concentrations can be measured by making proper adjustments in the procedure.

Interferences – The method is practically specific for beryllium since most heavy metals which would interfere are removed. Zn, however, may interfere slightly even though it is complexed with KCN. According to Walkley, in a 10 ml solution containing 0.01 μg Be, 100 μg of Zn will give a blank of 0.004 μg Be though 75 μg Zn gave no detectable blank.

Precision and accuracy – No good quantitative data are available. On 200 spiked samples on filter paper Welford and Harley reported 92 per cent recovery. Walkley reported average recovery of 110 per cent on 10 spiked samples on filter paper.

e. Tentative Method of Analysis for Manganese Content of Atmospheric Particulate Matter 12132-01-70T

Principle of the method – Manganese may be present in atmospheric particulate matter in the range 0.01–3.0 μg/m^3. For estimating microquantities of manganese, the periodate method is one of the most suitable wet chemical methods. By this method the collected manganese is dissolved in nitric and sulfuric acids and the resulting acid solution, free of chlorides, is oxidized with potassium periodate to convert the manganese to permanganate. The latter is determined in a spectrophotometer at 525 nm.

Range and sensitivity – Using a light path of 2 cm in the spectrophotometer, concentrations of 6–120 μg of manganese in 10 ml of solution represent the limiting range of the instrument.

A sample or an aliquot representing 20 m^3 of air will contain sufficient manganese for analysis if the air contains a minimum of 3.0 μg/m^3.

A longer light path or a larger sample or both may be used for lower concentrations.

Interferences – The maximum amounts of interfering elements which can be present in the sample without significantly altering results are as follows:

Element	Maximum
Beryllium	0.0001%
Bismuth	0.01%
Boron	0.005%
Chromium	0.01%
Copper	0.01%
Lead	0.03%
Iron	0.01%
Magnesium	0.01%
Nickel	0.02%
Silicon	0.05%
Tin	0.01%
Zinc	0.30%

Ferrous salts or other strong reducing agents will reduce the periodate and liberate iodine which will invalidate the results. The use of more HNO_3 in the digestion process should eliminate this interference.

If HCl is added to dissolve the sample, it must be completely removed by fuming with H_2SO_4 before the color is developed.

Precision and accuracy – No data are available in the literature for the range required. On the basis of mg/m^3, an error of ±7.7% was reported by the American Conference of Governmental Industrial Hygienists.

f. Tentative Method of Analysis for Molybdenum Content of Atmospheric Particulate Matter 12134-01-70T

Principle of the method – Molybdenum may exist in the atmosphere as the element, the oxide or as molybdates. Concentrations of molybdenum in ambient air range from below detectable amounts to maximum values in the order of 0.34 $\mu g/m^3$ of air. The satisfactory determination of molybdenum depends largely on the quantitative formation of a stable molybdenum (V) thiocyanate complex. This complex is formed, in the presence of iron and copper, by reduction of molybdenum VI with stannous chloride in a solution containing potassium thiocyanate. The thiocyanate complex is extracted with methyl isobutyl ketone. The absorbance of the extract is measured in a 10 cm cell at 500 nm using methyl isobutyl ketone in the reference cell. The absorbance is a linear function of molybdenum concentration, at least between 0 and 15 μg of molybdenum/25 ml of solvent.

Range and sensitivity – With a 10 cm light path cell the method covers the range from 0–15 μg molybdenum/25 ml of solvent.

The sensitivity being 1.1 μg of molybdenum/0.05 absorbance units, a concentration of 0.055 μg of molybdenum/m^3 of air can be measured if a 20 m^3 sample is taken for analysis.

Interferences – The only elements which cause interferences are rhenium, platinum, palladium, rhodium, selenium and tellurium.

The interferences caused by these elements are negligible in concentrations found in ambient air.

Precision and accuracy – Twenty solutions each containing 75 μg of molybdenum were extracted and the absorbances measured using a 2 cm cell. The absorbances of all 20 solutions were between 0.670 and 0.676. The method is accurate ±4 per cent.*

4. Methods Reviewed by the AIHA Analytical Chemistry Committee

Additional information relative to analytical methods may be found in the AIHA *Analytical Guides.* The Guides for cadmium, fluoride, uranium, and zirconium are reproduced for additional information by permission of the copyright owner.

a. Cadmium (Cd)

Note: Recommended cadmium atmospheric concentrations (8 hours); 0.1 mg cadmium oxide fume per cubic meter of air (ACGIH-TLV 1969).

References

1. **Saltzman, B. E.:** Colorimetric Microdetermination of Cadmium with Dithizone. *Anal. Chem., 25*:493 (1953).
2. **Sandell, E. B.:** *Colorimetric Determination of Traces of Metals.* Third Edition, p. 353, Interscience Publishers, Inc., N.Y. (1959).
3. American Conference of Governmental Industrial Hygienists: *Manual of Analytical Methods.* Determination of Cadmium in Air (1957).
4. **Elkins, H. B.:** *The Chemistry of Industrial Toxicology,* 2nd Edition, p. 308-309, John Wiley & Sons, Inc., N. Y. (1959).

Method of Collection

Solids may be collected from air by filtration through filter paper (Whatman No. 41 or equivalent) at a rate of 15 lpm (liters per minute) for a one-inch diameter paper, or through a molecular filter at a rate of 10 lpm or less for a one-inch diameter paper. The electrostatic precipitator may be used at a rate of 2 to 3 cfm (cubic feet per minute). The midget and standard impingers may be used for dust using 5% nitric acid (by volume) as the collecting medium at rates of 0.1 and 1 cfm respectively.

Principle

Cadmium forms a red complex with dithizone. Cadmium is extracted as dithizonate from a strongly alkaline solution containing CN by using a strong solution of dithizone in chloroform. The cadmium is returned to an aqueous layer by shaking the chloroform layer with 1% tartaric acid. The tartaric acid is made alkaline with NAOH-KCN solution and the cadmium is again extracted using a dilute dithizone solution. The red cadmium dithizonate is determined in a spectrophotometer at 518 nm.

Interferences

The method is remarkably free from interferences from other metals. By making suitable minor modifications, interferences can be essentially eliminated except in the case of thallium. A recovery of about 98% of cadmium is obtained when thallium is separated by extraction of thallic bromide from 0.3*N* HBr.

Sensitivity

In a 100-liter sample of air, 0.03 mg of cadmium per cubic meter of air can be determined. The molecular extinction coefficient is about 70,000 to 80,000 at 518 nm.

Special Reagents and Equipment

Spectrophotometer or colorimeter.

*From Katz, M., Ed., *Methods of Air Sampling and Analysis,* Intersociety Committee, Am. Public Health Assoc., Washington, D.C., 1972. With permission.

Committee Comment

Method is satisfactory for air and biological samples. Care must be exercised to prevent the final extract from contacting acid in stopcocks. Such contact with acid will result in partial reversion of the cadmium dithizonate to free dithizone. The final extract should be placed in a dark environment prior to the time that its absorptivity is determined, since the color of the extract may fade when it is exposed to sunlight or other strong light. Other methods involving polarography and atomic adsorption spectrophotometry have been developed.

A method designed specifically for the analyses of cadmium in urine has also been developed.

b. Fluoride (Manual Method)

Reference

Intersociety Committee: Tentative Method of Analysis for Fluoride Content of the Atmosphere (Manual Method) (12204-01-68T). *Health Lab. Sci.*, *6*(2):64-83 (April 1969).

Method of Collection

Samples may be collected by the following methods:

A. Dry collectors

1. Alkali fixative impregnated filters (for gaseous and particulate fluorides).
2. Alkali fixative-coated beads (for gaseous fluorides).
3. Membrane or glass fiber filters (for particulate fluorides).

B. Wet collectors

1. Dilute alkaline solution.
2. Distilled water.

Principle

After isolation of the fluoride by distillation, ion exchange or diffusion, the fluoride is determined by either direct titration or spectrophotometric measurement. In the titrimetric method the fluoride is titrated with standard thorium nitrate. In the spectrophotometric method the fluoride is reacted with a metal dye complex which results in a decrease (Zirconium-Eriochrome Cyanine R and Zirconium SPANDS reagents) or increase (Lanthanum-Alizarin Complexone reagent) in the absorbance of the solution. Zirconium-Eriochrome Cyanine R and Zirconium-SPANDS is read at 536 nm; Lanthanum-Alizarin Complexone is read at 622 nm.

Interferences

For the spectrophotometric measurement, aluminum, iron, phosphate and sulfate interfere and must be separated by the preliminary treatment. With the Lanthanum-Alizarin Complexone method, the acidity should not exceed the capacity of the buffer system to maintain a pH of 4.50±0.02.

Sensitivity

Zirconium-Eriochrome Cyanine R and Zirconium-SPANDS will determine 0.02 microgram fluoride per ml of solution.

Lanthanum-Alizarin Complexone will determine 0.015 microgram fluoride per ml of solution

Special Reagents and Equipment

Alizarin Complexone (1,2-dihydroxy-3-anthraquinonylamine-*N,N*-diacetic acid) available from Hopkins & Williams, Ltd., Chadwell Heath, Essex, England (Cat. No. 1369.4).
Eriochrome Cyanine R (Mordant Blue Color Index No. 43820).
Lanthanum Chloride, 99.9% assay, available from Kleber Laboratories, Burbank, Cal.
SPANDS (4,5-dihydroxy-3-(*p*-sulfophenylazo)-2-7-naphthalensulfonic acid trisodium salt), available from Eastman Organic Chemicals, Rochester, N.Y. Cat. No. 7309.
Zirconyl chloride octahydrate.
Spectrophotometer for the visible range.

c. Uranium

Note: Recommended maximum atmospheric concentration (8 hrs); soluble compounds: 0.05 mg per cu m (0.5 μg per liter); insoluble compounds: 0.25 mg per cu m (0.25 μg per liter) (ACGIH-TLV-1964).

Hydrogen Peroxide Method

References

1. **Grimaldi, F. S., I. May, M. H. Fletcher and J. Titcomb:** Collected Papers on Methods of Analysis for Uranium and Thorium. Geological Survey Bull. 1006, U.S. Govt. Printing Office (1954).
2. **Sandell, E. B.:** *Colorimetric Determination of Traces of Metals.* 3rd Ed., p. 915, Interscience Publishers, Inc., New York (1950).

Method of Collection

Samples may be taken on filter paper (Whatman No. 41 or equivalent) at 15 lpm (for 1" filter) or on molecular filters at 10 lpm or less (for 1" filter). The electrostatic precipitator may be used at 2 to 3 cfm. The midget and standard impingers may be used at the standard air flow rates with water as the collection medium.

Principle

In an alkaline medium containing peroxide, uranium gives a yellow color which can be measured to determine the uranium quantitatively. The sample is dissolved, the necessary separations are made, and the color is developed and measured.

Interferences

Vanadium, chromium, and molybdenum interfere but can be separated prior to the colorimetric determination. Uranium may be separated from interferences by ethyl acetate extraction of uranyl nitrate or by precipitation of quadrivalent uranium (with titanium as a carrier) with cupferron.

Sensitivity

With a 1 cm light path, an absorbance of 0.9 is obtained for a concentration of 0.2 mg per ml.

Special Reagents and Equipment

Spectrophotometer

Committee Comment

The method is applicable to air samples but it is not sensitive enough for use with biological samples.

Fluorometric, Sodium Fluoride Flux Method

References

1. **Grimaldi, F. S., I. May, M. H. Fletcher and J. Titcomb:** Collected Papers on Methods of Analysis for Uranium and Thorium. Geological Survey Bull. 1006, U.S. Govt. Printing Office (1954).
2. **Rodden, C. J., Editor:** *Analytical Chemistry of the Manhattan Project,* NNES. Division VIII, Vol. 1, McGraw-Hill Book Co., New York, N.Y. (1950).
3. **Price, G. R., R. J. Ferretti and S. Schwartz:** *Fluorophotometric Determination of Uranium. Anal. Chem.* 25: 322 (1953).

Method of Collection

Samples may be taken on filter paper (Whatman No. 41 or equivalent) at 15 lpm for 1" filter, or on molecular filters at 10 lpm or less for 1" filter. The electrostatic precipitator may be used at 2 to 3 cfm. The midget and standard impingers may be used at the standard air flow rates with water as the collecting medium.

Principle

When fused with sodium fluoride or other suitable flux, hexavalent uranium emits a visible fluorescent light if irradiated with a suitable ultraviolet light. Measurement of this fluorescent light permits a quantitative determination of the uranium. The ashed sample is dissolved in nitric acid. An aliquot is evaporated in a special platinum dish and fused with the flux (NaF or Na_2CO_3-K_2CO_3-NaF). The fluorescence of the melt is read in a fluorimeter.

Interferences

By proper filtration of the incident ultraviolet and emitted fluorescent lights, it is possible to reduce positive interferences to insignificance. Negative interferences include iron, platinum, chromium, and many other metals, but if the ratio of uranium to interferences is high enough, the aliquot can be made so small that the interferences are negligible. Commercial ores have a favorable ratio of uranium to interferences, while an iron-bearing biological sample such as blood has an unfavorable ratio. Thus an iron separation is required with the latter. Urine contains little iron and can be analyzed without separations, but precautions are required to avoid the solution of platinum from the dish. Negative interferences can be determined by a "spiking" technique.

Sensitivity

The sensitivity of the fluorimetric method varies greatly depending upon the instrument, purity and uniformity of the flux, and the reproducibility of the blank. Instrument sensitivity is about 10^{-10} to 10^{-12} gram uranium per division; a deflection of several divisions normally is significant. For an instrument having 10^{-10} gram U per division sensitivity, the range is from 10^{-7} to 10^{-10} gram uranium.

Special Reagents and Equipment

K_2CO_3-Na_2CO_3-NaF flux, if used, must be prepared by analyst.

Fluorimeter.

Committee Comment

The exact procedure will be determined by the fluorimeter and other equipment available. There are two general types of fluorimeters – "transmitting" and "reflecting." The interferences depend to some extent on the instrument and the flux (which is dictated by the instrument) used. Special platinum fusion dishes are convenient but matched platinum crucible lids can be used.

The method is applicable to air samples and to suitably prepared biological samples.

d. Zirconium

Note: Recommended maximum atmospheric concentration (8 hrs); 5 mg per cu m (5 μg per liter) (ACGIH-TLV-1964).

Chloranilic Acid Method

References

1. **Bricker, C. E. and R. Waterbury:** Separation and Determination of Microgram Quantities of Zirconium. *Anal. Chem.* 29: 558 (1957).
2. **Campbell, E. E.:** Determination of Zirconium in Air. *Amer. Ind. Hyg. Assoc. J.* 20: 281 (1959).

Method of Collection

Air is filtered through a 1 1/8" or 2 1/8" Whatman No. 41 filter paper held in a suitable adaptor. A sampling rate of 1 to 5 liters of air per minute is satisfactory.

Principle

The filter paper is wet ashed with nitric-sulfuric-perchloric acid digestion mixture and the zirconium isolated by precipitation as the p-bromomandelic acid complex. The amount of zirconium is determined colorimetrically with chloranilic acid in a spectrophotometer at 340 mμ.

Interferences

Virtually specific except for fluorides and phosphates which reduce the zirconium recovery. Hafnium may interfere because it is also precipitated with p-bromomandelic acid.

A slight change in the normality of the perchloric acid used and the effect of light on the prepared chloranilic acid solution will cause noticeable differences in analyses; therefore, it is advisable to carry standards through each set of determinations to adjust for these correction factors.

Sensitivity

1 to 10 μg zirconium per aliquot. With a 10 liter air sample, from 0.1 to 1 mg of zirconium per cubic meter of air may be determined, using 1 cm cuvettes.

Special Reagents and Equipment

p-Bromomandelic acid (Pilot Chemicals, Inc.)

Chloranilic acid (2,5-dichloro-3,6-dihydroxy-p-quinone) (Eastman Organic Chemicals P-4539)

Spectrophotometer *

5. *Sulfur – Colorimetric Determination in Organic Particulate*

A procedure that may be applied to sulfur-bearing particulates such as thioureas, thiocyanates, elemental sulfur, mercaptans, thiophenes, and thiocyclohexane by reduction to H_2S with Raney nickel followed by quantitation as methylene blue has been developed. The essential details follow:

Principle and Limitations

Organically bound sulfur and elemental sulfur present in the sample are reduced to sulfide with Raney nickel. After acidification with hydrochloric acid, hydrogen sulfide is evolved and collected in alkaline zinc acetate.

The amount of hydrogen sulfide present in the zinc acetate scrubber is determined colorimetrically through the formation of methylene blue from *N,N*-dimethyl-*p*-phenylenediamine at 670 mμ. Raney nickel does not quantitatively reduce inorganic sulfates and other compounds of hexavalent sulfur. For those compounds that are quantitatively reduced (e.g., divalent sulfur) and elemental sulfur the method is sensitive to concentrations as low as 0.01 ppm.

Precision and Accuracy

Although no detailed study is available, the precision based on quadruplicate determinations on one sample was ± 0.1 ppm at the 1 ppm level. The accuracy based on a comparison of known amounts of sulfide added to the scrubber and recovery of thiophene from known concentrations in toluene also was ± 0.1 ppm at the l-ppm level.

Apparatus

1. A spectrophotometer (Beckman DU-2 or its equivalent) is recommended.

2. Reaction and distillation assembly: the all-glass assembly closely resembles the Knorr carbon dioxide apparatus for the determination of CO_2 in carbonates as described by the American Association of Agricultural Chemists (AOAC) and recommended by the (ASTM) (D215) for the determination of SO_2 in paint pigments. A pressure-equalized dropping funnel is substituted for the conventional counterpart to permit nitrogen blanketing of the entire system. This apparatus can be easily assembled from Ace Mini-lab® components[90] as follows:

All components are fitted with standard taper 14/20 joints and polytetrafluoroethylene (PTFE) joint sleeves, which require no lubrication.

a. 100-ml side arm (for thermometer) flask (Catalog No. 9460).
b. Claissen Adapter (Catalog No. 9067).
c. Drop-leg Adapter (Catalog No. 9121) (seal off side arm). This adapter is inserted between the center neck of the Claissen Adapter and the dropping funnel to permit liquid addition under the surface of the flask contents. (Drop leg must be long enough to provide a liquid seal in the flask contents.)
d. Dropping funnel (Catalog No. 9493T) – pressure equalized, 60 mm. This unit is mounted on the center neck of the Claissen Adapter.
e. Hose Adapter (Catalog No. 9069) – to connect nitrogen supply to the dropping funnel.
f. Condenser (Catalog No. 9461) – to be mounted on the Claissen side arm.
g. Outlet Adapter (Catalog No. 9088) – to connect top of the condenser to the absorber (h) through a PTFE heat-shrinkable tube.[91]

*From Monkman, J. L., Ed., *Analytical Guides,* AIHA, Akron, O., 1965. With permission.

h. Midget Bubbler (Catalog No. 7532) – 25 ml, to absorb H_2S generated in the reaction flask (a).

i. Thermometer (10°C to 250°C) – insert through rubber sleeve on the reactum flask side arm – must clear stirrer bar (j).

j. Magnetic Stirrer (Catalog No. 3632) – hot plate, with PTFE spinbar (Catalog No. 3656) – 7/8 in.

Reagents

1. Raney nickel alloy, Raney Catalyst Co., Chattanooga, Tenn.

2. Raney nickel.

   a. Place 50 g of Raney nickel alloy in a 2-l beaker and slurry with 100 ml of distilled water.

   b. While stirring the mixture in the beaker, add 1 liter of 2.5*N* sodium hydroxide, *dropwise.* This operation must be carried out in a hood. The first 250 ml will cause vigorous reaction during the addition. As it becomes apparent that the reaction is subsiding, the rate of addition of sodium hydroxide may be increased.

   c. Continue stirring for 30 min after the addition of sodium hydroxide is completed and then allow the reaction mass to stand overnight.

   d. Decant the supernatant liquid and wash the nickel with several 300-ml portions of distilled water. Continue washing until the wash water gives no color when tested with phenolphthalein paper. *Caution:* It is essential that the Raney nickel be kept moist at all times. Due to the extremely fine particle size it can burst into flames if allowed to dry in air.

   e. Wash the Raney nickel with 3 successive 100-ml portions of isopropyl alcohol.

   f. Store the Raney nickel in a 4-oz jar under 100 ml of isopropyl alcohol. Each batch of Raney nickel should be checked by running the 1 ppm standard and comparing results to the calibration curve.

3. *N,N*-Dimethyl-*p*-phenylenediamine hydrochloride solution.

   a. Dilute 450 ml of concentrated hydrochloric acid to 1 liter with distilled water.

   b. Dissolve 1 g of *N,N*-dimethyl-*p*-phenylenediamine sulfate (Eastman White Label) in this solution.

4. Ferric chloride solution: dissolve 0.62 (± 0.01 g) of $FeCl_3 . 6H_2O$ (reagent grade) in a solution prepared by diluting 100 ml of concentrated hydrochloric acid to 1 liter with distilled water.

5. Sodium hydroxide solution, 3*N*: dissolve 120 g of CP sodium hydroxide in 1 liter of distilled water.

6. Zinc acetate solution: dissolve 10 g of CP zinc acetate in 1 liter of distilled water.

7. Nitrogen, oxygen free.

8. Hydrochloric acid, 7*N*: dilute 600 ml of 12*N* hydrochloric acid to 1 liter with distilled water.

9. Toluene, sulfur free: this reagent must be sulfur free as determined by this method of analysis. If sulfur-free toluene is not available, sulfur may be removed from the reagent by treatment with Raney nickel.

Procedure Hazards

Raney nickel is extremely hazardous. After the aluminum is dissolved from the Raney nickel alloy in caustic solution the remaining nickel is highly reactive. It must be kept moist at all times. Dry Raney nickel will ignite spontaneously in air. Clean spills with copious amounts of water. Waste Raney nickel should be dissolved in dilute hydrochloric acid.

Sodium hydroxide solution, 3*N* is corrosive and can cause blistering if spilled on the skin. In case of spills wash the affected area with copious amounts of water.

Procedure

1. Add from a buret 15 ml of 1% zinc acetate to the scrubber.

2. Add by pipet 5 ml of 3*N* sodium hydroxide to the scrubber. A white precipitate will form when sodium hydroxide is added.

This will not cause any problems in the determination.

3. Connect the scrubber to the apparatus.

4. Place the filter membrane or a toluene extract of the filter in the digestion flask.

5. Add 50 ml (pipet) of sulfur-free toluene that has been checked previously by this analysis.

6. Add approximately 250 mg of Raney nickel. Make this addition with a Scoopula®. This provides a large excess of nickel so there is no need for a weighing.

7. Place a magnetic stirring bar in the flask.

8. Connect the flask to the apparatus.

9. Add 15 ml of 7*N* HCl to the dropping funnel and connect the N_2 line.

10. Turn on the condenser water.

11. Turn on the N_2 (rate: 1 bubble/sec).

12. Turn on the magnetic stirrer.

13. Allow the reaction to proceed for 15 min at room temperature.

14. Turn on the heat and bring the reaction mixture to a boil.

15. Allow the reaction mixture to boil for 30 min.

16. Turn off the agitation, leaving the heat turned on.

17. Carefully open the dropping funnel stopcock and add all the hydrochloric acid, *dropwise,* to the reaction.

18. Allow the solution to boil for 15 min.

19. Disconnect the scrubber from the apparatus and place a 24/40 glass stopper on the scrubber.

20. Rinse the filter tube assembly with a portion of color-forming reagent as it is being added to the scrubber (step 22).

21. Turn off the heat and the nitrogen pressure.

22. Add 25 ml (pipet) of *N,N*-dimethyl-*p*-phenylenediamine reagent to the scrubber.

23. Add 0.5 ml (pipet) of ferric chloride solution to the scrubber.

24. Allow the scrubber to stand for 30 min.

25. Measure the absorbance of the solution on a spectrophotometer at 670 mμ in a 1-cm cell against a blank as a reference. The blank includes all steps except sample addition.

26. Read the micrograms of sulfur from the calibration curve.

27. Calculation:

$$\text{mg S/m}^3 = \frac{\mu\text{g S found}}{\text{sample volume x 1,000}}$$

where sample volume is in m^3.

Calibration

1. Prepare the following solutions in sulfur-free toluene:

   a. Stock solution – Dissolve 0.0263 g (±0.0001 g) of thiophene in 100 ml of sulfur-free toluene. This solution contains 100 μg/ml of sulfur.
   b. 1 μg/ml Standard – Pipet 10 ml of stock solution into a 1-liter volumetric flask. Dilute to the mark with sulfur-free toluene.
   c. 0.5 μg/ml Standard – In a 250-ml volumetric flask dilute 125 ml of 1 ppm standard to the mark with sulfur-free toluene.
   d. 0.2 μg/ml Standard – In a 250-ml volumetric flask dilute 50 ml of 1 ppm standard to the mark with sulfur-free toluene.

e. 0.1 μg/ml Standard – In a 250-ml volumetric flask dilute 25 ml of 1 ppm standard to the mark with sulfur-free toluene.

2. Analyze in duplicate, 1 ml of each of the 4 standard solutions.

3. Draw a calibration curve correlating the absorbance at 670 mμ with sulfur in micrograms.

C. REFERENCES FOR CHAPTER II

1. **Lippmann, M.,** *Air Sampling Instruments for Evaluation of Atmospheric Contaminants,* 4th ed., ACGIH Cincinnati, O., 1972.
2. **Katz, S. H. et al.,** Comparative tests of instruments for determining atmospheric dust, *U.S. Public Health Bull.,* 144 (1925).
3. **Littlefield, J. B. and Schrenk, H. H.,** Bureau of Mines Midget Impinger for Dust Sampling, R.I. 3360, Bureau of Mines, Department of the Interior, 1937.
4. **Stokinger, H. E.,** Threshold Limit Values for Chemical Substances and Physical Agents in the Workroom Environment with Intended Changes for 1972, Committee on Threshold Limits, ACGIH, Cincinnati, O., 1972.
5. Specifications for the Midget Impinger, R.I. 3387, Bureau of Mines, Department of the Interior, 1938.
6. **Dowling, T., Davis, R. B., and Linch, A. L.,** Lead in air analyzer, *Am. Ind. Hyg. Assoc. J.,* 19, 330, 1958.
7. Ace Glass, Inc., Vineland, N.J.
8. **Linch, A. L. and Corn, M.,** The standard midget impinger – design improvement and miniaturization, *Am. Ind. Hyg. Assoc. J.,* 26, 601, 1965.
9. **Drinker, P. and Hatch, T.,** *Industrial Dust,* 2nd ed., McGraw-Hill Book Co., New York, 1954.
10. **McAfee, J.,** All-glass midget impinger unit, *Ind. Eng. Chem.* (anal. ed.), 16, 346, 1944.
11. Efficiency of Impingers for Collecting Lead Dust and Fumes, R.I. 3401, Bureau of Mines, Department of the Interior, 1958.
12. **Kerrigan, J. V., Snajberk, K., and Anderson, E. S.,** Collection of sulfuric acid mist in the presence of a higher sulfur dioxide background, *Anal. Chem.,* 32, 1168, 1960.
13. **Davies, C. N. and Aylward, M.,** The trajectories of heavy, solid particles in a two-dimensional jet of ideal fluid impinging normally upon a plate, *Proc. Phys. Soc.,* 64, 889, 1951.
14. **Ranz, W. E. and Wong, J. B.,** Jet impactors for determining the particle size distribution of aerosols, *Am. Med. Assoc. Arch. Ind. Hyg. Occup. Med.,* 5, 464, 1952.
15. **Davies, C. N., Aylward, M., and Leagey, D.,** Impingement of dust from air jets, *Am. Med. Assoc. Arch. Ind. Hyg. Occup. Med.,* 4, 354, 1951.
16. **Williams, C. R.,** A method of counting samples taken with the impinger, *J. Ind. Hyg. Toxicol.,* 21, 226, 1939.
17. **Brown, C. E. and Shrenk, H. H.,** Relation Between and Precision of Dust Counts (Light and Dark Field). Some Simultaneous Impinger, Midget Impinger, Electric Precipitator and Filter Paper Samples, R.I. 4568, Bureau of Mines, Department of the Interior, 1949.
18. **Chapman, H. M. and Ruhf, C.,** Dust counting reliability, *Am. Ind. Hyg. Assoc. Quart.,* 16, 3, 1955.
19. **Jacobson, M., Terry, S. L., and Ambrosia, D. A.,** Evaluation of some parameters affecting the collection and analysis of midget impinger samples, *Am. Ind. Hyg. Assoc. J.,* 31, 442, 1970.
20. **Stern, A. C., Ed.,** *Air Pollution, Vol. 2, Analysis, Monitoring and Surveying,* 2nd ed., Academic Press, New York, 1968.
21. **Gussman, R. A., Dennis, R., and Silverman, L.,** Notes on the design and leak testing of sampling filter holders, *Am. Ind. Hyg. Assoc. J.,* 23, 480, 1962.
22. Aerosol Analysis Monitors, Catalog No. MAWG 037-AO, Type AA, Catalog MC/1:2A360, Millipore Corp., Bedford, Mass., 1972.

22a. Mighty-Mite®, Model 440A, Personnel Monitor, National Environmental Instruments, Inc., P.O. Box 590, Pilgrim Station, Warwick, R.I.

23. **Gehman, C.,** Microporous membrane technology. I. Historical development and applications, *Anal. Chem.,* 37, 29A, 1965.
24. "Nuclepore" Membrane Filters and Filter Holders for Microanalysis, General Electric Co., Schenectady, N.Y.

24a. 37-mm Visual Metricel VM-1 (5-μ pore size), Catalog No. 3600-101, National Environmental Instruments, Inc., P.O. Box 590, Pilgrim Station, Warwick, R.I.

25. *The Industrial Environment, Its Evaluation and Control,* Publication No. 614:B-5-12, Public Health Service, Department of Health, Education and Welfare, Washington, D.C., 1965.

25a. **Renshaw, F. M., Bachman, J. M., and Pierce, J. O.,** The use of midget impingers and membrane filters for determining particle counts, *Am. Ind. Hyg. Assoc. J.,* 30, 113, 1969.

26. Guides to Clean Room Operation and How to Size and Count Air-Borne Particulate Contamination, Abstracted from U.S. Air Force, Technical Order 00-25-203, 1963.

27. Dust Topics, Vol. 3, No. 1, 18, Gelman Instrument Co., Ann Arbor, Mich., 1966.

28. **Yaffe, C. D., Paulus, H. J., and Jones, H. H.,** *Am. Ind. Hyg. Assoc. Quart.,* 15, 206, 1954.

29. *Procedure for the Determination of Particulate Contamination of Air in Dust Controlled Spaces by the Particle Count Method, Aerospace Recommended Practice,* Soc. Automotive Engineers, Inc., New York, 1962, 743.

30. **Key, M. M., Ed.,** Criteria for a Recommended Standard – Occupational Exposure to Asbestos, HSM 72-10267, Natl. Inst. Occup. Safety and Health, Health Services and Mental Health Admin., Public Health Service, Department of Health, Education and Welfare, Washington, D.C., 1972; *Fed. Register,* 37(No. 110), 1318, June 7, 1972, chap. 17, part 1910.

31. **Dreessen, W. C., Dallavalle, J. M., Edwards, J. I., Miller, J. W., and Sayers, R. R.,** A study of asbestosis in the asbestos textile industry, *U.S. Public Health Bull.,* 241, 1938.

32. **Ayer, H. E. and Lynch, J. R.,** Motes and fibers in the air of asbestos processing plants and hygienic criteria for airborne asbestos, *Proceedings of an International Symposium Organized by the British Occupational Hygiene Society,* 1965, 511.

33. **Lynch, J. R. and Ayer, H. E.,** Measurement of asbestos exposure, *J. Occup. Med.,* 10, 21, 1968.

34. **Lynch, J. R. and Ayer, H. E.,** Measurement of dust exposure in the asbestos textile industry, *Am. Ind. Hyg. Assoc. J.,* 27, 431, 1966.

35. **Keenan, R. G. and Lynch, J. R.,** Techniques for the detection, identification and analysis of fibers, *Am. Ind. Hyg. Assoc. J.,* 31, 587, 1970.

36. **Edwards, G. H., and Lynch, J. R.,** The method used by the Public Health Service for enumeration of asbestos dust on membrane filters, *Ann. Occup. Hyg.,* 11, 1, 1968.

37. **Lynch, J. R., Ayer, H. E., and Johnson, D. L.,** The Measurement of Exposure to Airborne Mineral Fibers, presented at the Am. Ind. Hyg. Assoc. Conf., Denver, 1969.

38. **Lane, R. E. et al.,** Hygiene standard for chrysotile asbestos dust, *Ann. Occup. Hyg.,* 11, 47, 1968.

39. Asbestos – OSHA Sampling Data Sheet #2 (revised), Occup. Safety and Health Admin., Department of Labor, June 27, 1972.

40. Unico Model C115 and Micronair Sampler, National Environmental Instruments, Inc., P.O. Box 590, Pilgrim Station, Warwick, R.I.

41. Crystalline Silica – OSHA Sampling Data Sheet #3, Occup. Safety and Health Admin., Department of Labor, December 21, 1972.

42. **Talvitie, N. A.,** Determination of free silica. Gravimetric and spectrophilometric procedures applicable to air-borne and settled dust, *Am. Ind. Hyg. Assoc. J.,* 25, 169, 1964.

43. **Talvitie, N. A.,** Determination of quartz in the presence of silicates using phosphoric acid, *Anal. Chem.,* 23, 626, 1951.

44. **Talvitie, N. A. and Hyslop, F.,** Colorimetric determination of siliceous atmospheric contaminants, *Am. Ind. Hyg. Assoc. J.,* 19, 54, 1958.

45. Ayer, H. E., The proposed ACGIH mass limits for quartz. Review and evaluation, *Am. Ind. Hyg. Assoc. J.,* 30, 117, 1969.

46. **Ohman, H.,** Methods of sampling and analyzing health-hazardous silica dust (in Swedish), *Arbetsmedicinska Institute Stockholm, Sweden,* Al-Rapport No. 3:8, 1968.

47. **Kingsley, K.,** Rapid quartz analysis by X-ray spectrometry, *Am. Ind. Hyg. Assoc. J.,* 11, 185, 1950.

48. **Russ, J. C.,** Elemental X-Ray Analysis of Materials – EXAM Methods, Edax International Inc., P.O. Box 135, Prairie View, Ill.

48a. Selas Flotronics Membranes – Silver, Flotronics Division, Sales Corp. of America, Spring House, Pa., 1973.

49. X-Ray Fluorescence Analysis, Inax Instruments Ltd., P.O. Box 6044 Station J., Ottawa, Can.

49a. **Bianconi, W. O. and Thomas, F. W.** Reproducibility of aerosol photometer, midget impinger and membrane. Filter counts for limestone and cool dusts, *Am. Ind. Hyg. Assoc. J.,* 26, 362, 1965.

49b. **Blanco, A. J. and Hoidale, G. B.,** Microspectrophotometic technique for obtaining the infrared spectrum of microgram quantities of atmospheric dust, *Environment* (St. Louis), 2, 328, 1968.

49c. **Friegieri, P., Trucco, R., Anzani, P., and Caretta, E.,** Spectroscopic analysis of elements present in air-borne materials (in Italian), *Chem. Inc.* (Milan), 54, 12, 1972.

49d. **Bowman, H. R., Conway, J. G., and Asard, F.,** Atmospheric lead and bromine concentration in Berkeley, California (1963–1970), *Environ. Sci. Technol.,* 6, 558, 1972.

49e. **Keane, J. R. and Fisher, E. M. R.,** Analysis of trace elements in air-borne particulates by neutron activation and gamma-ray spectrometry, *Atmos. Environ.,* 2, 603, 1968.

49f. **Gray, D., McKown, D. M., Kay, M., Eichor, M., and Vogt, J. R.,** Determination of trace element levels in atmospheric pollutants by instrumental neutron activation analysis, *IEEE Trans. Nucl. Sci.,* 19, 194, 1972.

50. **Lippmann, M.**, "Respirable" dust sampling, *Am. Ind. Hyg. Assoc. J.*, 31 138, 1970.
51. **Hatch, T. F. and Gross, P.**, *Pulmonary Deposition and Retention of Inhaled Aerosols,* Academic Press, New York, 1964.
52. AIHA Aerosol Technology Committee, Guide for respirable mass sampling, *Am. Ind. Hyg. Assoc. J.*, 31, 133, 1970.
53. **Weibel, E. R.**, *Morphometry of the Human Lung,* Academic Press, New York, 1963.
54. **Morrow, P. E., Gibb, F. R., and Johnson, L.**, Clearance of insoluble dust from the lower respiratory tract, *Health Phys.*, 10, 543, 1964.
55. **Davies, C. N.**, Dust sampling and lung disease, *Br. J. Ind. Med.*, 9, 120, 1952.
56. **Orenstein, A. J., Ed.**, *Proc. Pneumoconiosis Conf., Johannesburg, 1959,* J. and A. Churchill, Ltd., London, 1960.
57. Threshold Limit Values of Air-Borne Contaminants for 1968, Committee on Threshold Limits, ACGIH, Cincinnati, O., 1968.
58. **Sutton, G. W. and Reno, S. J.**, Respirable Mass Concentrations Equivalent to Impinger Count Data, presented at 1968 Annual Meeting of AIHA, St. Louis, Mo., May 1968.
59. *Fed. Register,* 34(No. 96), 7946, May 20, 1969.
60. **Brown, J. H., Cook, K. M., Ney, F. G., and Hatch, T.**, Influence of particle size upon the retention of particulate matter in the human lung, *Am. J. Public Health,* 40, 450, 1950.
61. Report of Task Group on Lung Dynamics to ICRP Committee. 2. Deposition and retention models for internal dosimetry of the human respiratory tract, *Health Phys.*, 12, 173, 1966.
62. **Mercer, T. T.**, Air Sampling Problems Associated with the Proposed Lung Model, presented at the 12th Annu. Bioassay and Anal. Chem. Meeting, Gatlinsburg, Tenn., October 13, 1966.
63. **Watson, H. H.**, Dust sampling to simulate the human lung, *Br. J. Ind. Med.*, 10, 93, 1953.
64. **Higgins, R. I. and Dewell, P.**, A Gravimetric Size-Selecting Personal Dust Sampler, BCIRA Rep. 908, British Cast Iron Research Assoc., Alvechurch, March 1968.
65. **Knight, G. and Lichti, E. K.**, Comparison of Cyclone and Horizontal Elutriator Size Selectors, presented at the Annual Meeting of AIHA, Denver, May 1969.
66. **Lippmann, M. and Harris, W. B.**, Size-selective samplers for estimating respirable dust concentrations, *Health Phys.*, 8, 155, 1962.
67. **Hyatt, E. C., Schulte, H. F., Jensen, C. R., Mitchell, R. N., and Ferran, G. H.**, A study of two-stage air samplers designed to simulate the upper and lower respiratory tract, in *Proceedings of the 13th International Congress of Occupational Health,* New York, 1961.
68. **Sutton, G. W.**, Calibration of Aerodynamic Size vs. Efficiency for a Miniature Cyclone, presented at the Annual Meeting of AIHA, Pittsburgh, May 1966.
69. **Knuth, R. H.**, Recalibration of size-selective samplers, *Am. Ind. Hyg. Assoc. J.*, 30, 379, 1969.
70. **Tomb, T. F. and Raymond, L. D.**, Evaluation of the Collecting Characteristics of Horizontal Elutriator and of 10 mm Nylon Cyclone Gravimetric Dust Samplers, presented at the Annual Meeting of AIHA, Denver, May 1969.
71. **Timbrell, V.**, The terminal velocity and size of airborne dust particles, *Br. J. Appl. Phys.*, Suppl. 3, 86, 1954.
72. **Ettinger, H. J. and Royer, G. W.**, Calibration of a Two-Stage Air Sampler, LA-4234, Los Alamos Scientific Laboratory, August 1969.
73. **Thomas, J. W. and Knuth, R.**, Settling velocity and density of monodisperse aerosols, *Am. Ind. Hyg. Assoc. J.*, 28, 173, 1967.
74. **Roesler, J. F., Stevenson, H. J. R., and Nader, J. S.**, Size distribution of sulfate aerosols in the ambient air, *J. Air Pollut. Control Assoc.*, 15, 576, 1965.
74a. **Donaldson, H. M., Hiser, R. A., and Schwenzfeier, C. W.**, Realistic air sampling of beryllium production facilities, *Am. Ind. Hyg. Assoc. J.*, 25, 69, 1964.
74b. **Hyatt, E. C.**, Techniques for measuring radioactive dusts, in *Radiological Health and Safety in Mining and Milling of Nuclear Materials,* Vol. I, Int. Atomic Energy Agency, Vienna, 1964.
74c. **Schulte, H. F.**, Personal Air Sampling and Multiple Stage Sampling, presented at the ENEA Symposium on Radiation Dose Measurements, Stockholm, June 12-16, 1967.
74d. **Fischoff, R. L.**, The relationship between and the importance of the dimensions of uranium particles dispersed in air and the excretion of uranium in the urine, *Am. Ind. Hyg. Assoc. J.*, 26, 26, 1965.
75. Dorr-Oliver 10 mm Cyclone, Mine Safety Appliances Co., 201 N. Braddock Ave., Pittsburgh.
76. Dorr-Oliver 10 mm Cyclone, Dorr-Oliver, Inc., Stanford, Conn.
77. **Andersen, A. A.**, A sampler for respiratory health hazard assessment, *Am. Ind. Hyg. Assoc. J.*, 27, 160, 1966.
78. "Andersen Mini-Sampler," Research Appliance Co., Allison Park, Pa.
78a. **Lamonica, J. A. and Treaftis, H. N.**, The Effect of Pulsation Damping on Respirable Dust Collected By Coal Mine Dust Personal Samplers, R. I. 7636, Bureau of Mines, Department of the Interior, 1972.
79. **Ayer, H. E.**, Sampling methods for oil mist in industry, *Am. Ind. Hyg. Assoc. J.*, 25, 151, 1964.
79a. **Lamonica, J. A. and Treaftis, H. N.**, Investigation of Pulsation Dampers for Personal Respirable Dust Samplers, R. I. 7545, Bureau of Mines, Department of the Interior, 1971.
80. MSA Gravimetric Dust Sampling Kit, Catalog No. 0810-l-A, Mine Safety Appliances Co., 201 Ninth Braddock Ave., Pittsburgh, 1969.

81. **Jacobson, M. and Lamonica, J. A.,** Personal Respirable Dust Sampler, Tech. Progr. Rep. 17, Bureau of Mines, Department of the Interior, 1969.
82. Mandatory Health Standards – Surface Work Areas of Underground Coal Mines and Surface Coal Mines, *Fed. Register,* 37, 6368, March 28, 1972.
83. Micronair Gravimetric Dust Sampler, Catalog No. 3900-10; with Sampling Head (Catalog No. 3900-90), National Environmental Instruments, Inc., P.O. Box 590, Pilgrim Station, Warwick, R.I.
84. Specifications for Coal Dust Personal Monitor, in *Code of Federal Regulations,* January 11, 1970, chap. 1, subchap. O, title 3,O, part 74, *Fed. Register,* 35(No. 48).
85. Cotton Dust, OSHA Sampling Data Sheet #5, Occup. Safety and Health Admin. Standards, 29 CFR-1910.93, (b) Table G-1 of August 13, 1971.
86. **Pate, J. B. and Tabor, E. C.,** Analytical aspects of the use of glass fiber filters for the collection and analysis of atmospheric particulate matter, *Am. Ind. Hyg. Assoc. J.,* 23, 145, 1962.
87. **Morse, K. M.,** Problems in the gravimetric measurement of respirable coal mine dust, *J. Occup. Med.,* 12, 400, 1970.
88. Lead and Its Inorganic Compounds (except lead arsenate) – OSHA Sampling Data Sheet #4, Occup. Safety and Health Admin., Department of Labor, December 21, 1972.
89. **Katz, M., Ed.,** Analysis for inorganic lead content of the atmosphere (12128-01-71T), in *Methods of Air Sampling and Analysis,* Intersociety Committee Am. Public Health Assoc., Washington, D.C., 1972.
90. "Ace Mini-lab" Glassware, *Catalog No. 600,* Ace Glass, Inc., Vineland, N.J., 1972.
91. Pope "Flo-Tite" Tubing, Pope Scientific, Inc., 13600 W. Reichert Ave., Menomonee Falls, Wisc., 1972.
92. **Fogo, J. K. and Popowsky, M.,** Spectrographic determination of hydrogen sulfide, *Anal. Chem.,* 21, 732, 1949.
93. **Linch, A. L.,** *Biological Monitoring for Industrial Chemical Exposure Control,* CRC Press, Cleveland, in press, 1973.
94. **Linch, A. L., Wiest, E. G., and Carter, M. D.,** Evaluation of tetraalkyl lead exposure by personnel monitor surveys, *Am. Ind. Hyg. Assoc. J.,* 31, 170, 1970.
95. **Quino, E. A.,** Field method for the determination of inorganic lead fumes in air, *Am. Ind. Hyg. Assoc. J.,* 20, 134, 1959.
96. Unico Lead-In-Air Detector, National Environmental Instruments, Inc., P.O. Box 590, Pilgrim Station, Warwick, R.I., 1960.

D. GENERAL REFERENCES FOR CHAPTER II

Air Pollution Manual, Parts I and II, AIHA, Akron, O., 1971.

Bloor, W. A. and Dinsdale, A., A long-period gravimetric dust sampler, *Ann. Occup. Hyg.,* 9, 29, 1966.

Brown, J. H., Cook, K. M., Ney, F. G., and Hatch, T., Influence of particle size upon the retention of particulate matter in the human lung, *Am. J. Public Health,* 40, 450, 1950.

Casarett, L. J., Some physical and physiological factors controlling the fate of inhaled substances. II. Retention, *Health Phys.,* 2, 379, 1960.

Cralley, L. V., Cralley, L. J., and Clayton, G. D., *Industrial Hygiene Highlights,* Industrial Health Foundation, Inc., Pittsburgh, 1961.

Davies, C. N., *Aerosol Science,* Academic Press, New York, 1966.

Davies, C. N., Deposition and retention of dust in the human respiratory tract, *Ann. Occup. Hyg.,* 7, 169, 1964.

Donaldson, H. M., Hiser, R. A., and Schwenzfeier, C. W., Realistic air sampling of beryllium production facilities, *Am. Ind. Hyg. Assoc. J.,* 25, 69, 1964.

Fischoff, R. L., The relationship between and the importance of the dimensions of uranium particles dispersed in air and the excretion of uranium in the urine, *Am. Ind. Hyg. Assoc. J.,* 26, 26, 1965.

Hamilton, R. J. and Walton, W. H., The Selective Sampling of Respirable Dust, in *Proceedings of the International Symposium on Inhaled Particles and Vapors,* Oxford, 1960, 465.

Hatch, T. F. and Gross, P. *Pulmonary Deposition and Retention of Inhaled Aerosols,* Academic Press, New York, 1964.

Hyatt, E. C., A Study of Two-Stage Air Samplers Designed to Simulate the Upper and Lower Respiratory Tract, LA-2440, OTS, Department of Commerce, Washington, D.C., 1960.

Katz, M., Ed., *Methods of Air Sampling and Analysis,* Intersociety Committee, Am. Public Health Assoc., Washington, D.C., 1972.

Leithe, W., *The Analysis of Air Pollutants,* Ann Arbor Science, Ann Arbor, Mich., 1970.

Lippmann, M., *Air Sampling Instruments for Evaluation of Atmospheric Contaminants,* 4th ed., ACGIH, Cincinnati, O., 1972.

Lippmann, M. and Harris, W. B., Size-selective samplers for estimating respirable dust concentrations, *Health Physics,* 8, 155, 1962.

May, K. R. and Druett, H. A., The pre-impinger. A selective aerosol sampler, *Br. J. Ind. Med.,* 10, 142, 1953.

McCrane, W. C. and Delly, J. G., *The Particle Atlas,* 2nd ed., Ann Arbor Science, Ann Arbor, Mich., 1972.

Mitchell, R. I., Retention of aerosol particles in the respiratory tract – a review, *Am. Rev. Resp. Dis.,* 82, 627, 1960.

Mitchell, R. N., Aerosol sampling and the importance of particle size, in *Air Sampling Instruments,* 2nd ed., ACGIH, Cincinnati, O., 1962.
Morrow, P. E., Evaluation of inhalation hazards based upon the respirable dust concept and the philosophy and application of selective sampling, *Am. Ind. Hyg. Assoc. J.,* 25, 213, 1964.
Morrow, P. E., Some physical and physiological factors controlling the fate of inhaled substances. I. Deposition, *Health Phys.,* 2, 366, 1960.
Olishifski, J. B. and McElroy, F. E., Eds., *Fundamentals of Industrial Hygiene,* Natl. Safety Council, Chicago, 1971.
Recommended Practices for Collection by Filtration and Determination of Mass, Number and Optical Sizing of Atmospheric Particulates, ASTM Designation D2009-65, ASTM Standards – Industrial Water Atmospheric Analysis Part 23:854, Philadelphia, 1967.
Report of Task Group on Lung Dynamics to ICRP Committee. 2. Deposition and retention models for internal dosimetry of the human respiratory tract, *Health Phys.,* 12, 173, 1966.
Shanty, F. and Hemeon, W. C. L., The inhalability of outdoor dust in relation to air sampling network, *Air Pollut. Control Assoc. J.,* 13, 211, 1963.
Stern, A. C., Ed., *Air Pollution, Vol. 2, Analysis, Monitoring and Surveying,* 2nd ed., Academic Press, New York, 1968.
Wright, B. M., A size selecting sampler for air-borne dust, *Br. J. Ind. Med.,* 11, 284, 1954.

Chapter III

BATTERY POWERED AIR SAMPLERS

A. SPECIFICATIONS

Ideally, motor driven self-contained personnel monitors should collect sufficient contaminant sample to produce a significant response in the analytical system without encumbrance, discomfort, or annoyance for the wearer. From a practical standpoint, this ideal can be approached but not fully attained. The noise level, bulk, and weight must be reduced to the lowest limit attainable by current technology if acceptance by the workmen is to be expected. These objectives can be attained by:

1. Miniaturization of the mechanical and electrical components.
2. Maximization of electrical energy storage capacity.
3. Minimization of parasitic (frictional, hysteresis) power losses.
4. Minimization of vibration by choice of construction materials and precision machining of moving parts.
5. Minimization of the sample size by increasing the sensitivity of the analytical procedure.

Application of currently available technology, mechanical components, and electrical circuitry can produce an acceptable, operable personnel monitor for aerosols, gases, and vapors based on the following specifications.

a. Weight: Less than 550 g. One instrument manufacturer has produced a sampler that weighs less than 350 g (45 x 70 x 100 mm).

b. Size: Thickness, less than 50 mm (2 in.); width, less than 75 mm (3 in.); length, less than 150 mm (6 in.).

c. Sampling time: Minimum 8 hr (4 hr acceptable in many industrial applications if further miniaturization can be attained thereby).

d. Sampling rate: 2.8 l/min at 305 mm (12 in.) water pressure differential or 0.6 l/min at 180 mm (7 in.) water with 0.61-mm nozzle in the microimpinger.[2]

e. Parastic power losses: Less than 20% (some diaphragm pumps consume up to 40% of the available power in diaphragm flexing – hysteresis loss).

f. Battery: Sealed, rechargeable, incorporates maximum charge consistent with size and weight (silver-cadmium cell), rapid charge rate (preferably less than 16 hr for complete recharge), is chemically stable, mechanically rugged, and has broad operating temperature range (−30 to 60°C) and current limiting resistors to protect against accidental short circuit destruction of the cells. The battery or battery pack should be easily removed for independent charging when the sampler is in use with a spare battery.

g. Battery charger should not generate hazardous heat or hydrogen on overcharge (controlled voltage).

h. Pump should operate at constant speed, be constructed from corrosion and solvent resistant materials, and be self-lubricating (polytetrafluoroethylene plastics).

i. Flow rate control should be based on a needle valve by-pass arrangement rather than a rheostat motor speed control. Flow adjustment should be provided by a tool inserted into a recessed and capped valve stem, or accessible only from within the case to prevent tampering or accidental maladjustment.

j. On-off switch should be recessed and capped to prevent accidental starting or stopping.

k. Substantial inlet and outlet connectors should be designed to provide both pressure and vacuum. All internal connections and joints in the airflow system should be sealed to eliminate air leakage in the sampler system.

l. No protrusions, except the belt clip, on either the back or face of the instrument.

m. Flow rate meter: A rotameter, either dual or single tube with sufficient range to cover the operating capabilities of the pump (0.5 to 3 l/min) and with accuracy of ± 5% of the true value (calibration will be required).

n. Airflow pulsation: The oscillations or ripple produced by the reciprocating pump action in the airflow through the collection system should be less than 2% of the average steady state flow (2.8 l/min ± 1% = 2,800±28 ml/min). This pulsation damping is critical for collection systems that depend upon impingement or cyclonic action to classify and retain aerosols.

o. Total volume readout meter: A pump stroke counting device is preferred. A shaft rotation counter, or timer, would be considered as a second choice only if constant speed pumping is provided. This total volume meter is a basic necessity required to eliminate nonproductive attention time consumed by rotameter reading, flow rate adjustments, and time keeping in the field to obtain a reasonably accurate estimation of sample size.

p. All component parts should be readily accessible and demountable to facilitate inspection, adjustment, maintenance, and repair in the field.

q. Explosion hazard integrity: The air sampler unit must be certified by an appropriate agency before operation in an explosion hazard area is attempted (Factory Mutual for intrinsically safe operation in Class 1, Group D, hazardous (flammable) areas, or U.S. Bureau of Mines and U.S. Bureau of Occupational Health and Safety approval for use in coal mines).

r. Receptacle to auxiliary battery connection (banana plug).

B. POWER SOURCE

For quantitative personnel monitoring, an air mover to draw the air sample through the collecting system must be provided with an energy source. Power for the pump is limited by current technology to batteries designed to energize a small electric motor. Preferably, these batteries should be rechargeable, mechanically rugged, and endure long storage life over a broad temperature range in the charged state. The silver-cadmium cell and the nickel-cadmium battery provide the most consistent service at a nearly constant voltage.

The nickel-cadmium is the most universally used battery in small cordless electrical appliances where its chemical and mechanical ruggedness are necessary. Many units have survived more than 20 years of continuous use, and others have been returned to service without obvious ill effects after more than 20 years of idleness. This battery does not require periodic recharging or other maintenance during periods of idleness at normal temperatures, although charged cells slowly lose their charge in storage. Sealed cells of the type used in personnel monitors and portable instruments may be operated in any position without electrolyte spillage, neither require nor permit maintenance other than recharging, and are commercially available in a great many sizes and shapes.

The nickel-cadmium battery is characterized by delivery of very high current output on demand and of unusually uniform voltage during most of the discharge cycle (85 to 90% discharged) at normal current rates, especially in sealed construction, and can be recharged at voltages only slightly higher than those delivered on discharge without evolution of hazardous hydrogen gas. Oxygen is generated when overcharging occurs, but sealed units are so designed as to consume this oxygen internally, thereby preventing gas pressure build-up. Although the electrode reactions are complex in detail, the overall cell reactions may be summarized:

$$\underset{\text{Positive}}{2\ \text{NiO hydrate}} + \underset{\text{Negative}}{\text{Cd (OH)}_2} \ \underset{\text{discharge}}{\overset{\text{Charge}}{\underset{\longleftarrow}{\overset{\longrightarrow}{\text{NaOH}}}}} \ \underset{\text{Positive}}{2\ \text{NiO (OH) hydrate}} + \underset{\text{Negative}}{\text{Cd}}$$

The electrolyte concentration is not significantly changed during these cycles; therefore, the electrolyte may be considered only as a transfer medium for –OH ions between the plates. Furthermore, the freezing point and voltage drop are essentially independent of the charge condition of the battery. The active materials are extraordinarily stable in contact with and almost totally insoluble in the electrolyte. Consequently, parasitic, electrochemically irreversible changes (including corrosion of the battery components) do not present any problems.

Although the nickel-cadmium battery can be charged and discharged over hundreds of cycles of deep (complete) discharge, thousands of cycles can be expected if the battery is not completely discharged (80 to 85%) before the next recharge cycle. The "nominal" discharge voltage is 1.2 to

1.3 V per cell. Series circuits can be employed to deliver voltage in any multiple of this value (6 V = 1.2 x 5). Nominal charging voltage is limited to 1.4 to 1.6 V per cell to minimize oxygen evolution from overcharging. Sealed cells are usually charged by the constant-current method through a series resistance or limiting device at about 1 A for each 10 Ahr of rated capacity for a time sufficient to deliver to the battery about 110 to 150% of the number of Ahr previously withdrawn by load and self-discharge. Under these conditions the charging current does not spontaneously decline to low levels when the cells are fully charged. Consequently, prolonged overcharge can cause substantial gas evolution and overheating of the cells. Ambient temperature effects must be compensated, and correction charts should be consulted for operation under temperature extremes.[1] The temperature range for both use and storage covers a very wide range (−28.89 to 60°C). For low and moderate discharge rates, the average nominal capacity of the sealed-cell battery is 1.6 $Whr/in.^3$ and 15 Whr/lb.[1]

The silver-cadmium battery offers a greater capacity on an equal size basis and a somewhat wider permissible temperature range than its nickel-cadmium counterpart, but has not been readily available outside of military applications. The useful operating temperature range includes −28.89 to 73.89°C and the capacity is rated at 24 to 33 Whr/lb and 1.9 to 2.7 $Whr/in.^3$ The nominal voltage is slightly lower at 1.1 V, and the recommended charging voltage is somewhat higher at 1.7-V maximum than the nickel-cadmium limits. The cell reactions are

$$\underset{\text{Positive}}{Ag} + \underset{\text{Negative}}{CdO} \underset{\text{Discharge}}{\overset{\text{Charge}}{\underset{\longleftarrow}{\overset{\longrightarrow}{\scriptstyle KOH}}}} \underset{\text{Positive}}{AgO} + \underset{\text{Negative}}{Cd}$$

Other chemical and physical characteristics of the silver-cadmium cell are quite similar to the more widely used nickel-cadmium counterpart.[2]

C. PISTON PUMPS

1. Micronair® – Unico

Two models are available from National Environmental Instruments, Inc., Warwick, R.I. Both have the same sealed (potted) pump-motor power unit. The more compact model, the Micronair®, was designed to meet U.S. Bureau of Mines and U.S. Bureau of Occupational Safety and Health specifications for establishing compliance with the Federal Coal Mine Health and Safety Act of 1969 (Figure III.1). With the gravimetric sampling assembly – which is composed of a cyclone separator to reject nonrespirable dust particles, and a tangentially oriented filter – the pump will deliver 1.6, 1.8, or 2.0 l/min airflow rates over a 100-mm water pressure drop (ΔP) for at least 8 hr with not more than ±0.1 l/min variation in flow rate. The microimpinger[2] also can be used with this unit.

FIGURE III.1. Dust collection with the Micronair® permissible air sampling pump. (Courtesy of National Environmental Instruments, Inc., Pilgrim Station, Warwick, R.I.)

Controls:

Switch: Pump unit has an on-off switch on outside of case. The switch is protected against dust entering the mechanism and against accidental stopping or starting by the use of a screw-on dust cover.

Flow rate adjustment: To prevent accidental change of adjustment, the flow rate adjuster is recessed and requires use of an adjusting tool.

Battery characteristics: Power supply is a removable nickel-cadmium rechargeable battery located in pump case. Battery capacity is 3.75 V, 1 Ahr. Battery pack contains current limiting resistors to protect against accidental short circuit destruction of cells. Miniature jack connection provides for charging battery in or out of case.

Battery performance: With fully charged battery, pump is capable of operating for more than 10 hr at a flow rate of 2.0 l/min against resistance of 4 in. water measured at inlet of pump.

Pulsation: Piston-type pump has minimum frequency of 20 Hz.

Belt clips: Pump unit is provided with spring belt clip for suspension from the operator's belt.

Dimensions: The overall dimensions of the pump unit are 156 x 83 x 64 mm and 510 g.

Construction and protection features: The case is constructed from a high impact resistant ABS plastic. Dust entry into the case is prevented by means of a vinyl gasket, and by exhausting the pump into the case to maintain a slight positive internal pressure.

Flow rate indicator: A flowmeter is provided as an integral part of pump unit, calibrated within 5% at 2.0, 1.8, and 1.6 l/min.

Flow rate range: Pump is capable of operating continuously for 8 hr over a range of 1.5 to 2.5 l/min and is adjustable over this range by means of its control valve.

Battery charger

Power supply: 117-V 60-Hz power line.

Connector: Cord and polarized connector to charging inlet on pump or battery case.

Safety: Supplied with fuse and grounded power plug. Not susceptible to damage when operated without a battery on charge.

Charging rates: Charger will operate at either a 16-hr or 64-hr charge rate. A safety feature ensures that it will not overcharge a discharged battery in 16 hr when operating at the 16-hr charge rate or in 88 hr when operating at the 64-hr charge rate.

A multiple unit (gang) charger also is available commercially from the same supplier (Figure III.2) for charging up to five Micronair units simultaneously.

FIGURE III.2. Multiple charging unit for Micronair® pumps.

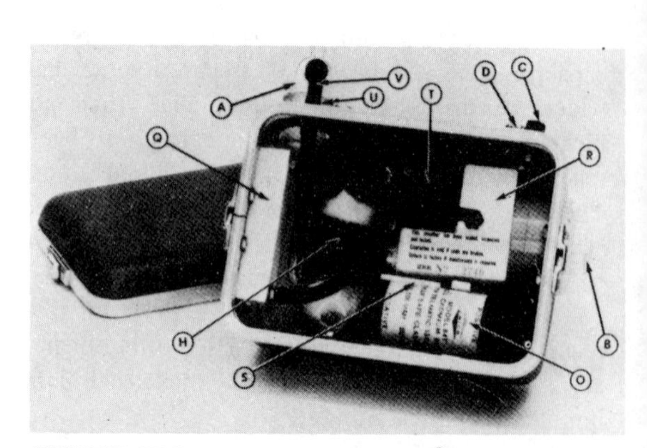

FIGURE III.3. The Unico Telmatic® air sampler. A. Standard midget impinger-scrubber. B. Timer. C. Potentiometer on-off and adjust. D. Carrying handle. H. Charcoal trap. O. 8.4-V 500-mAhr battery. Q. Flo-Set® rotameter. R. Air pump. S. Battery clip. T. Built-in charger module. U. Airflow feed-through. V. Rubber tubing connects side arm of inpinger to feed-through. (Courtesy of National Environmental Instruments, Inc., Pilgrim Station, Warwick, R.I.)

2. Telmatic® – Unico

The second model is actually a handheld sampler for "walk-around" surveys, or mobile fixed-station sampling (Figure III.3). The sampling rate is controlled by a rheostat in the electrical circuit, and an automatic timer to terminate unattended sampling is included. The sampler is one component of the Telmatic 150® analyzer kit which is available as a complete self-contained "suitcase laboratory" for the analysis of

TEL/TML, diisocyanates, and inorganic lead in air. The pump is powered by an 8.4-V 0.5-Ahr rechargeable nickel-cadmium battery which provides 5 hr of continuous sampling with the standard midget impinger at 2.8 l/min and 305 mm water Δ P. The sampler unit has been approved by Factory Mutual as intrinsically safe for Class 1, Group D, hazardous (flammable) areas. Flow rate can be adjusted over a 1 to 5 l/min range. The unit weighs just over 1 kg.

D. DIAPHRAGM PUMPS

Present-day designs can be traced to the original pump described during 1960 in England by Sherwood and Greenhalgh for collecting radioactive dust.[3] The pumping train consisted of a 6-V constant speed motor that drove a simple diaphragm pump [22-mm diameter polytetrafluoroethylene (PTFE) diaphragm] at about 1,700 rpm and consumed power at a 21-mA rate under no load conditions. The motor speed remained constant within a ±2% limit in the range 4.8 to 7.5 V. A second generation unit incorporated a running-time indicator, had a greater battery capacity (4 mercury-type with an operational life of 100 hr vs. 30 hr for the original), and weighed 900 g (original weighed 450 g). The flow rate was very nearly linear over the range 600 ml/min at 100 mm water Δ P to 400 ml/min at 305 mm Δ P. Current drain varied from 185 to 235 mW over this range.

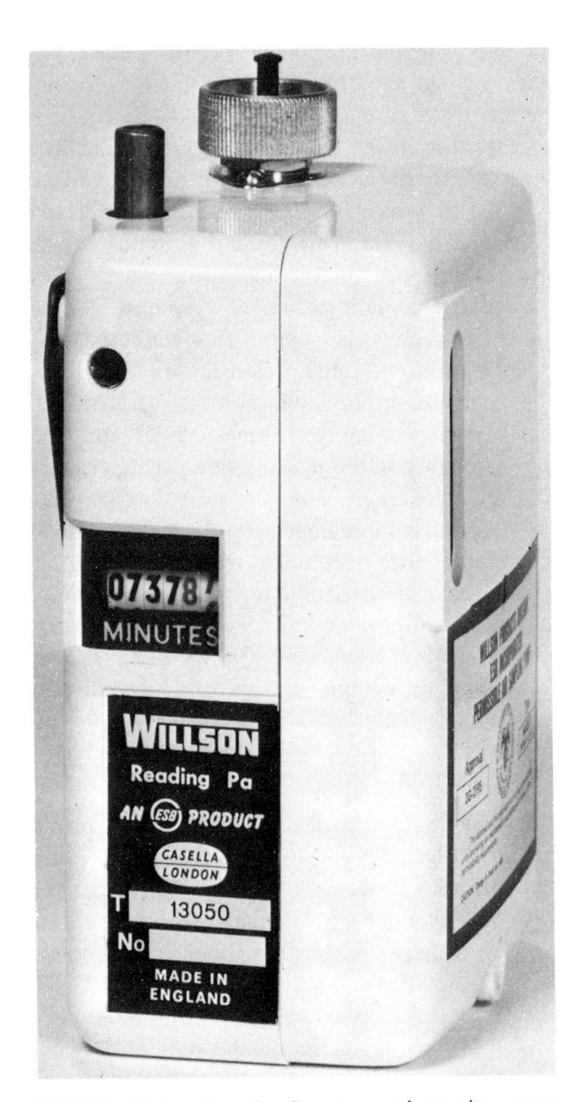

FIGURE III.4. The Casella personnel monitor pump with total volume readout scaler.

1. Casella Personal Sampler

Commercial production and promotion of this air sampler was undertaken by C. F. Casella & Company, Limited[4] and is currently available in the United States as the Mark II personal air sampler[5] (Figure III.4). Operated by a 6.4-V rechargeable nickel-cadmium battery (0.9 Ahr) the pump provides 8 to 10 hr continuous service with a current drain of 90 mA and 2 l/min flow rate against a 450-mm water Δ P.[6]

The speed-governed DC motor drives through a reduction gear, an eccentric to which is coupled a flexible diaphragm. Together with two flap valves and a counter, the motor and diaphragm unit are mounted onto a one-piece die-cast chassis. The gear shaft runs in PTFE/lead bearings for a long service-free life, while gears are stainless steel and plastic.

The case is molded in a white, tough plastic with round corners for comfortable pocket wear. It is divided into two compartments, one having the motor/pump unit, and the other a battery easily removable if required but normally charged in place. A window shows minutes run on the gear-driven counter. The lid of the case is held shut with a screw which can be sealed to prevent unauthorized opening.

The unit is started by inserting a three-pin plug into a socket. This gives a readily visible indication that the unit is on and proof against accidental switching off. The time counter provides an accurate estimate of the volume sampled and thus the concentration of contaminant.

An adjustment allows the user to adjust the

flow rate to match the collector characteristics by means of an air "bleed" accessible from the exterior of the case (recessed, slotted screw head) for adjusting flow rate. A flowmeter can be connected to the collector for setting purposes. Given the flow rate and the time run, an accurate estimate of the volume aspirated gives the mean concentration for the sampling period.

A special pulsation damper ("smoother") is required to deliver the smooth flow necessary for respirable dust sampling with the cyclone collection system. For a mean flow rate of 1.9 l/min, corresponding to a mean inlet of 210 cm/sec velocity without a pulsation damper, the inlet velocity varied from 40 to 1,000 cm/sec. When the pulsation damper was used at the same mean flow rate, the inlet velocity pulsations settled down to a maximum ripple of 210 $\pm$ 8 cm/sec. The maximum capacity appears to be 3 l/min.[5] The dimensions are 57 x 79 x 127 mm and 600 g. Bureau of Mines intrinsic safety approval has been obtained.

2. Misco® Lapel Sampler

A similar unit was manufactured in California by the Electro-Neutronics Division of the International Chemical and Nuclear Corporation, which was acquired by Microchemical Specialties Company (Misco) in 1970.[7] Their lapel sampler measures 30 x 90 x 110 mm, weighs 570 g, and delivers a 2 l/min flow rate at a 630-mm water Δ P or 2.8 l/min at a 35-mm water Δ P. The rechargeable battery capacity is rated at 0.5 Ahr and will deliver 8 hr of sampling at 2.5 l/min with a 28.6-mm Whatman® No. 41 filter paper before in-place recharging becomes necessary. The reserve capacity permits adjustment in flow rate to maintain 2.5 l/min as filter media loading progresses. An integral timer, or revolution counter readout, and pulse dampener are not included.

3. Air Chek® – SKC

SKC, Incorporated[8] offers yet another variation which features a 45 x 89 x 165 mm lightweight stainless steel case, ball bearings, total weight of 1 kg, "zero-leak" valves, voltage regulator, 3-V, in-place, rechargeable battery, and is available in three models – 1, 2, and 2,000 ml/min flow rates. No digital readout, either timer, revolution counter, or total volume, is available.

4. Monitair® – MSA

In 1965 the Mine Safety Appliance Company introduced their Monitair sampler.[9] The original model (currently available) combines the pump, battery, control valves, and rotameter in a pressure-molded, high-impact plastic case that measures 51 x 102 x 107 mm and weighs (entire unit) 540 g (Figure III.5). The 6-V rechargeable battery provides power for 7 hr of sampling (pressure differential not stated). However, with the midget impinger at the standard flow rate of 2.8 l/min and 305 mm water Δ P, the sampling time is limited to 4 hr. The unit is available in two capacities – 0 to 1 l/min and 0 to 4.7 l/min – which are convertible by interchanging flow tubes. Valve-in-valve construction permits efficient pump operation through dual flow control for minimum battery drain on either high or low resistance sampling. A fitting on the pump exhaust supplies compressed air for pressure applications. A 16-hr recharge cycle replaces 90% of the power consumed during 7 hr of sampling. A built-in rotameter is included, but no timing, or total volume readout, facility is furnished. Flow rate is controlled by the dual valve-within-valve assembly. The larger diameter valve stem controls the sample flow directly while the smaller inner valve stem controls by-pass air, which enters through the core of the stem. The by-pass permits efficient pump and rotameter operation by compensation for excessive flow resistance in the sampling system (Δ P).

Recently, the Monitair sampler design was revised to provide 8 hr of continuous sampling at 1.7, or 1.4 l/min for gravimetric sampling with 2.0, 1.7, or 1.4 l/min for gravimetric sampling with the cyclone-filter assembly designed for cool mine dust collection. The dimensions were increased somewhat to accommodate the larger battery (length from 107 mm to 127 mm). The sampling pump is Bureau of Mines approved (Figure III.6).

5. Unico® C115 Personnel Air Sampler

Greater pumping capacity has been attained by increasing the overall size and weight of the assembly to accommodate a larger battery in the Models C110 and C115 personnel samplers produced by the Bendix Corporation and distributed by National Environmental Instruments, Incorporated.[10] Constant flow for 8 hr at 2.8 l/min $\pm$ 2% and 305 mm water Δ P

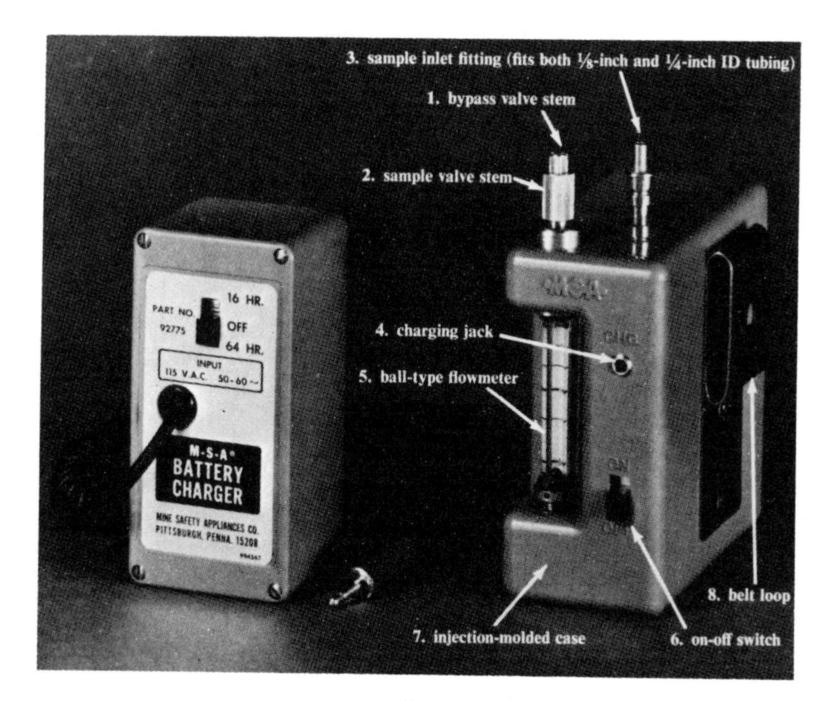

FIGURE III.5. The MSA Monitair® sampling pump and battery charger. (Courtesy of the Mine Safety Appliances Co., Pittsburgh.)

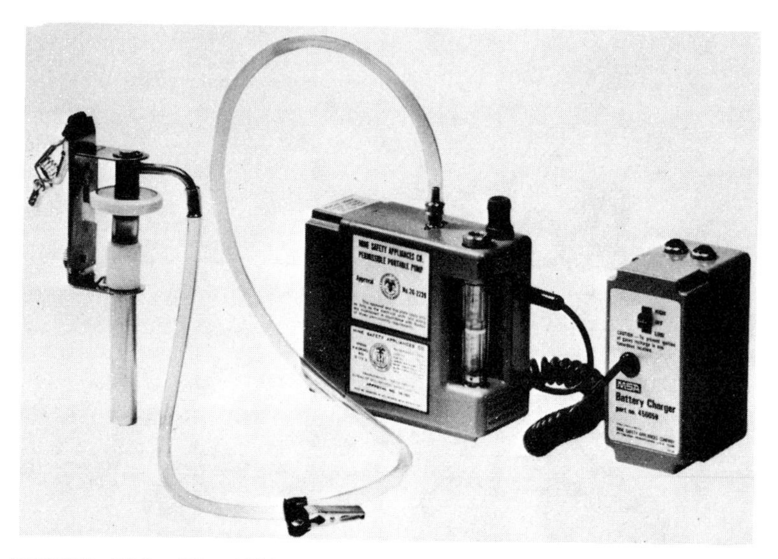

FIGURE III.6. The MSA permissible portable pump and respirable dust sampling unit. (Courtesy of the Mine Safety Appliances Co., Pittsburgh.)

(midget impinger) or 14 hr at 2 l/min and 100 mm water ΔP is supplied by a 6-V, 2-Ahr rechargeable nickel-cadmium battery, which can be charged in place or removed for external charging, and supplies constant voltage to the motor through a regulator. The maximum rate is 3.5 l/min through a 430-mm water pressure drop resistance (Figure III.7). A built-in battery voltage meter provides a quick check of battery condition. The battery is fitted with a resistor unit that prevents overheating in case of accidental shorting, and the unit has Bureau of Mines approval for intrinsically safe operation in explosive atmospheres, and Environmental Health Service Bureau of Occupational Safety and Health approval. A 0.5 to 5 l/min recessed rotameter and a pulsation eliminator are included as standard accessories. Fittings are provided for both pressure and vacuum. Flow rate is controlled by an adjustable by-pass valve, and the controls (rate and on-off switch) are recessed and capped. All parts are readily accessible for field servicing (Figure III.8). The sampling assembly is housed in a 63 x 140 x 165 mm ABS plastic case, weighs 1,360 g, and is supplied with a detachable belt clip.

6. Sipin Personal Sampler Pump

After completion of the writing of this book, a personnel monitor that does record the total volume of air sample drawn through the sampling

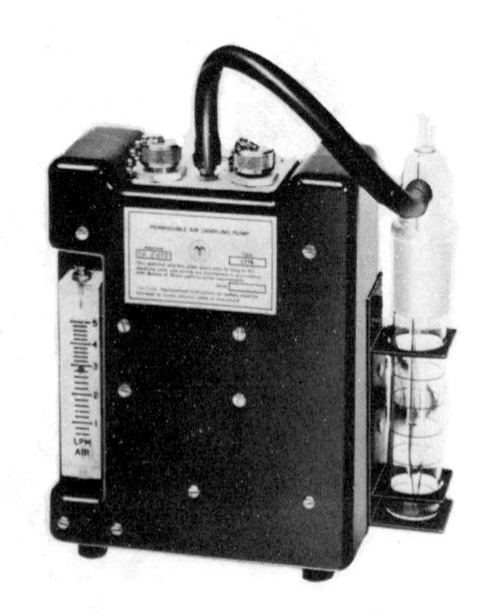

FIGURE III.7. The Unico® Model C115 personnel monitor pump. (Courtesy of National Environmental Instruments, Inc., Pilgrim Station, Warwick, R.I.)

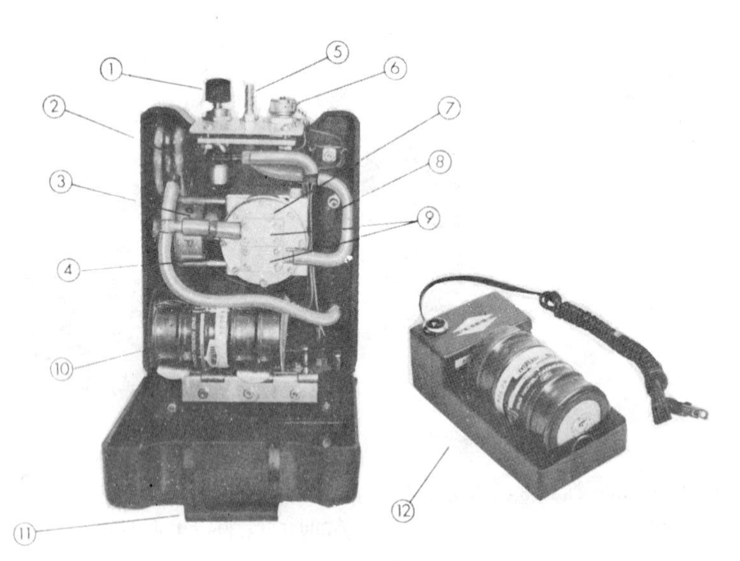

1. FLOW CONTROL VALVE
2. FLOW EQUALIZER (SNUBBER)
3. HI-TORQUE PRECISION ELECTRIC MOTOR
4. DELRIN GEAR-TRAIN ASSEMBLY
5. INLET PORT
6. COVERED ON-OFF SWITCH
7. PRECISION DIAPHRAGM PUMP
8. FLOWMETER
9. INTAKE/EXHAUST VALVE HOUSINGS
10. 6.0 VOLT 2 AMP-HOUR REACHARGEABLE NICKEL CADMIUM BATTERY
11. BELT LOOP
12. 110 VOLT 60 CYCLE BATTERY CHARGER

FIGURE III.8. The Unico® C110 personnel monitor pump construction details. (Courtesy of National Environmental Instruments, Inc., Pilgrim Station, R.I.)

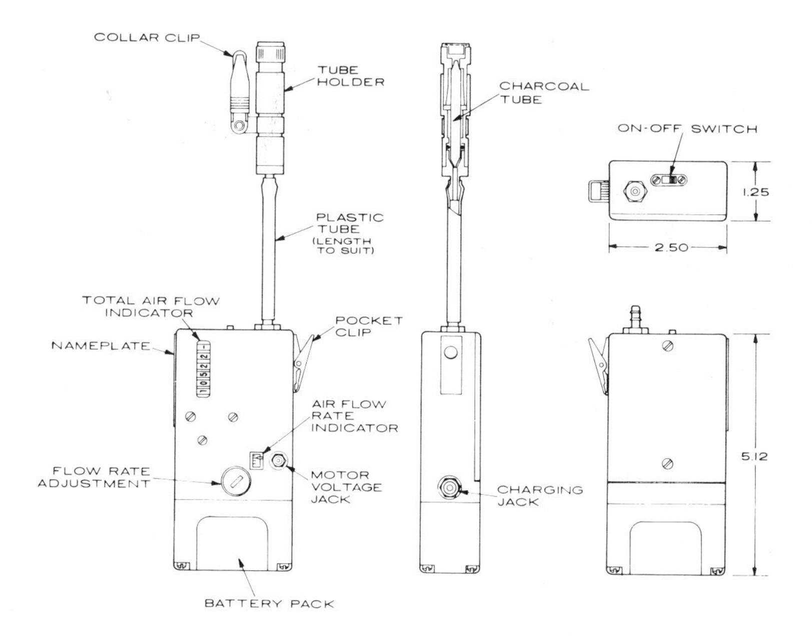

FIGURE III.9. The Sipin personnel monitor pump with charcoal sampling tube and total volume readout scaler. (Courtesy of the Anatole J. Sipin Co., New York.)

system was announced. This development was considered to be of sufficient importance to add as an insertion at this point.

The Sipin Personal Sampler Pump has been developed under sponsorship of the National Institute for Occupational Safety and Health to monitor the concentrations of harmful air pollutants in industrial environments.

This equipment fills a present need for a miniature pump to continuously draw an air sample from a worker's breathing zone through a sorbent tube for the duration of a working shift, and to measure the total volume of air that has been pumped.

The . . . Pump has dimensions of 1-1/4″ in depth, 2-1/2″ in width and 5-1/8″ in length; and the weight is less than 12 oz. It is carried in the breast pocket of the worker's shirt or coat or in a pouch clipped to the worker's belt. The pump is connected to a charcoal tube holder. More than 8 hours of continuous operation can be provided before recharge of the battery . . . . The flow rate is adjustable, and it is calibrated. Total airflow passing through the pump during its operating period is indicated.

Sipin Personal Sampler Pump Specifications

Flow Rate: Model SP-1 – 50 to 200 cc/min., Adjustable.
Model SP-2 – 25 to 100 cc/min., Adjustable.
Pressure Drop: Maximum of 2.5 inches of water through the charcoal tube at a flow rate of 200 cc/min.
Power: Rechargeable nickel-cadmium batteries.
Duration of Pump Operation: Eight hours at constant speed without recharging batteries.
Dimensions: 1-1/2″ x 2-1/2″ x 5-1/8″.
Weight: 12 ounces.
Total Flow Indication: Pump stroke indication from 0 to 99,999.
Charcoal Tube: The charcoal tube is protected by a separate holder clipped to the worker's collar or lapel.
Flow Adjustment: The flow adjustment is recessed and screwdriver operated.
On-Off Switch: The switch is inaccessible when carried.
Accuracy: Pump stroke indication is linearly related to total airflow within ±5%.
Flow Regulation: Flow rate change is not more than 10% of initial value over duration of operation.
Qualification: Prototype unit (model SP-1) has been tested and accepted by the National Institute for Occupational Safety and Health.*

The schematic drawings are shown in Figure III.9, an internal view of the pump in Figure III.10, in place on a workman in Figure III.11, relationship of flow rate to motor drive speed in Figure III.12, and relationship of flow rate vs. motor terminal voltage (contacts accessible on exterior of the case) in Figure III.13. By the time this copy goes to press, a new model with a

*From Sipin, A. J., The Sipin Personal Sampler Pump, Anatole J. Sipin Co., 386 Park Ave. S., New York. With permission.

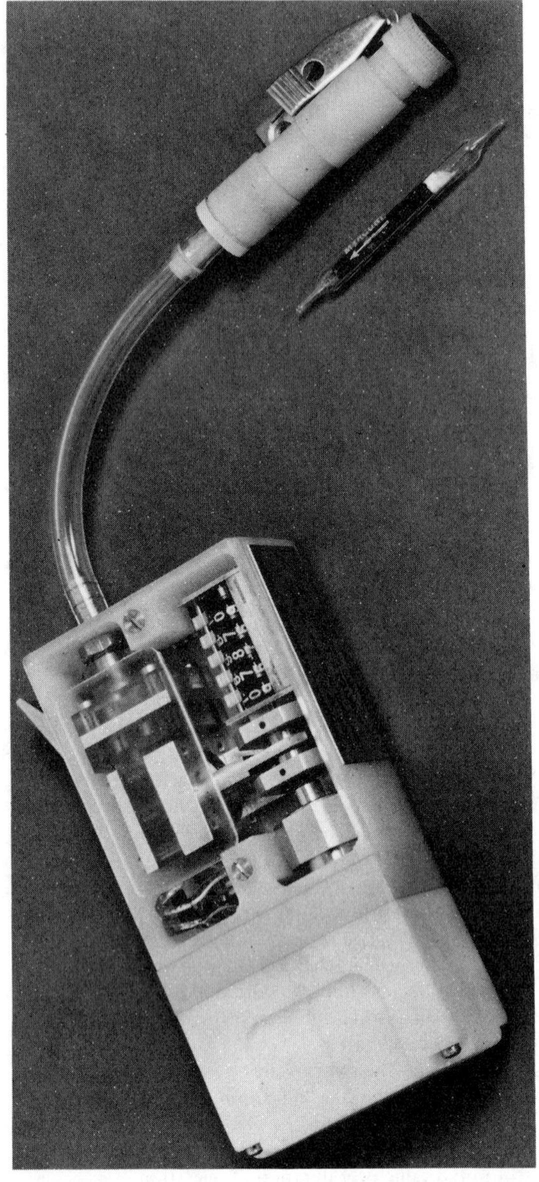

FIGURE III.10. The Sipin personnel monitor pump – cutaway view. (Courtesy of the Anatole J. Sipin Co., New York.)

FIGURE III.11. The Sipin personnel monitor pump in place. (Courtesy of the Anatole J. Sipin Co., New York.)

six-place digital readout that will permit a full 8-hr sampling period at 200 ml/min will be available.

7. Other Miniature Pumps

Salient features of these six samplers described are summarized in Table III.1.

a. Solonoid Type

An electrically driven miniature pump which has not been applied to personnel monitoring equipment offers some rather unique advantages. Pumping is accomplished through reciprocal displacement (solonoid principle) of a flexible diaphragm by a piston attached to an electromagnetic coil. The coil is energized by a 150-cps square wave current pulse that causes movement in a permanent magnetic field. The regulated constant-amplitude current pulse is generated by a self-contained, 1-C digital electronic circuit from a 6-V DC input. Vacuum to 500 mm water and a 1 l/min rate can be attained with the smallest unit, which measures 51 mm on a side (cubical shape) and weighs 170 g. Current drain is rated at 260 mA. A slightly larger unit has a 2 l/min output.

b. Brushless DC Motor

Brushless DC motors which are solid state commutated are available in a range of miniature sizes, but require higher voltages (18 to 28 V). Pumping characteristics are good: 2 to 2.8 l/min at 660 mm water ΔP, and the weight is not excessive (230 g). However, further miniaturization would be desirable (38 x 72 x 97 mm).[12] Centrifugal blower models are available also in miniature sizes.

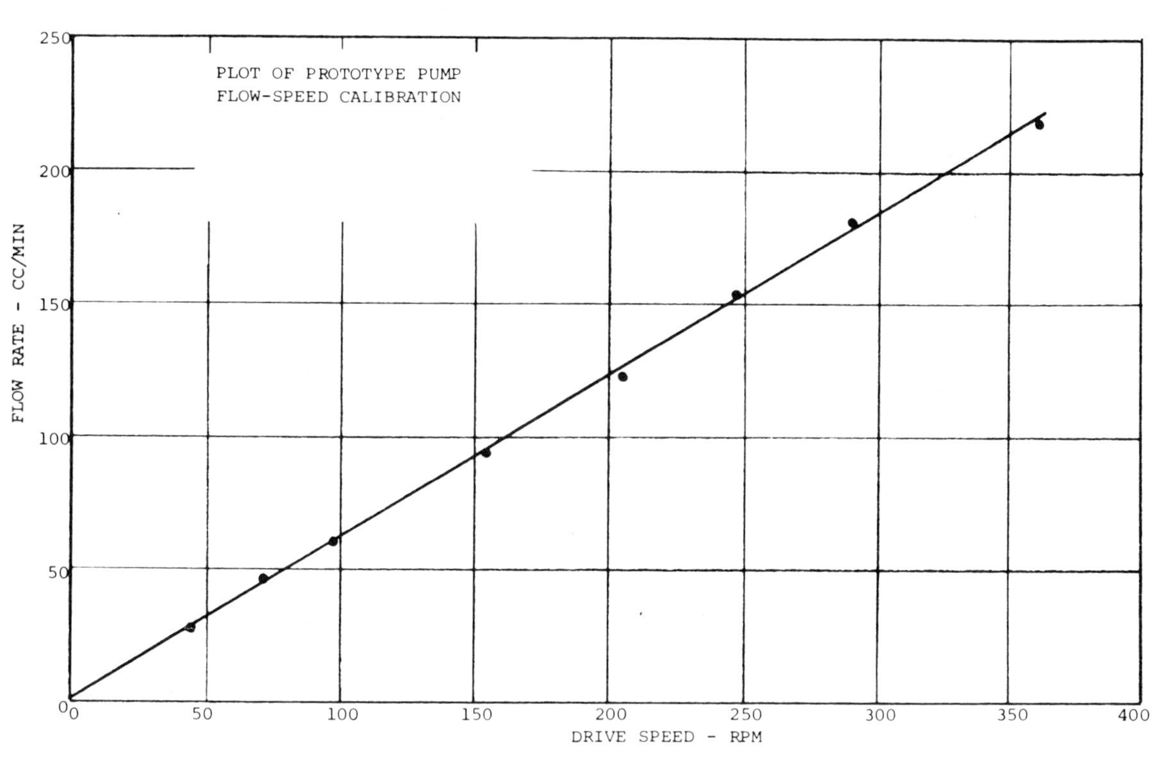

FIGURE III.12. The Sipin personnel monitor pump performance – plot of prototype pump flow-speed calibration. (Courtesy of the Anatole J. Sipin Co., New York.)

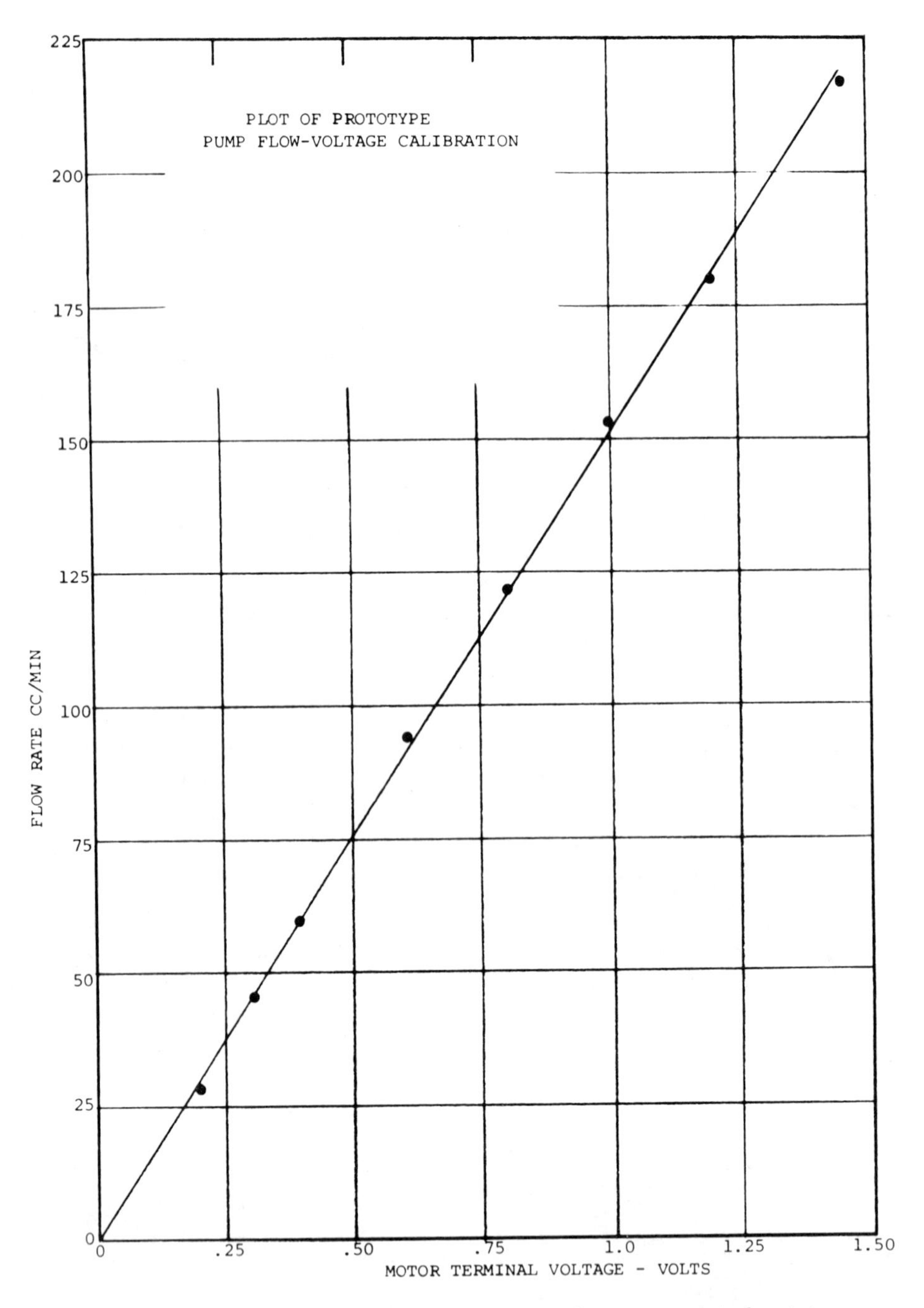

FIGURE III.13. The Sipin personnel monitor pump performance – plot of prototype pump flow-voltage calibration. (Courtesy of the Anatole J. Sipin Co., New York.)

TABLE III.1

Battery Powered Air Sampler Summary (Diaphragm Pumps Unless Noted)

Supplier–Model	Weight (g)	Dimensions (mm) T	W	L	Battery V	Ahr	Sampling capacity Rate (l/min)	ΔP (mm)	(hr)
Specification	550*	50*	75*	150*			2.8	305	8
Micronair® Unico (piston)	510	64	83	156	3.8	1.0	2.0	100	8
Telmatic® Unico (piston)	1000	100	120	185	8.4	0.5	2.8	305	5
Casella Mark II	600	57	79	127	6.2	0.9	2.0	450	8+
Misco®	570	30	90	110	?	0.5	2.0	630	8
SKC	1000	45	89	165	3.0	?	2.0	?	?
Monitair® (MSA)	540	51	102	107	6.0	?	2.8	305	4
Revised Monitair	740	51	102	127	6.0	?	2.0	?	8
Unico C115	1360	63	140	165	6.0	2.0	2.8	305	8
							2.0	100	14
RAC® 440-B (rotary; Battery 3400)	2500	89	185	254	12.0	?	13.0	68	1
							1.8	270	
Staplex® Models BN and BS (rotary)	3650	?	?	?	?	?	14.0	100	8

Supplier–Model	Electric short control	Recessed controls	Pressure connection	Rotameter	Pulse damper	Timer	Total volume readout	Accessible parts	Removable battery	Battery plug	Explosion safe	Constant-speed pump	Flow rate control	Reference
Specification	+	+	+	+	+	–	+	+	+	+	+	+	+	1a
Micronair® Unico (piston)	+	–	–	+	–	–	–	+	+	+	+	–	+	10
Telmatic® Unico (piston)	+	–	–	+	–	+	–	+	+	–	+	–	+	10
Casella Mark II	–	+	–		+	+	–	–	–	+	+	+	+	4,5
Misco®	–	–	–	–	–	–	–	?	?	?	?	–	+	7
SKC	–	–	–	–	–	–	–	?	?	+	?	+	–	8
Monitair® (MSA)	–	–	+	+	–	–	–	–	–	+	+	–	+	9
Revised Monitair	–	–	+	+	?	–	–	–	–	+	+	–	+	9
Unico C115	+	–	+	+	+	–	–	+	+	+	+	+	+	10
RAC® 440-B (rotary; Battery 3400)	–	–	+	+	–	–	–	+	+	+	–	–	+	13
Staplex® Models BN and BS (rotary)	–	–	?	–	+	–	–	?	?	?	–	–	?	14

*Maximum

ΔP = Pressure differential in mm water
T = Thickness or depth
W = Width
L = Length

E. ROTARY PUMPS – CARBON VANE, OIL-LESS

Although no electrically driven rotary pump has been miniaturized sufficiently to be considered as a personnel monitor, current models are small enough to qualify for "carry around" or walk-through surveys. Variations on the original Gast pump include a direct 12-V DC motor directly coupled to a miniature carbon vane pump, a rotameter, and an airflow regulator valve which provides an outside bleed to prevent motor overheating at low flow rates (Figure III.14). This assembly mounted in an aluminum carrying case weighs 2,500 g without the lead acid battery, which weighs another 3,400 g. Although the pumping rate and vacuum attained is high (13 l/min at 68 mm water ΔP), the available power lasts for only 1 hr. Pump vane and motor brush lives are reported to be 5,000 hr and 300 hr, respectively (see Table III.1 for further details).[13]

Another variation extends the operating cycle to 8 hr by drawing power from a silver-cadmium battery (Model BS) which has a 1- to 2-year service life expectancy. The sampler also is available with nickel-cadmium batteries (Model BN). Other than the weight, 3,650 g, no dimensions were given in the available technical literature. The sampling rate appears to be 14 l/min for 8 hrs, but no ΔP information was supplied, although one could assume 100 mm/water since only a specific filter was offered for air sampling.[14]

Undoubtedly this type of pump could be further miniaturized, and PTFE resin substituted for the carbon vanes and pump case to produce a pulse-free flow. These improvements also would overcome another major weakness, relatively short pump service life, without sacrificing entirely the good qualities: (1) high flow rate, (2) minimal ripple in the flow, (3) high vacuum capacity, (4) valves not required, (5) self-lubricating bearings and vanes, and (6) potentially higher power efficiency.

FIGURE III.14. The Research Appliance Company portable air sampling unit. (Courtesy of the Research Appliance Company, Allison Park, Pa.)

F. THE PERSONNEL MONITOR

By combining the advantages offered by the microimpinger[14] (Chapter I, Section B.1.b), the membrane filter[15] (Chapter II, Section B) and the battery-powered miniature sampling pump, a "universal" sampler that collects both the particulate fraction on a filter and the gas phase fraction in a liquid reagent system was developed. This assembly (Figures III.15 and III.16) was

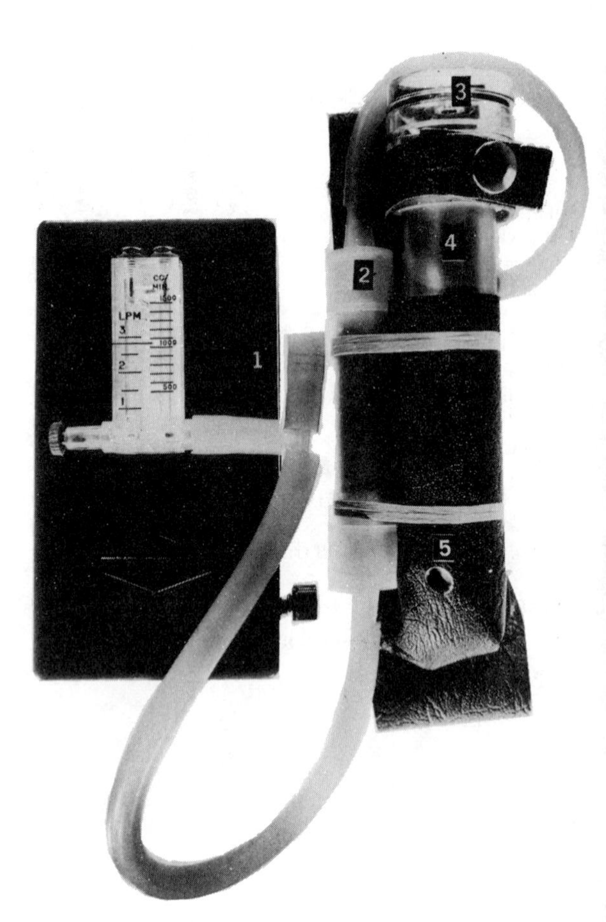

FIGURE III.15. Assembled personnel monitoring sampler.

employed in the evaluation of tetraalkyl lead exposure in a manufacturing operation during 1968,[16] and to determine the source of methylene-*o*-dichloroaniline exposure incurred by production workers during 1969.[17] Details relative to these two biological monitoring studies will be found in Reference 18.

As noted in Figure III.15, the assembly consists of the battery-powered pump (1), a capillary by-pass at (1) which allows sufficient air to pass through the rotameter to attain a stable calibration reference point, charcoal trap (2) to prevent corrosive vapors from the microimpinger (iodine monochloride and HCl in Reference 16) from reaching the rotameter and pump, filter holder (3), aluminum shield (4) to protect the glassware from damage, and plastic-coated cloth carrying case (4). Further detail is shown in the exploded view in Figure III.16. A typical field operation is shown in Figure III.17 (methylene-*o*-chloroaniline project[17]). Although the pump unit shown is now obsolete, several satisfactory replacements are available (Figures III.1 and III.4 to III.6).[5,9,10] A second generation sampler (Figure I.46) which recently has been developed further miniaturizes the collecting assembly without sacrificing any of the advantages.

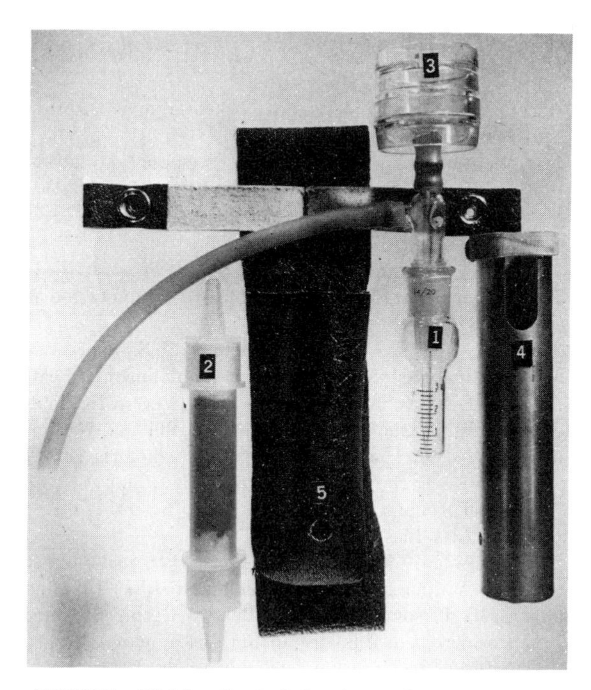

FIGURE III.16. Exploded view of the personnel monitoring sampler.

FIGURE III.17. The personnel monitoring sampler in place for field survey.

G. REFERENCES FOR CHAPTER III

1. Characteristics and Uses of Nickel-Cadium Batteries, The International Nickel Co., Inc., 67 Wall Street, New York, 1966.

1a. Mighty-Mite®, Model 440A, Personnel Monitor, National Environmental Instruments, Inc. (formerly Union Industrial Equipment Corp., Fall River, Mass.), P.O. Box 590, Pilgrim Station, Warwick, R.I.

1b. **Linch, A. L. and Corn, M.,** The standard midget impinger – design improvement and miniaturization, *Am. Ind. Hyg. Assoc. J.,* 26, 601, 1965.

1c. **Kupel, R. E. and White, L. D.,** Report on a modified charcoal tube, *Am. Ind. Hyg. Assoc. J.,* 32, 456, 1971.

2. **Campbell, E. E. and Ide, H. M.,** Air sampling and analysis with microcolumns of silica gel, *Am. Ind. Hyg. Assoc. J.,* 27, 323, 1966.

3. **Sherwood, R. J. and Greenhalgh, D. M. S.,** A personal air sampler, *Ann. Occup. Hyg.,* 2, 127, 1960.

4. Personal Samplers of Airborne Contaminants, Leaflet 930-S-A1 and Supplementary Information Sheet to Leaflet 930-S-AR, C. F. Casella & Co., Ltd., Regent House, Britannia Walk, London, N.I., 1967.

5. Willson Products Division, ESB Inc., 2nd and Washington Streets, Reading, Pa.

6. **Higgins, R. I. and Dervell, P.,** A Gravimetric Size-Selecting Personal Dust Sampler, BCIRA Rep. 908, British Cast Iron Research Assoc., 1968 (available from C. F. Casella & Co., Ltd.).

7. Lapel Sampler, Model 4000, Misco Scientific Co., 1825 Eastshore Highway, Berkeley, Cal. (formerly Flowmaster Products, Inc. – Filtronics®).

8. Life Guard® Model 222, Air-Chek® Personal Pump, SKC, Inc., P.O. Box 8538, Pittsburgh.

9. M-S-A Monitaire Sampler Bulletin No. 0811-14, Mine Safety Appliances Co., 201 N. Braddock Ave., Pittsburgh.

10. C115 Personnel Monitor, Bendix Corp., Baltimore; distributed by National Environmental Instruments, Inc. (formerly Unico Environmental Instruments), Box 590, Pilgrim Station, Warwick, R.I.

10a. **Sipin, A. J.,** The Sipin Personal Sampler Pump, Anatole J. Sipin Co., 386 Park Ave. S., New York.

11. Gas Sampling Pump, AS-100, Atmospheric Sciences Inc., 343 Second St., Suite L, Los Altos, Cal.

12. Miniature Brushless D.C. Diaphragm Blower, Brailsford & Co., Inc., Milton Road, Rye, N.Y.

13. Midget Air Sampler, Model 440B (formerly Gast AD-440-3 Air Sampler), Research Appliance Co., Craighead Rd., Allison Park, Pa.

14. **Linch, A. L.,** The spill-proof microimpinger, *Am. Ind. Hyg. Assoc. J.,* 28, 497, 1967.

15. Aerosol Analysis Monitors, Catalog No. MAWG 037-A0, type AA, and Catalog No. MC/1:2A360, Millipore Corp., Bedford, Mass., 1972.

16. **Linch, A. L., Wiest, E. G., and Carter, M. D.,** Evaluation of tetraalkyl lead exposure for personnel monitor surveys, *Am. Ind. Hyg. Assoc. J.,* 31, 170, 1970.

17. **Linch, A. L., O'Connor, G. B., Barnes, J. R., Killian, A. S., and Neeld, W. E.,** Methylene-bis-ortho-chloroaniline (MOCA®): evaluation of hazards and exposure control, *Am. Ind. Hyg. Assoc. J.,* 32, 802, 1971.

18. **Linch, A. L.,** *Biological Monitoring for Industrial Health Control,* CRC Press, Cleveland, in press, 1973.

Chapter IV

QUALITY CONTROL FOR SAMPLING AND LABORATORY ANALYSIS

A. INTRODUCTION

The measurement of physical entities such as length, volume, weight, electromagnetic radiation, and time involves uncertainties that cannot be eliminated entirely, but when recognized can be reduced to tolerable limits by meticulous attention to detail and close control of the significant variables. Other errors, often unrecognized, are introduced by undesirable physicochemical effects and by interferences in chemical reaction systems. In many cases, absolute values are not directly attainable; therefore, standards from which the desired result can be derived by comparison must be established. Again, errors are inherent in the measurement system. Although the uncertainties cannot be reduced to zero, methods which provide reliable estimates of the probable true value and the range of measurement error are available.

Numbers are employed either to enumerate objects or to delineate quantities. If 16 air samples are taken simultaneously at different locations in a warehouse where gasoline-powered fork lift trucks are in motion, the count would be the same regardless of who counted them, when the count was made, or how the count was made. However, if each individual sample were analyzed for carbon monoxide, 16 different results undoubtedly would be obtained. Furthermore, if replicate determinations were made on each sample, a range of carbon monoxide concentrations would be found.[1] The procedure, which required weighing, volumetric measurements, and reading instruments, or length of stain estimates, included regions of uncertainty. The accumulated errors govern the accuracy of the final results, and combined with the variability of the atmosphere sampled influence the reliability of the final result.

B. DETERMINATE ERROR

1. Detection

Experimental errors are classified as determinate or indeterminate. A 15 count of the warehouse air samples would be a determinate error quickly disclosed by recount. An indeterminate instrumental error would be encountered when the carbon monoxide content was determined by gas chromatography, infrared, or a colorimetric technique.

For example, if the estimation of carbon monoxide concentration was made with a length of stain detector tube, and a 6.5 mm stain length equivalent to 57 ppm was recorded by the observer, whereas the true stain length was 6.0 mm and equivalent to 50 ppm, the observational error would have been $(57 - 50) \times 100 \div 50 = 14\%$.

All analytical methods are subject to errors. The determinate ones contribute constant error or bias while the indeterminate ones produce random fluctuations in the data. The concepts of accuracy and precision as applied to the detection and control of error have been clearly defined and should be used exactly.

a. Accuracy

Accuracy relates the amount of an element or compound recovered by the analytical procedure to the amount actually present. For results to be accurate, the analysis must yield values close to the true value.

b. Precision

Precision is a measure of the method's variability when repeatedly applied to a homogeneous sample under controlled conditions without regard to the magnitude of displacement from the true value as the result of systematic or constant indeterminate errors which are present during the entire series of measurements. Stated conversely, precision is the degree of agreement among results obtained by repeated measurements of "checks" on a single sample under a given set of conditions.[1a]

A concept of the difference between accuracy and precision can be visualized by the pattern formed by shots aimed at a target, as shown in Figure IV.1. From the scatter of four shots, one can see that a high degree of precision can be attained without accuracy, and accuracy without precision is possible. The ultimate goal is, of course, accuracy with precision, target 4. (See also American Society for Testing and Materials (ASTM) Designation D-1129-68 for definitions.[3])

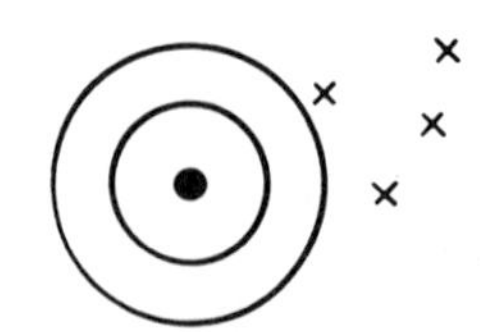

IMPRECISE AND INACCURATE

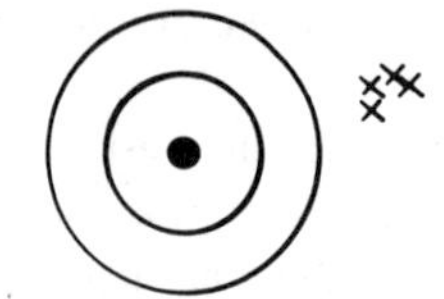

PRECISE BUT INACCURATE

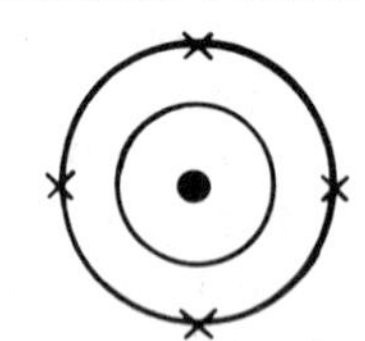

ACCURATE BUT IMPRECISE

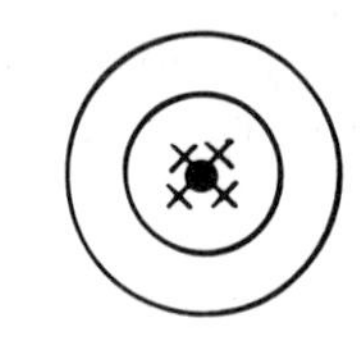

PRECISE AND ACCURATE

FIGURE IV.1. Precision and accuracy.

c. Mean Error

Mean error is the average difference, with regard to sign, between the test results and the true result. It is also equal to the difference between the average of a series of test results and the true result.

d. Relative Error

Relative error is the mean error of a series of test results expressed as a percentage of the true result.

e. Determinate Error and Accuracy

The terms "determinate" error, "assignable" error, and "systematic" error are synonymous. A determinate error contributes constant error or bias to results which may agree precisely among themselves.

Sources of determinate error – A method may be capable of reproducing results to a high degree of precision, but only a fraction of the component sought may be recovered. A precise analysis may be in error due to inadequate standardization of solutions, inaccurate volumetric measurements, inaccurate balance weights, improperly calibrated instruments, or personal bias (e.g., color estimation). Method errors that are inherent in the procedure are the most serious and most difficult to detect and correct. The contribution from interferences is discussed in Section B.2.c, "Chemical Interferences." Personal errors other than inherent visual acuity deficiencies (e.g., color judgment) include consistent carelessness, lack of knowledge, and personal bias, and are exemplified by calculation errors, use of contaminated or improper reagents, nonrepresentative sampling, and poorly calibrated standards and instruments.

Effects of determinate error – Determinate error falls into two categories:

Additive – Has a constant value regardless of the amount of the constituent sought in the sample. A plot of the analytical value vs. the theoretical value (Figure IV.2) will disclose an intercept somewhere other than zero.

Proportional – Is a determinate error that changes magnitude according to the amount of constituent present in the sample. A plot of the analytical value vs. the theoretical value (Figure IV.3) not only fails to pass through zero, but discloses a curvilinear rather than a linear function.

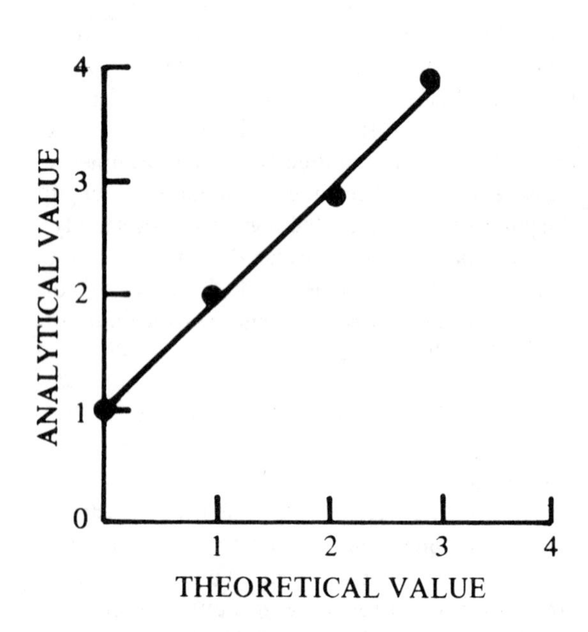

FIGURE IV.2. Additive error.

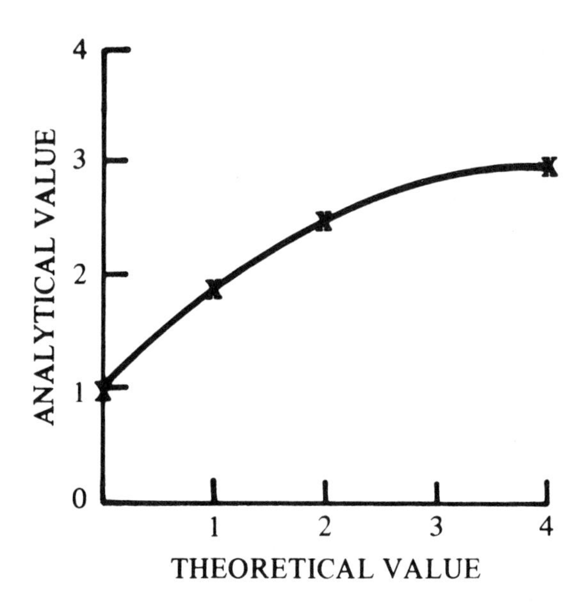

FIGURE IV.3. Proportional error.

f. Recovery or "Spiked" Sample Procedures

"Spiked" samples provide a technique for the detection of determinate errors. A recovery procedure does not provide a correction factor to adjust the results of an analysis; however, a basis for evaluating the applicability of a particular method to any given sample and analytical quality control can be derived from the results.

A recovery should be performed at the same time as the sample analysis, but need not be run for confirmation of method integrity on a routine basis with samples whose general composition is known or when using a method whose applicability is well established. However, the principle applied on a routine basis provides an excellent quality control program.

In brief, the recovery technique requires application of the analytical method to a reagent blank; to a series of known standards covering the anticipated range of concentration of the sample; to the sample itself, in at least duplicate; and to spiked samples, prepared by adding known quantities of the substance sought to separate aliquots of the sample itself, which are equal in size to the unspiked sample taken for analysis. The substance sought should be added in sufficient quantity to exceed in magnitude the limits of analytical error but the total in the sample should not exceed the range of the standards selected.

The results are first corrected for reagent influence by subtracting the reagent blank from each standard, sample, and spiked sample results. From a plot of the known standard values on graph paper the amount of sought substance in the sample itself is determined. This quantity is subtracted from each of the spiked determinations, and the remainder divided by the known amount originally added and then multiplied by 100 to provide the percentage recovery. Table IV.1 illustrates an application of this technique to the analysis of blood for lead content.

This recovery technique may be applied to a colorimetric, flame photometric, fluorometric, polarographic, atomic absorption spectrophotometric, and liquid phase and gas phase chromatographic analysis, and in a simplified form to titrimetric, gravimetric, and other types of analyses.

Specifications for percentage recoveries required for acceptance of analytical results usually are determined by the state of the art and the final disposition of the results. Recoveries of substances within the calibrated range of the method, of course, may be very high or very low and approach 100% as the relative magnitude of the errors of the method diminishes and as the upper limit of the calibration range is approached. In general, procedures for trace substances that have relatively large inherent errors due to operation near the limits of sensitivity deliver recoveries that would be considered very poor by the classical analytical chemist and yet, from the practical viewpoint of usefulness, may be quite acceptable. Poor recovery may reflect interferences present in the sample, excessive manipulative losses, or the method's technical inadequacy in the range of application. The recovery error range becomes increasingly greater as the limit of the sensitivity of the method is approached (Table IV.1 – 2 μg spike). The limit of sensitivity may be considered the point beyond which indeterminate error is a greater quantity than the desired result.

It must be stressed, however, that the judicious use of recovery methods for the evaluation of analytical procedures and their applicability to particular circumstances is an invaluable aid to the analyst in both routine analysis and research investigations.

g. Control Charts

Trends and shifts in control chart responses also may indicate determinant error. The standard deviation for the analysis is calculated from spiked samples and control limits (usually ± 3 standard

TABLE IV.1

Lead In Blood Analysis

Analyst: DJM

μg Pb added	Optical density	μg Pb found		% Recovery
		Total	Recovered	
None – blank	0.0969	–	–	–
5-μg Calibration point	0.2596	–	–	–
None	0.1427	1.6	–	–
None	0.1337	1.3	–	–
None	0.1397	1.4	–	–
None	0.1397	1.4	–	–
Average	0.1389	1.4	–	–
2.0	0.1805	2.9	1.5	75
4.0	0.2636	5.4	4.0	100
6.0	0.3372	7.8	6.4	107
8.0	0.3925	9.4	8.0	100
10.0	0.4437	11.4	10.0	100
Total 30.0	–	36.9	29.9	96

Basis: 10 g blood from blood bank pool ashed and lead determined by double extraction, mixed color, dithizone procedure.
Analyst: DJM
Calculation of mean error:
Mean error = 36.9 – (30.0 + 5 x 1.4) = 0.1 μg for entire set
= 2.9 – (2.0 + 1.4) = 0.5 μg for 2-μg spike
Calculation of relative error:
Relative error = (0.1 x 100)/37.0 = 0.27% for entire set
= (0.5 x 100)/3.4 = 14.7% for 2-μg spike

deviations). (For calculation of standard deviation see Section C.1.a, "Standard Deviation," and for in-depth discussion of control limits see Reference 3.) In some cases such as biological oxygen demand (BOD) and pesticide samples, spiking to resemble actual conditions is not possible. However, techniques for detecting bias under these conditions have been developed.[2]

Control charts may be prepared even for samples which cannot be spiked or for which the recovery technique is impractical. A reference value is obtained from the average of a series of replicate determinations performed on a composite or pooled sample which has been stabilized to maintain a constant concentration during the control period (e.g., nitric acid in urine). An example has been prepared from a lead in blood study (Figure IV.4). Although these data were drawn from the same blood pool used to illustrate the application of the spiking technique for quality control, the consecutive aliquot analyses plotted as a control chart furnish additional information. The control limits were reduced to ± 2 standard deviations to further sharpen the trends. The effect of personal bias is shown by KD's vs. JD's performance.

h. Change in Methodology

Analysis of a sample for a particular constituent by two or more methods that are entirely unrelated in principle may aid in the resolution of indeterminate error.

In Table IV.2, an interlaboratory evaluation of three different methods for the determination of lead concentration in ashed urine specimens (mixed color dithizone, atomic absorption, and polarography) is summarized. If the highly specific polarographic method is selected as the primary standard, then the dithizone procedure is subject to a + 7.1 μg/l bias as compared with a + 3.6 μg/l bias in the atomic absorption method for lead.

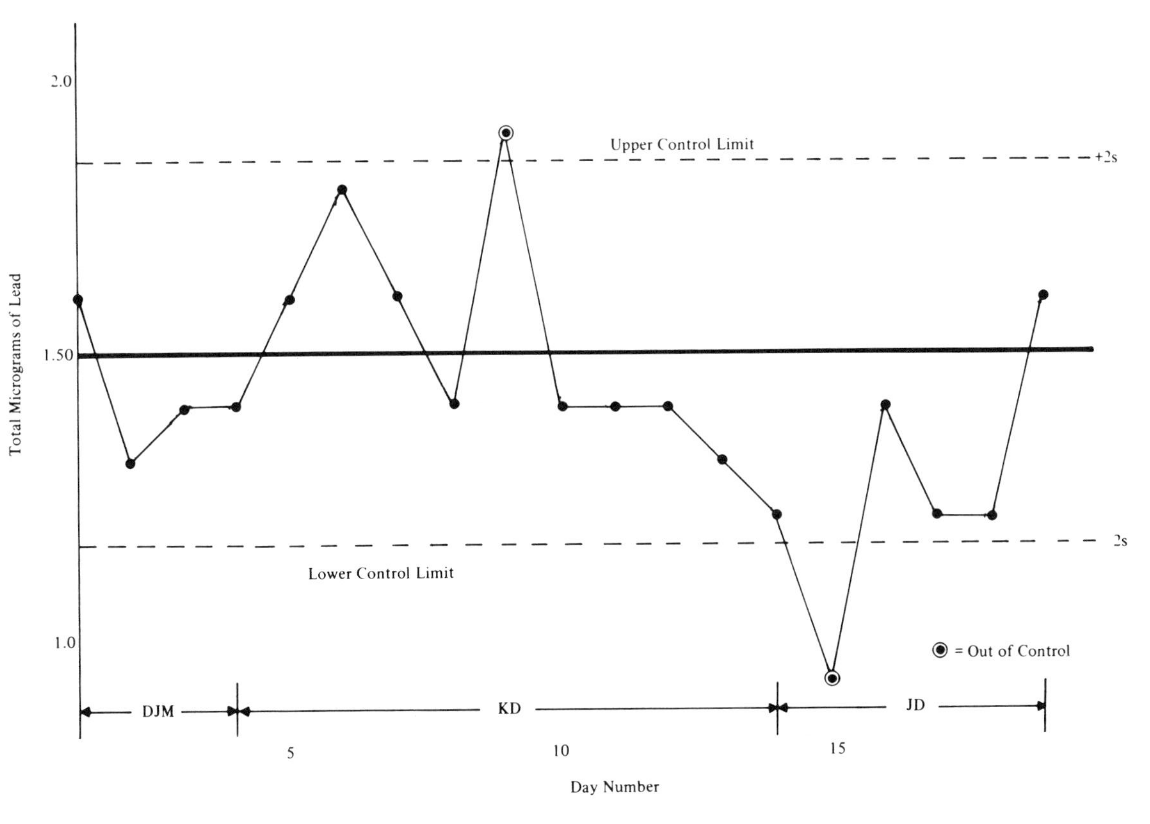

FIGURE IV.4. Lead in blood control chart.

TABLE IV.2

An Interlaboratory Study of the Determination Error in the Dithizone Procedure for the Determination of Lead in Nine Urine Specimens

Polarographic method	Mixed color found	Dithizone difference	Atomic absorption	
			Found	Difference
10	25	15	10	0
14	28	14	22	8
12	12	0	16	4
15	20	5	16	1
21	20	1	22	1
22	30	8	24	2
27	40	13	36	9
19	22	3	22	3
12	22	10	16	4
Mean	–	+7.1	–	3.6

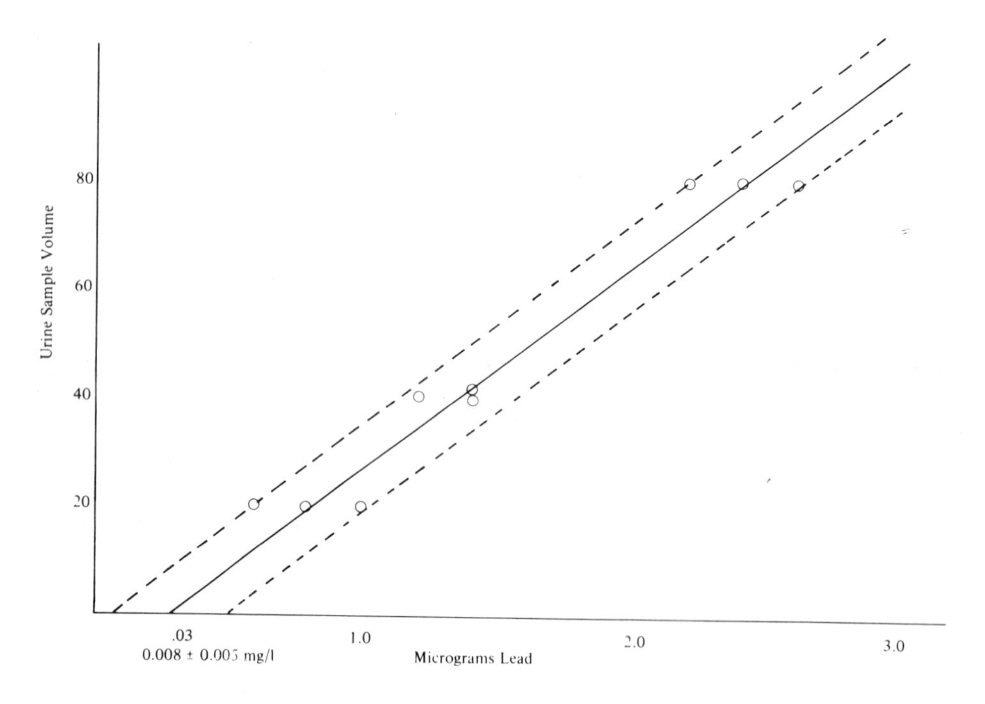

FIGURE IV.5. Effect of sample size on determination of lead in urine.

i. Effect of Sample Size

If the determinate error is additive, the magnitude may be estimated by plotting the analytical results vs. a range of sample volumes or weights. If the error has a constant value regardless of the amount of the component sought, then a straight line fitted to the plotted points will pass through the origin. The effect of urine volume on the analysis for lead is shown in Figure IV.5.

2. Correction

a. Physical

Determinate errors should not be tolerated unless careful evaluation of the magnitude of the controlling variables and consideration of alternative routes produce no practical solution for the problem. In many cases error can be reduced to tolerable levels by quantitating the magnitude over the operating range and developing either a corrective manipulation directly in the procedure or a mathematical correction in the final calculation. Temperature coefficients (parameter change per degree) are widely applied to both physical and chemical measurements. For example, the stain length produced by carbon monoxide in the detector tubes previously cited for illustration is temperature dependent as well as air sampling rate and CO concentration dependent. Therefore, when these tubes are used outside the median temperature range, a correction must be applied to the observed stain length (Table IV.3).[1b]

As a general rule most instruments exhibit maximum reliability over the center 70% of their range (midpoint ±35%). As the extreme to either side is approached the response and reading errors become increasingly greater. For example, optical density measurements in colorimetric analysis should be confined to the range 0.045 to 0.80 by concentration adjustment or cell path choice.

Extrapolation to limits outside of the range of response established for the analytical method or instrumental measurement may introduce large errors, as many chemical and physical responses are linear only over a relatively narrow band in their total response capability. In absorption spectrophotometric measurements, Beer's law relating optical density to concentration may not be linear outside of rather narrow limits in some instances (e.g., colorimetric determination of formaldehyde at high dilution by the chromatropic acid method).

The well-worn axiom that the reliability of an analytical result can be no better than the quality of the sample submitted for analysis can be appropriately applied to air sampling. The source of the contaminant, air flow direction, and

TABLE IV.3

Temperature Correction for Kitagawa® Carbon Monoxide Detector Tube No. 100

Chart readings (ppm)	Correct Concentration (ppm) 0°C (32°F)	10°C (50°F)	20°C (68°F)	30°C (86°F)	40°C (104°F)
1,000	800	900	1,000	1,060	1,140
900	720	810	900	950	1,030
800	640	720	800	840	910
700	570	640	700	740	790
600	490	550	600	630	680
500	410	470	500	520	560
400	340	380	400	420	440
300	260	290	300	310	320
200	180	200	200	200	210
100	100	100	100	100	100

velocity, whether due to wind, forced circulation, or thermal gradients; density of the contaminant; intensity of sunlight; time of day; presence of obstructions such as trees, buildings, partitions, machinery, etc. which act as baffles to produce turbulence; humidity; and half-life of the contaminant determine the concentration at any given time and location. That the concentration can vary by several orders of magnitude within a relatively short radius from the point of reference and within short time intervals has been confirmed repeatedly.

Forced circulation effects – circulating fans, air conditioners, hot air heating systems (whether convection or forced draft), filtration units, hoods, fume ducts and vents, etc., thermal gradients that set up convection currents and eddies, drafts from outside air leakage through structural voids, open doors, and windows, roof ventilators, open stacks, etc. – often are controlling factors. Smoke tubes (NH_4Cl) are quite useful in delineating air flow patterns for planning a sample survey.

The density of the pollutant in many cases will counteract diffusion processes and establish stratification. The phenomenon prevails when certain highly toxic and irritant gases (phosgene, mustard gas, lewisite, chlorine, etc.) are encountered. Natural convective circulation and diffusion in confined spaces such as silos, coal mines, wells, pits, tanks, etc. may be insufficient to maintain a respirable atmosphere (above 15% oxygen).

The location of sampling sites, whether on a temporary or "grab" sample basis or on a fixed station basis for continuous monitoring over long periods of time, is often critical and will determine the validity of the conclusions based on the results. In industrial atmospheres the variability may be sufficiently extreme to vitiate the results from fixed-station monitors regardless of the care taken in locating the samplers. In an alkyl lead manufacturing facility, for example, no relationship could be established between exposure indicated by urinary lead excretion analyses and numerous fixed-station samplers. Correlation was not established until battery-powered monitors in the form of a small filter-microimpinger assembly were worn to collect the airborne lead compounds in the breathing zone during the entire 8-hr shift by workmen assigned to these areas.[4]

Even greater care is required to establish uniform mixtures of pollutants in air for use as calibration mixtures, whether the system be dynamic or static in principle. Passage of gases through lengths of small-diameter tubing ensures effective turbulent mixing. Other systems employ some form of baffled chamber in the mixing line.[5] Rotating paddles or even loose pieces of sheet metal or plastic (PTFE) activated by shaking have been used in containers for mixing static mixtures. A more effective alternative mixes the contaminant with the dilution air as it is released into an evacuated container. Diffusion alone cannot be relied upon to produce a homogeneous mixture in most cases. Homogeneity can be attained directly by constant rate vapor release from permeation tubes in a dynamic flow system.[6,7]

b. Internal Standards

The internal standard technique is used primarily for emission spectrograph, polarographic, and chromatographic (liquid or vapor phase) procedures. This technique enables the analyst to compensate for electronic and mechanical fluctuations within the instrument.

In brief, the internal standard method involves the addition to the sample of known amounts of a substance to which the instrument will respond in a manner similar to its response to the contaminant in the system. The ratio of the internal standard response to the contaminant response determines the concentration of contaminant in the sample. Conditions during analyses will affect the internal standard and the contaminant identically and thereby compensate for any changes. The internal standard should be chemically similar to the contaminant, approximate the concentration anticipated for the contaminant, and be of the purest attainable quality.

A detailed discussion of the sources of physical error, magnitude of their effects, and suggestions for minimizing their contribution to determinate bias and error will be found in Reference 7.

c. Chemical Interferences

The term "interference" relates to the effects of dissolved or suspended materials on analytical procedures. A reliable analytical procedure must delineate and minimize anticipated interferences.

The investigator must be aware of possible interferences and be prepared to use an alternate or modified procedure to avoid intolerable errors. One expedient method for avoiding interference is provided by analyzing a smaller initial aliquot which may suppress or eliminate the effect of the interfering element through dilution. The concentration of the substance sought is likewise reduced; therefore, the aliquot must contain more than the minimum detectable concentration. When the results display a consistently increasing or decreasing pattern by dilution, then interference is indicated (see Section B.1.i, "Effect of Sample Size").

In general, an interfering substance may produce one of three effects:

1. React with the reagents in the same manner as the component being sought (positive interference).
2. React with the component being sought to prevent complete isolation (negative interference).
3. Combine with the reagents to prevent further reaction with the component being sought (negative interference).

The sampling and analytical technique employed for the surveillance of airborne toluene diisocyanate (TDI) in the manufacturing environment furnishes a good example in which all three factors may be encountered. The TDI vapor is absorbed and quantitatively hydrolyzed in an aqueous acetic acid-HCl mixture to toluene diamine (MTD) which then is diazotized by the addition of sodium nitrite. The excess nitrous acid is destroyed with sulfamic acid and the diazotized MTD coupled with *N*-(1-naphthyl)ethylenediamine to produce a bluish red azo dye.[8] In the phosgenation section of the operations, the starting material (MTD) may coexist with TDI in the atmosphere sampled. If so, then a positive interference will occur, as the method cannot distinguish between free MTD and MTD from the hydrolyzed TDI.

This problem can be resolved by collecting simultaneously a second sample in ethanol. The TDI reacts with ethanol to produce urethane derivatives, which do not produce color in the coupling stage of the analytical procedure. The MTD is quantitated by the same diazotization and coupling procedure after boiling off the ethanol from the acidified scrubber solution. Then the difference represents the TDI fraction in the air sampled.

On the other hand, if the relative humidity is high or alcohol vapors are present, negative interference will reduce the TDI recovered as a result of formation of the carbanilide (dimer) or the urethane derivative, which will not produce color in the final coupling stage. Alternative methods have not been developed for these conditions. If high concentrations of phenol are absorbed, then a negative interference will arise from side reactions with the nitrous acid required to diazotize the MTD. This loss can be avoided by testing for excess nitrous acid in the diazotization stage and adding additional sodium nitrite reagent if a deficiency is indicated.

An estimate of the magnitude of an interference may be obtained through the recovery procedure (see Section B.1.f, "Recovery or 'Spiked' Sample Procedures").

If recoveries of known quantities exceed 100%,

a positive interference is present (Condition 1). If the results are below 100%, a negative interference is indicated (Condition 2 or 3; see Reference 7 for details).

When treatment of the sample for removal of an interfering substance is necessary, the following approaches may solve the problem.

Distillation of the sample leaving the interference behind – With reference to the example cited for correction of interferences that react in the same manner as the component being sought, the ethanol must be boiled off before the sodium nitrite reagent is added, as even traces of ethanol will interfere in the coupling reaction.[9] The distilland must be strongly acid to prevent steam distillation of the MTD. Although a reverse application of the distillation technique, the principle is the same.

Removal of the interference by ion exchange resins – The isolation of Δ-aminolevulinic acid (ALA) from urine on cation exchange resin to eliminate interference from urea and porphobilinogen serves as a good illustration of this principle.[10] Elevated ALA levels have been proposed as an early warning system for lead intoxication.

Addition of complexing agents – The use of the cyanide and citrate ions in a strongly ammoniacal solution to prevent heavy metal ions other than bismuth, stannous tin, and thallium from reacting with dithizone in the colorimetric procedure for the determination of lead in trace quantities is a classic example of this technique.[11]

Extraction into organic solvents – Returning to the analysis for lead in urine we find a recent reference to the use of a small volume of methylisobutyl-ketone to extract and concentrate the pyrrolidine dithiocarbamate complex with lead for atomic absorption analysis. Cyanide is added as a masking agent and only large quantities of bismuth and cadmium cause interference, which can be controlled by ashing or reducing the sample size.[12]

Ashing – In the dithizone procedure for the analysis of traces of lead in urine, if (ethylenedinitrilo)-tetraacetic acid used in the treatment of lead intoxication is present, the specimen must be burned to a white ash to avoid the sequestering effect of this compound, which is not removed by iodine oxidation in the rapid screening procedure.[13]

pH adjustment – A good example of the effect of acidity is furnished by addition of sodium carbonate to the color coupling stage of the procedure for toluenediisocyanate (TDI) in air as a modification that permits use of the same system for methylene-bis-(4-phenylisocyanate) (MDI). The added alkaline buffer reduces the time for complete color development from 2 hr to 15 min.[8]

Different reaction rates – In the foregoing example a differential analysis in cases where both TDI and MDI would be expected to be present can be carried out by dividing the sample into two aliquots before adding the sodium carbonate reagent. The sodium carbonate is then added to one aliquot and the color density determined after 15 min. The unbuffered aliquot is read between 3 and 10 min. The difference will be a good approximation of the MDI fraction of the diisocyanates collected.

Change of temperature – In the preferred procedure for the analysis of tetraalkyllead compounds in the atmosphere, the iodine monochloride collecting reagent oxidizes off only two of the alkyl groups to produce dialkyl lead salts at room temperature. To convert to inorganic lead ions for analysis by the dithizone procedure, the mixture must be heated to 50°C for a minimum of 10 min. The method can be calibrated for dialkyl lead dithizonates, but this alternative is much less expedient and is subject to error from inorganic lead that also may be present.[14]

The source of the interferences should be identified and eliminated if possible. The cause may be found in the following locations:

1. Present at the sampling site.
2. Introduced during sample collection.
3. Developed in sample storage.
4. Originated in the laboratory analysis procedure.

These four sources of error may be avoided normally by applying adequate sampling and analysis techniques.

Physical, chemical, and biological interferences may be outlined as follows to assist the analyst in recognizing and correcting deviant effects.

1. Physical
   a. Heat
      1.1. Chemical equilibrium may be temperature sensitive.

2.1. Side reactions may become significant as the temperature increases.
3.1. Rate of reaction is temperature dependent. Serious errors may be encountered at low temperatures. An example is iodometric oxidation of tetraethyl lead to inorganic lead iodide.

b. Light
1.1. Photodecomposition – generation of iodine in potassium or sodium iodide solutions.
2.1. Photooxidation – generation of yellow color in perfluoroisobutylene reagent by ultraviolet rays (positive error).
3.1. Fading of detector reagent colors (lead and mercury dithizonates – negative error).

c. Time
1.1. Half-life of the component analyzed:
aa. In the atmosphere sampled.
bb. In storage after collection and before analysis (deterioration rate).
2.1. Fading of dyes in colorimetric procedures after maximum optical density attainment.
3.1. Deterioration of reagents.
4.1. Time required for a chemical reacting system to reach equilibrium.
5.1. Optimum sampling rate. Collection efficiency in a liquid reagent or on a solid support is rate dependent. Change in sample rate through length of stain detectors without recalibration will produce serious errors.

d. Contamination
1.1. Sampling equipment during use in high concentration areas (aniline in rubber tubing).
2.1. Extraneous dust and dirt.
aa. Failure to keep collection equipment and reagents stoppered.
bb. Cross-contamination in recharging liquid reagents or changing filters, especially in the field.
cc. Electrolytes in conductivity and pH meters.
3.1. Impure reagents – high background blanks.
4.1. Biological
aa. Algal growth
bb. Insects
cc. Waste products (animal and insects)
5.1. Background "noise" in electronic circuits.

2. Chemical
a. Humidity
1.1. Reaction of the contaminant with H_2O (e.g., TDI, $COCl_2$, etc.).
2.1. Reaction of the collection medium with H_2O (e.g., anhydrous reagents for $COCl_2$).
3.1. Dilution of the collection medium (e.g., H_2SO_4 reagent for formaldehyde).

b. pH control in aqueous systems
1.1. pH-sensitive collection system
2.1. pH-sensitive color development
3.1. Buffered vs. unbuffered systems

c. Interferences
1.1. Negative (e.g., NH_3 in Hg detector tubes) – redox cancellations (e.g., SO_2 in the colorimetric H_2S procedure and ozone in the colorimetric SO_2 procedure).
2.1. Positive
aa. Different shade or color produced (e.g., dithizone + oxidizing agents).
bb. Reaction same – result increased (e.g., ozone in nitrogen oxides system). Color intensified by interference (e.g., formaldehyde in acrolein procedure).

d. Concentration effects
1.1. Adjust concentrations of reagents to attain maximum effect (e.g., color density) – ratios often are not stoichiometric.
2.1. Beer's law may not be applicable at high or very low concentrations, i.e., attempting to operate outside the

optimum range for which the method was developed.

3.1. Effect on reaction rate and equilibrium (e.g., formaldehyde + fuchsin colorimetric reagent).

4.1. Optimum sample size. Attempts to increase sensitivity by increasing sample size may produce serious errors (e.g., repetitive sample aliquots through the benzene length of stain detector tube requires recalibration after second aliquot).

e. Catalytic effects

1.1. Decomposition on contact with tubing and container surfaces (e.g., O_3 and H_2O_2 on metal surfaces).

2.1. Promotion of undesirable side reactions.

3.1. Failure to react in the absence of a catalyst.

4.1. Surface reactions – porous glass bubblers and glass bead packed absorbers.

5.1. Inhibition of reactions (e.g., brucine alkaloid prevents oxidation of SO_2 by atmospheric O_2).

C. INDETERMINATE ERROR

1. Statistical Evaluation

a. Standard Deviation (σ)

When indeterminate or experimental errors occur in a random fashion, the observed results (x) will be distributed at random around the average ($\bar{x}$). The average usually is referred to as the arithmetic mean. The sum of all results divided by the number of results (n) equals the mean. A median refers to the result that lies exactly in the center of the results tabulated in order of ascending or descending magnitude. The result which occurs most frequently is designated the mode.

Given an infinite number of observations, a graph of the relative frequency of occurrence plotted against magnitude will describe a bell-shaped curve known as the Gaussian or normal curve (Figure IV.6). However, if the results do not occur in a random fashion the curve may be flattened (no peak), skewed (unsymmetrical), narrowed, or exhibit more than one peak (multimodal). In these cases the arithmetic mean will be misleading, and unreliable conclusions with respect to deviation ranges (σ) will be drawn from the data. A typical graph illustrating skew, multimodes, and a narrow peak is shown in Figure IV.7.

In any event, the investigator should confirm the normalcy of the data at hand by plotting magnitude vs. frequency before proceeding with the calculation of standard deviation, which is the most universally applied, fundamental tool of statistics. The normal curve (Figure IV.6) is completely defined by two statistically fundamental parameters: (1) the mean (arithmetic average), $\bar{x}$, of n observations, and (2) the standard deviation, σ, which determines the width or spread of the curve on each side of the mean. The relationship is further defined by

$$\sigma = (\Sigma(x - \bar{x})^2/n - 1)^{1/2}$$

when x = observed result, $\bar{x}$ = mean of all results, n = number of determinations, and Σ = sum of $(x - x)^2$.

The distribution of results within any given range about the mean is a function of σ. The proportion of the total observations that reside within $\bar{x} \pm 1\sigma$, $\bar{x} \pm 2\sigma$, and $\bar{x} \pm 3\sigma$ has been thoroughly established and is delineated in Figure IV.1. Although these limits do not define exactly any finite sample collected from a normal group, the agreement with the normal limits improves as n increases. As an example, suppose an analyst were to analyze a composite urine specimen 1,000 times for lead content. He could reasonably expect that 50 results would exceed $\bar{x} \pm 2\sigma$ and only 3 results would exceed $\bar{x} \pm 3\sigma$. However, the corollary condition presents a more useful application. In the preceding example, the analyst has found $\bar{x}$ to be 0.045 mg/l with $\sigma = \pm$ 0.005 mg/l. Any result that would fall outside the range 0.035 to 0.055 mg/l (0.045 $\pm 2\sigma$) would be questionable, as the normal distribution curve indicates this should occur only 5 times in 100 determinations. This concept provides the basis for tests of significance.

b. The t Distribution (Student's t)

In the following development of statistical methods the term "sample" refers to a group of observations or analytical results rather than a single finite fraction of the entity that has been observed or analyzed. The statistician regards ten

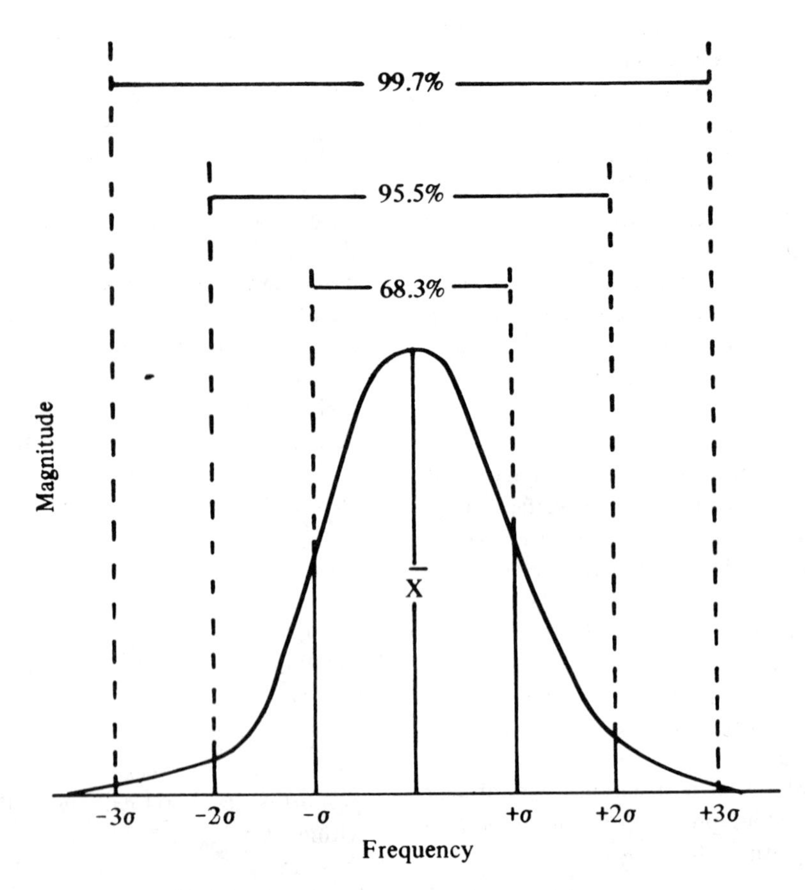

FIGURE IV.6. Gaussian or normal curve of frequencies.

replicate determinates of the sulfur dioxide content of an ambient atmosphere as a small sample, whereas the analytical chemist would consider analysis of a single 10-ml aliquot of that atmosphere as a small sample.

The concept of normal distribution was developed from large bodies of data and does not necessarily apply to small numbers of determinations. This weakness was recognized by W. S. Gosset, who published under the pseudonym "Student." Unless a large sample is used, the true σ and the true $\bar{x}$ cannot be established. However, the theoretical distribution of the average of a small sample ($\bar{x}_t$) drawn from a normal distribution can be derived by replacing σ with the sample standard deviation (s) and using a new distribution which is independent of σ. Then $t = \frac{(\bar{x}_t - \bar{x})n^{1/2}}{s}$ and is dependent only on the size of the sample (n). The curve for t is more peaked in the center and has higher tails than the normal distribution. For numbers greater than 30 the normal distribution is used.[15]

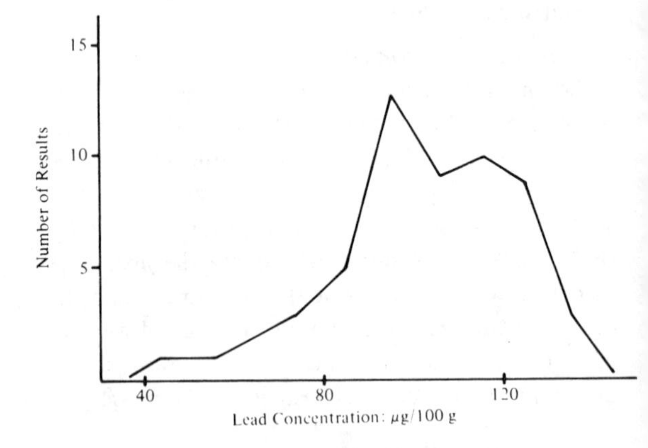

FIGURE IV.7. Frequency distribution of lead in blood — analytical results. Data are from 16 laboratories.[18]

c. Confidence Limits

When the distribution is normal, the confidence limits for any given parameter can be determined when σ has been established for a particular large sample with size n. When this sample yields a mean of $\bar{x}_t$, we are 95% confident that the true mean value for the population will be found within the limits $\bar{x}_t \pm 2\ \sigma/n^{1/2}$. These values are known as the 95% confidence limits (C.L.), and provide an estimate of the mean. They can also forecast the number of observations needed to ensure precision within prescribed limits.

The standard deviation may not be known, or may be estimated from a small number n of determinations. Then the 95% confidence limit for the mean for n can be calculated from the expression $\bar{x}_t \pm t_{.025}\ \sigma/n^{1/2}$, where t (Student) compensates for the tendency to underestimate variability and is equivalent to[16]

n	$t_{.025}$
2	12.71
3	4.30
4	3.18
5	2.78
10	2.26
∞	1.96

d. Range

The difference between the maximum and minimum of n results (range) also is related closely to σ. The range (R) for n results will exceed σ multiplied by a factor d_n only 5% of the time when normal distribution of errors prevails.

Values for d_n:

n	d_n
2	2.77
3	3.32
4	3.63
5	3.86
6	4.03

Since the practice of analyzing replicate (usually duplicate) samples is a general practice, application of these estimated limits can provide detection of faulty technique, large sampling errors, inaccurate standardization and calibration, and personal judgment and other determinate errors. However, resolution of the question, whether the error occurred in sampling or in analysis, can be done more confidently when single determinations on each of three samples rather than duplicate determinations on each of two samples are made. This approach also reduces the amount of analytical work required.[2] Additional information relative to the evaluation of the precision of analytical methods will be found in the ASTM standards.[17]

e. Rejection of Questionable Results

The question whether or not to reject results that deviate greatly from $\bar{x}$ in a series of otherwise normal (closely agreeing) results frequently arises. On a theoretical basis, no result should be rejected, as one or more determinate errors that render the entire series doubtful may be involved. Tests that are known to involve mistakes should not be reported as analyzed. A mathematical basis for rejection of "outliers" from experimental data may be found in statistics text books.[15]

f. Correlated Variables – Regression Analysis

A major objective in scientific investigations is the determination of the effect that one variable exerts on another. For example, a quantity of sample (x) is reacted with a reagent to produce a result (y). The quantity x represents the independent variable over which the investigator can exert control.

The dependent variable (y) is the direct response to changes made in x, and varies in a random fashion about the true value. If the relationship is linear, the equation for a straight line will describe the effect of changes in x on the response y: $y = a + bx$, in which a is the intercept with the y axis and b is the slope of the line (the change in y per unit change in x). In chemical analysis a is a measure of constant error arising from a colorimetric determination, trace impurity, blank, or other determinate source. The slope b may be controlled by reaction rate, equilibrium shift, or the resolution of the method. The term "regression analysis" is applied to this statistical tool.

A typical application is exhibited in Figure IV.8, which relates the concentration of lead in blood to the standard deviation of the method.[18] For this relationship, $y = 0.0022 + 0.054x$. Additional useful information can be obtained by certain transformations and shortcuts.[2,15,16,19]

g. Fitting Data to a Straight Line by Least Squares Method

Frequently the results when plotted do not clearly define the true position or slope of a straight regression line (Figure IV.8). In this case the method of least squares will fit a straight line that is not dependent on the investigator's judg-

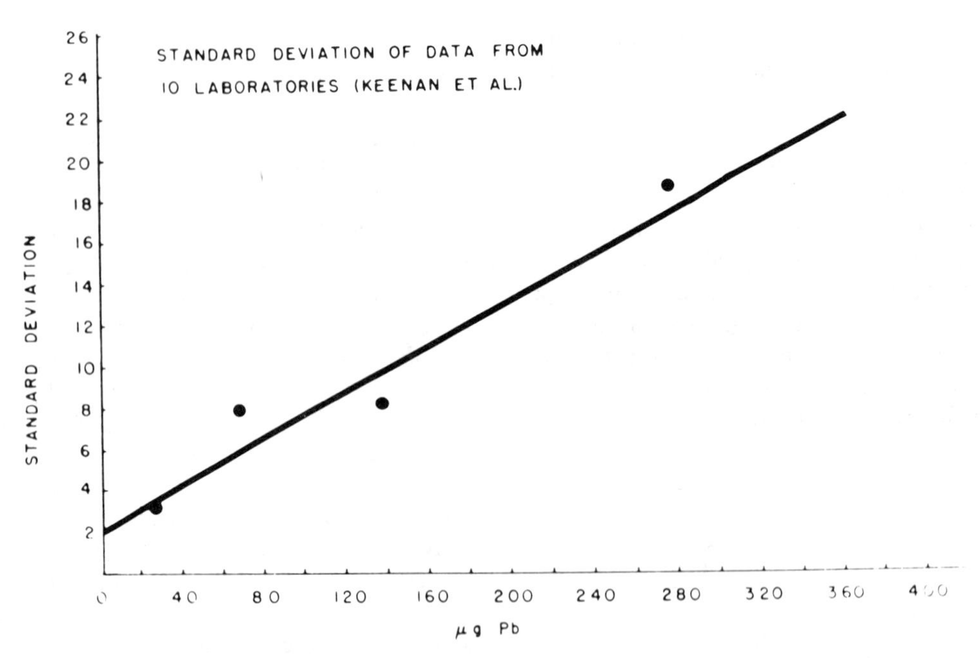

FIGURE IV.8. Fitting data to a straight regression line – standard deviation of data.

ment of the scatter of points. Additional functions are substituted in the equation for a straight line:

$$y = a + bx \qquad a = \frac{\Sigma y(\Sigma x^2) - (\Sigma xy)(\Sigma x)}{n(\Sigma x^2) - (\Sigma x)^2}$$

$$b = \frac{n(\Sigma xy) - \Sigma x(\Sigma y)}{n(\Sigma x^2) - (\Sigma x)^2}$$

where n = number of sets of x and y and Σ = sum of x, y, x^2, and xy included in n. After a and b have been calculated, substitute three convenient values for x in the equation and plot the points that will fall in a straight line if no errors are made in the calculations. The actual data should be plotted also to disclose the scatter on each side of the regression line. Examples and additional explanation are available in standard statistical texts.[2,15,16] In many cases, visual inspection will locate the regression line with reasonable assurance, especially if the constant a is known to be zero (Figure IV.9).[4]

2. Graphic Analysis for Correlations

Useful shortcuts may be elected to determine whether a significant relationship exists between x and y factors in the equation for a straight line (y = a + bx). The data are plotted on linear cross section paper and a straight line drawn by inspection through the points with an equal number on each side or fitted by the least squares method previously described (see Reference 19 for quick solution). If the intercept a must be zero, the fitting is greatly simplified. Then on each side equidistant from this line parallel lines are drawn corresponding to the established deviation (σ) of the analytical procedure, the points falling inside of the band formed by the ± σ lines are tallied up, and the correlation (conformance = number within band x 100 per total points plotted) calculated. This technique is illustrated in Figure IV. 9, which was used to relate urinary lead excretion to the airborne lead concentration obtained by personnel monitor surveys.[4] In this case more than one TLV was involved, so the TLV coefficient (TLVC) transformation was used for estimation of total lead exposure; $\text{TLVC} = \frac{\text{alkyl Pb found}}{\text{TLV}} + \frac{\text{inorganic Pb found}}{\text{TLV}}$. A plot of the monthly coefficients vs. corresponding average urinary excretion disclosed only a 69% conformance, whereas a plot of the previous month's TLVC's vs. the current month's average urinary excretion gave a 78% conformance. Furthermore, inspection of the chart indicated most of the "outliers" were contributed by the furnace crew. Deletion of this group raised conformance to 86% for the balance of the operation.[4] Correlations above 80% are considered quite good (see also Reference 19).

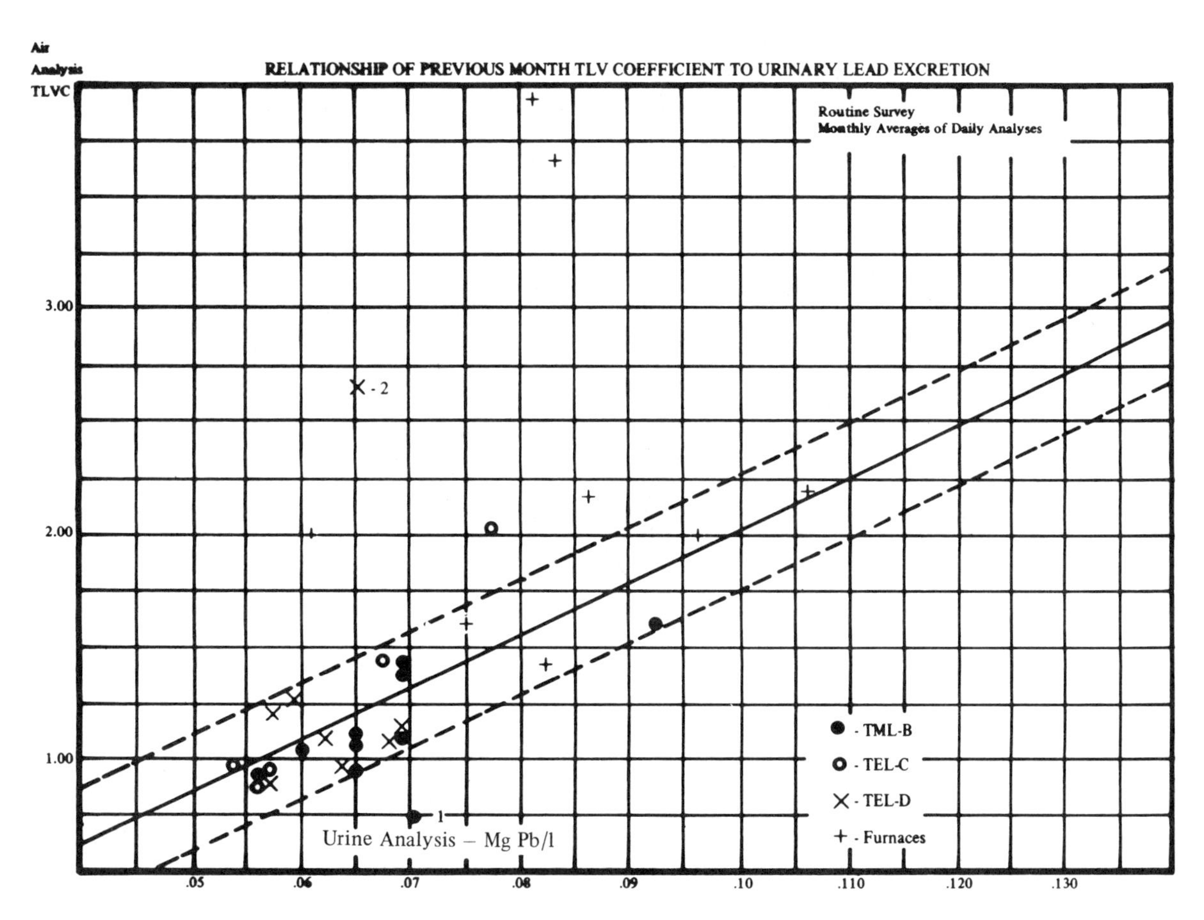

FIGURE IV.9. Relationship of previous month TLV coefficient to a urinary lead excretion regression curve. (Reproduced from Linch, A. L., Wiest, E. G., and Carter, M. D., *Am. Ind. Hyg. Assoc. J.*, 31, 170, 1970. With permission.)

Curvilinear functions can be accommodated, especially if a log normal[15] function is involved, and a plot of the data on semi-log paper yields a straight line.[9] Log-log paper also is available for plotting complex functions.

A combination of curvilinear and bar charts in some cases will reveal correlations not readily detected by mathematical processes. The data derived from an industrial cyanosis control program[20] illustrate an application that revealed a rather significant relationship between abnormal blood specimens and the frequency of cyanosis cases on a long-term basis (Figure IV. 10). In fact, one trend line could be fitted to both variables, and the predicted ultimate improvement was attained in 1966 when abnormal blood specimens dropped below 2% and the cyanosis cases below 4.[21]

Grouping data on a graph and approximating relationships by the quadrant sum test (rapid corner test for association) can provide useful results with a minimum expenditure of time.[19,22]

In those cases where application of mathematical tools is tedious or completely impractical, a system of ranking is sometimes applicable to the restoration of order out of chaos. Referring again to the cyanosis control program, a relationship between causative agent structure and biochemical potential for producing cyanosis and anemia was needed. Ten common factors (categories) for each of the 13 compounds under study had been recognized. The 13 compounds were ranked in each category in reverse order of activity (No. 1 most active, No. 13 least active) and the sum of the rankings obtained for each compound. These sums then were divided by the number of categories used in the total ranking to obtain the "score." The scores were then arranged in increasing numerical order in columnar form. The most potent cyanogenic and anemiagenic com-

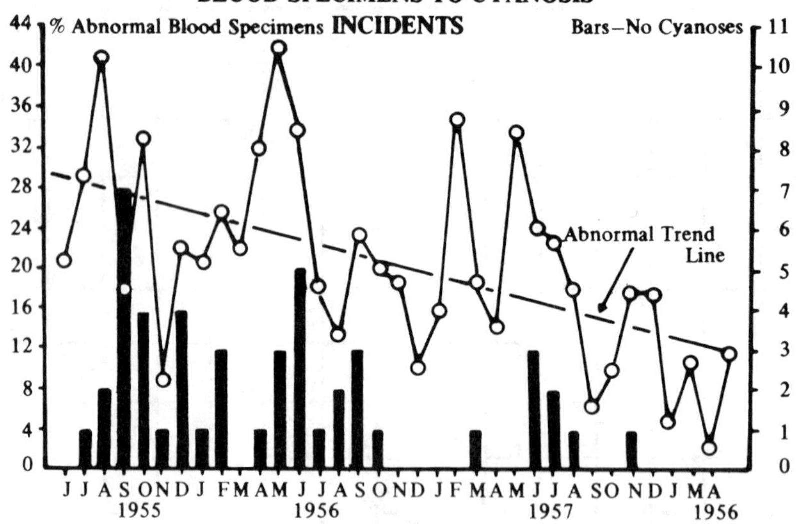

FIGURE IV.10. Relationship of abnormal blood specimens to cyanosis incidents. (Reproduced from Wetherhold, J. M., Linch, A. L., and Charsha, R. C., *Am. Ind. Hyg. Assoc. J.*, 20, 396, 1956. With permission.)

pounds then appeared at the top of the table and the least potent at the bottom.[21]

3. Routine Analysis Control

a. Internal

In addition to the use of internal standards, recovery procedures, and statistical evaluation of routine results, the laboratory should subscribe to a reference sample service to confirm precision and accuracy within acceptable limits. Apparatus should be calibrated directly or by comparison with National Bureau of Standards (NBS) certified equipment or its equivalent, reagents should meet or exceed ACS standards, and calibration standards should be prepared from AR (analytical reagent) grade chemicals[23] and standardized with NBS standards if available. To illustrate, in a laboratory engaged in an exposure control program based on biological monitoring by trace analysis of blood and urine for lead content, at least two calibration points, blanks, and a recovery should be included in each batch analyzed by the dithizone procedure. In addition, the wavelength integrity and optical density response of the spectrophotometer should be checked and adjusted if necessary by calibration with NBS cobalt acetate standard solution. Until standard deviation for the analytical procedure has been established within acceptable limits, replicate determinations should be made on at least two samples in each batch (either aliquot each sample or analyze duplicate samples) and thereafter with a frequency sufficient to ensure continued operation within these limits.

1. The Control Chart

Control charts, which are probably the most widely recognized application of statistics, provide "instant" quality control status when plotted daily or within another interval sufficiently short to disclose trends without undue oscillations from overrefinement of the data. Examples selected from a lead surveillance program illustrate the value of control charts. Figure IV. 11 for analytical control is based on recoveries of known quantities of lead added to blood. From this curve several conclusions may be drawn:

1. Background ("natural") lead concentrations lay very close to the ultimate sensitivity of the method (35 ± 5 μg).
2. The variability of the background lead concentration exerts a relatively strong controlling effect on the recovery.
3. Although only a short period is covered, a downward trend is noticeable.
4. A control limit set at 98% ± 5% probably is more realistic.

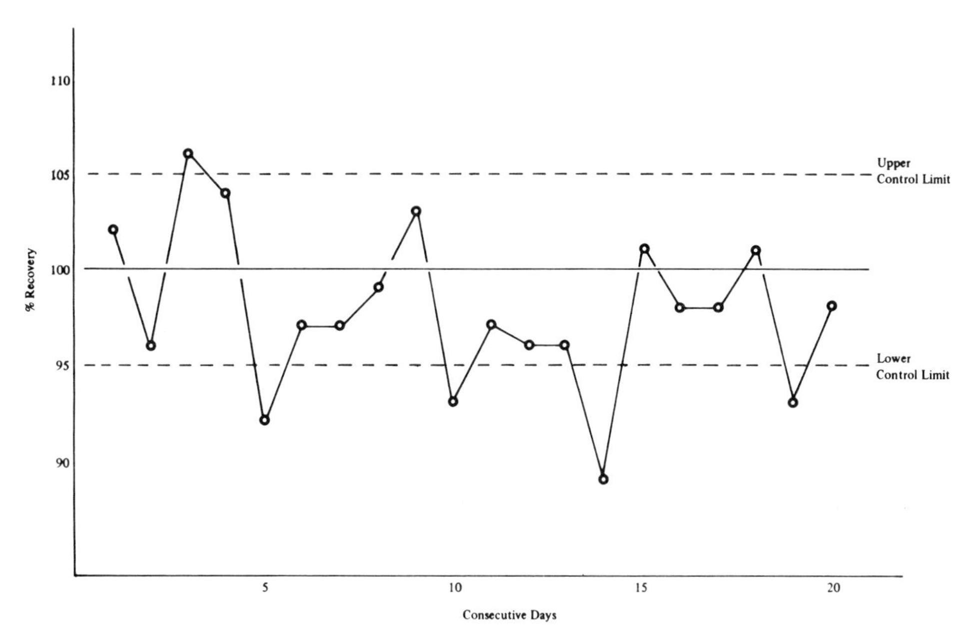

FIGURE IV.11. Control chart – recovery of lead from blood.

The same technique was applied to the evaluation of the quality of an exposure control program. A 1-year section from the control chart is presented in Figure IV. 12. The graph provided several significant conclusions upon which action was initiated:

1. An alleged bias in the technicians' performances was ruled out, as each had about the same number of peaks and valleys during the period (each technician in turn analyzed all of the urine specimens for the entire week plotted).
2. Trend lines which were drawn in by inspection disclosed a much closer correlation with production rate than with an alleged seasonal (temperature) cycle.
3. The peaks in the short-term oscillations were connected with particular rotating shift crews who engaged in "dirty" work habits that were corrected from time to time.
4. No correlation could be established with fixed-station air analysis data.

These examples are but two applications of a very extensive specialty within the field of statistics; therefore, the reader is referred to standard texts for additional information on refinements and procedures for extracting significant information from control charts.[15,24] On the basis of its raw simplicity, amount of information available for a minimum expenditure of time and effort, graphic presentation, and ease of comprehension, the control chart cannot be overrecommended.

b. Interlaboratory Reference Systems

Participation in interlaboratory studies, whether by subscription from a certified commercial laboratory supplying such a service or from a voluntary program initiated by a group of laboratories in an attempt to improve analytical integrity,[18] is highly recommended. Evaluation of the analytical method as well as evaluation of the individual laboratory's performance can be derived by specialized statistical methods applied to the data collected from such a study. However, inasmuch as most investigators will not be called upon to conduct or evaluate interlaboratory surveys, the reader is referred to the literature in the event such specialized information is needed.[18,24-26] In the absence of such programs, the investigator or laboratory supervisor should make every effort to locate colleagues engaged in similar sampling and analytical activity and arrange exchange of standards, techniques, and samples to establish integrity and advance the art.

D. QUALITY CONTROL PROGRAMS

1. Accreditation Requirements

A summary of Sections B, "Determinate Error,"

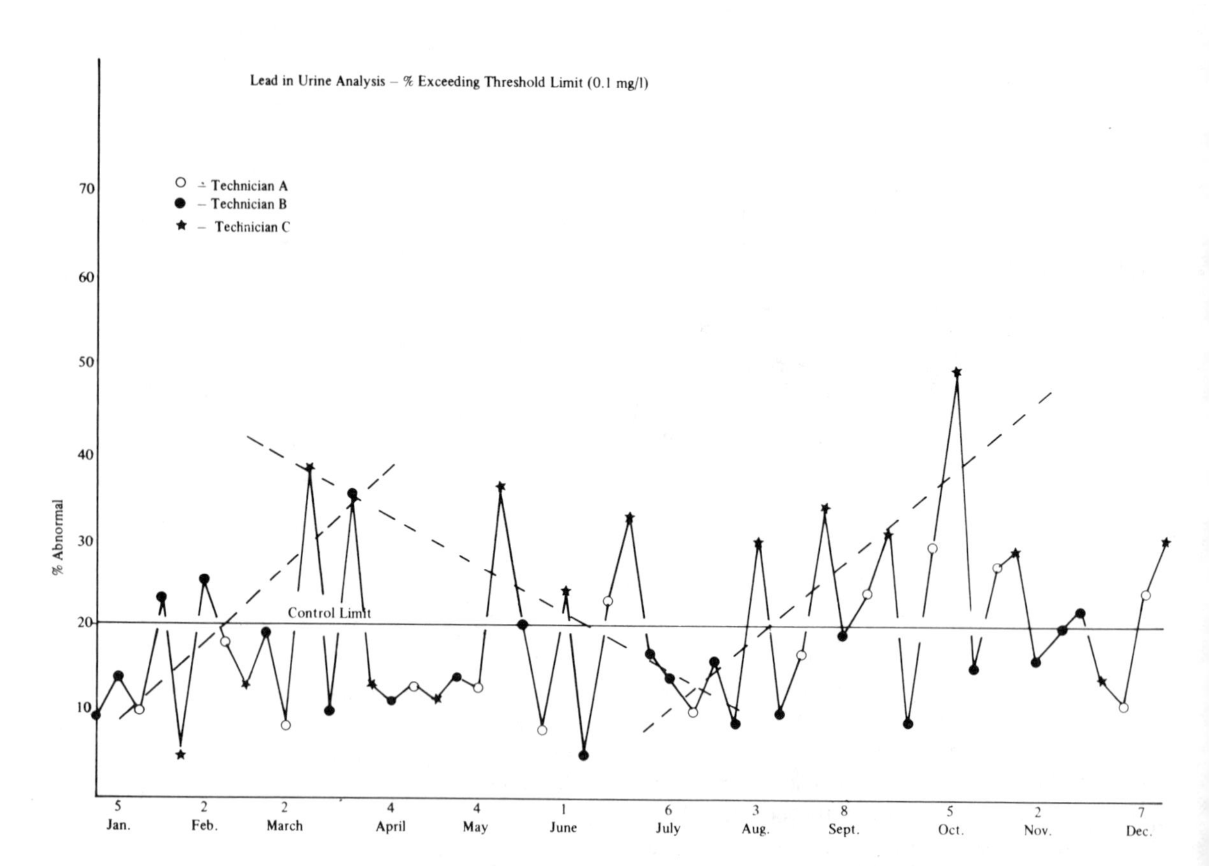

FIGURE IV.12. Control chart – lead in urine analysis.

and C, "Indeterminate Error," provides the foundation for construction of a quality control (QC) program which must be established before certification as an Industrial Hygiene Laboratory can be considered. Guidelines established by the American Industrial Hygiene Association for Accreditation of Industrial Hygiene Analytical Laboratories delineate the minimum requirements as follows:

Quality Control and Equipment

Routine quality control procedures shall be an integral part of the laboratory procedures and functions. These shall include:

1. Routinely introduced samples of known content along with other samples for analyses.
2. Routine checking, calibrating, and maintaining in good working order of equipment and instruments.
3. Routine checking of procedures and reagents.
4. Good housekeeping, cleanliness of work areas, and general orderliness.
5. Proficiency testing – The following criteria shall be used in the proficiency testing of industrial hygiene analytical laboratories accredited by the American Industrial Association.
   a. Reference laboratories – Five or more laboratories shall be designated as reference laboratories by the American Industrial Hygiene Association based on the appraisal of competence of the laboratories by the Association. The reference laboratories may be judged competent:

      a. in all industrial hygiene analyses
      b. specific industrial hygiene analyses.

   Proficiency samples shall be sent to designated reference laboratories for analyses. These data will be used for grading analytical data received from laboratories seeking or maintaining accreditation by the Association.
   b. Method of grading – Laboratories shall be graded on the basis of their ability to perform analyses within specified limits determined by the reference laboratories. Satisfactory performance shall be the reporting of results within two standard deviations of the mean value

obtained by the reference laboratories. Exception shall be made in cases where too few laboratories are in existence, a new procedure has not been adequately tested, or the range in variation from reference laboratories is too great to apply this method of grading.

c. Number and suitability of samples – Samples shall be either environmental materials, biological fluids, or tissues or synthetic mixtures approximating these. They shall be packaged, as nearly as possible, in an identical manner and the containers will be chosen so as to avoid exchange of the test material between the samples and container. Samples shall be analyzed by each participating laboratory in sufficient number and at proper intervals for the results to form an adequate basis for accreditation in the opinion of AIHA.
d. Frequency of samples – Samples shall be submitted to each laboratory quarterly.
e. Satisfactory performance – Satisfactory performance is a considered scientific judgment and is not to be judged exclusively by any inflexible set of criteria. The judgment shall be made, however, on the basis of the results submitted by the laboratories and a statistical estimation of whether the results obtained are probably representative of analytical competence considering inherent variables in the method.

6. Records – The industrial hygiene analytical laboratory shall maintain records and files proper and adequate for the services given. These shall include:

   a. The proper identification and numbering of incoming samples.
   b. An adequate and systematic numbering system relating laboratory samples to incoming samples.
   c. An adequate record system on internal logistics of each sample including date of incoming sample, analysis and procedures, and reporting of data.
   d. A records checking system of the calibration and standardization of equipment and of internal control samples.

2. Techniques

A basic quality control program requires close attention to:

1. Evaluation of the equipment performance and calibration to assure accuracy. As applied to colorimetric procedures, the choices in order of increasing reliability are

   a. Precalibrated curves – use of a curve supplied with the instrument or prepared in the laboratory for an indefinite period is expedient, but may be invalid as instrument or procedure deviations are not detected.
   b. Precalibrated curves occasionally compared with a standard – these are somewhat better, but only the instrument response is checked.
   c. Precalibrated curves with a standard for each determination – changes in instrument response are immediately detected, and application of correction factors will improve the accuracy. However, this approach does not evaluate the procedure.
   d. No precalibration curves: use of standards with each determination – analytical results may be either read from a plotted calibration curve or by direct calculation. Although the exact state of the instrumentation can be confirmed, no knowledge of overall analytical performance is obtained.
   e. Known control added to the standard – introduction of a known control in addition to the standard applied in d above will provide a check on the entire analytical procedure (see Section B.1.f, "Recovery or 'Spiked' Sample Procedures"). When the control simulates or duplicates the environmental sample, this combination will deliver the best accuracy, especially if a calibration curve is prepared for each series of samples (Table IV.1).

2. Evaluation of the present state of the art. Is the best analytical method used? (See Section B.1.h, "Change in Methodology.")

3. Evaluation of the expected range of normal analytical results. A standard deviation is calculated (see Sections C.1.a, "Standard Deviation," and C.1.d, "Range") and assigned on the basis of

satisfactory precision for a surveillance program. The mean range R between duplicate analyses then may be calculated from the standard deviation of the analysis.

4. Evaluation of the precision of the analytical procedure. Duplicate analyses are employed for the determination and control of precision within the laboratory and between laboratories. Approximately 20% of the routine samples with a minimum of 20 samples should be analyzed in duplicate to establish internal reproducibility.
5. Establishment of control charts. The control chart provides a tool for distinguishing the pattern of indeterminate (stable) variation from the determinate (assignable cause) variation. This technique displays the test data from a process or method in a form that graphically compares the variability of all test results with the average or expected variability of small groups of data – in effect, a graphical analysis of variance, and a comparison of the "within group" variability vs. the "between group" variability (see Section D.3, "Application").
6. Providing appropriate log sheets and procedures for sample control in the laboratory. Samples are shown in Figures IV.13 and IV.14 and Table IV.1.
7. Correlation of quantitatively related data to provide confirmation of accuracy, and evaluation of quality control analyses as produced. A standard or a repeatedly analyzed control, if available, should be included periodically for long-term accuracy control. The control chart technique is directly applicable, and appropriate control limits can be established by

Procedure – Ash ________

BLOOD
ANALYSIS URINE SPOT

Date Collected ________
Date Started ________
Date Competed ________

IDENTIFICATION									LEAD CONTENT		
NAME	NO.	LOCATION		NO.	SP. GRAV.	VOL. / WT.	POT TRANS	Pb in SAMPLE	Mg/Liter	Mg/100	
				Blank							
				Known Calib.							
				Cobalt Std.							
				Re-covery	1.0						
					1.0						
					1.0						
					1.0						
					1.0						
					1.0						
					1.0						
					1.0						
					1.0						
					1.0						
					1.0						
					1.0						
					1.0						
					1.0						
					1.0						
					1.0						
					1.0						
					1.0						
					1.0						

FIGURE IV.13. Analytical report form – blood and urine analysis.

URINE–CYANOSIS WORK SHEET

Instrument Identification –

Cobalt Standard Reading – %T =

Calibration Point Readings – %T =

Note: Blank internally compensated: Sample read against the blank

Area:

BUILDING EXPOSURE

Procedure No. ______ Analyzed As ______ Date ______

Name	Area Code	Number	Voided		Specific Gravity	Vol. ml.	Cell Size	Pct Trans			Remarks
			Date	Time					mmg/50 ml	mg/l.	
					1.0						
					1.0						
					1.0						
					1.0						
					1.0						
					1.0						
					1.0						
					1.0						
					1.0						
					1.0						
					1.0						
					1.0						
					1.0						
					1.0						
					1.0						
					1.0						
					1.0						
					1.0						
					1.0						
					1.0						
					1.0						

PERSONAL AND CONFIDENTIAL

FIGURE IV.14. Analytical report form – urine-cyanosis work sheet.

arbitrarily subgrouping the accumulated results or by using appropriate estimates of precision from an evaluation of the procedure.

3. Application

In order for quality control to provide a means of separating the determinate from indeterminate sources of variation, the analytical method must clearly emphasize those details that should be controlled to minimize variability. A check list would include:

1. Sampling procedures
2. Preservation of the sample
3. Aliquoting methods
4. Dilution techniques
5. Chemical or physical separations and purifications
6. Instrumental procedures
7. Calculation and reporting results

The next step to be considered is the application of control charts for evaluation and control of these unit operations. Decisions relative to the basis for construction of a chart are required:

1. Choose the method of measurement.
2. Select the objective:

   a. Precision (Figure IV.4) or accuracy evaluation (Figure IV.11)

   b. Observation of test results or the range of results

   c. Measurable quality characteristics (Figure IV.4), fraction defective (Figure IV.11), or number defects per unit (Figure IV.12)
3. Select the variable to be measured (from the check list).
4. Establish the basis of a subgroup, if used:

   a. Size – a subgroup size of n = 4 is frequently recommended. The chance that small changes in the process average remain undetected decreases as the statistical sample size increases.

b. Frequency of subgroup sampling – changes are detected more quickly as the sampling frequency is increased.

5. Control limits (CL) can be calculated, but judgment must be exercised in determining whether or not the values obtained satisfy criteria established for the method, i.e., does the deviation range fall within limits consistent with the solution or control of the problem? After the mean ($\overline{X}$) of the individual results (X) and the mean of the range (R) of replicate result differences have been calculated, then CL can be calculated from data established for this purpose (Table IV.4).[6]

TABLE IV.4

Factors for Computing Control Chart Lines

Observations in subgroup (n)	Factor A_2	Factor d_2	Factor D_4	Factor D_3
2	1.88	1.13	3.27	0
3	1.02	1.69	2.58	0
4	0.73	2.06	2.28	0
5	0.58	2.33	2.12	0
6	0.48	2.53	2.00	0
7	0.42	2.70	1.92	0.08
8	0.37	2.85	1.86	0.14

Grand mean $(\overline{\overline{X}}) = \overline{X}/k$
CL's on mean $= \overline{\overline{X}} \pm A_2$
Range $(\overline{R}) = R/k$, or $d_2\sigma$
Upper control limit (UCL) on range $= D_4\overline{R}$
Lower control limit (LCL) on range $= D_3\overline{R}$

Where k = number of subgroups, A_2, D_4, and D_3 are obtained from Table IV.4. $\overline{R}$ may be calculated directly from the data, or from the standard deviation (σ) using factor d_2. The lower control limit for R is 0 when $n \leqslant 6$.

The calculated CL's include approximately the entire data under "in control" conditions, and, therefore, are equivalent to $\pm 3\sigma$ limits, which are commonly used in place of the more laborious calculation. Warning limits (WL) set at $\pm 2\sigma$ limits (95%) of normal distribution included serve a very useful function in quality control (see Figures IV.4 and IV.11). The upper warning limit (UWL) can be calculated by:

$$UWL = \overline{R} + 2\sigma_R = 2.51\,\overline{R}$$

4. Construction of Control Charts

a. Precision Control Charts

The use of range (R) in place of standard deviation (σ) is justified for limited sets of data, $n \leqslant 10$, since R is approximately as efficient and is easier to calculate. The average range ($\overline{R}$) can be calculated from accumulated results, or from a known or selected σ ($d_2\sigma$). Lower control limit (LCL) = 0 when $n \leqslant 6$.

The steps employed in the construction of a precision control chart for an automatic analyzer illustrate the technique (Table IV.5):

1. Calculate R for each set of side-by-side duplicate analyses of identical aliquots.

2. Calculate $\overline{R}$ from the sum of R values divided by the number (n) of sets of duplicates.

3. Calculate the upper control limit (UCL) for the range:

$$UCL = D_4\overline{R}$$

Since the analyses are in duplicates, $D_4 = 3.27$ (from Table IV.4).

4. Calculate the upper warning limit (UWL):

$$UWL = \overline{R} + 2\,\sigma_R = \overline{R} \pm 2/3(D_4\overline{R}) = 2.51\ R$$

(D_4 from Table IV.4) which corresponds to the 95% confidence limits.

5. Chart $\overline{R}$, UWL, and UCL on an appropriate scale, which will permit addition of new results as obtained, as shown in Figure IV.15 and Table IV.5.

6. Plot results and take action on out-of-control points.

b. Accuracy Control Charts – Mean or Nominal Value Basis

$\overline{X}$ charts simplify and render more exact the calculation of control limits (CL), since the data distribution which conforms to the normal curve can be completely specified by $\overline{X}$ and σ. Stepwise construction of an accuracy control chart for the automatic analyzer, based on duplicate sets of results obtained from consecutive analysis of knowns, serves as an example (Table IV.6):

1. Calculate $\overline{X}$ for each duplicate set.

2. Group the $\overline{X}$ values into a consistent reference scale (in groups by orders of magnitude for the full range of known concentrations).

TABLE IV.5

Precision (Duplicates) Data

Date	Data			R
9/69	# 8	25.1	24.9	0.2
	#16	25.0	24.5	0.5
	#24	10.9	10.6	0.3
10/69	# 7	12.6	12.4	0.2
	#16	26.9	26.2	0.7
	#24	4.7	5.1	0.4
2/70	# 6	9.2	8.9	0.3
	#12	13.2	13.1	0.1
	#16	16.2	16.3	0.1
	#22	8.8	8.8	0.0
4/70	# 6	14.9	14.9	0.0
	#12	17.2	18.1	0.9
	#18	21.9	22.2	0.3
5/70	# 6	34.8	32.6	2.2
	#12	37.8	37.4	0.4
6/70	# 6	40.8	39.8	1.0
	#10	46.0	43.5	2.5
	#17	40.8	41.2	0.4
	#24	38.1	36.1	2.0
7/70	# 6	12.2	12.5	0.3
	#12	25.4	26.9*	1.5
	#18	20.4	19.8	0.6

$\overline{R}$ = 14.9/22 = 0.68
UCL = 3.27 x 0.68 = 2.2
UWL = 2.51 x 0.68 = 1.7

*Bad spike at top of peak

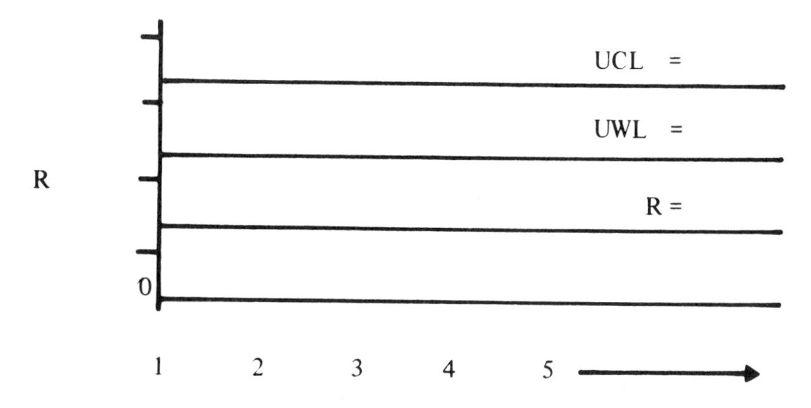

Order of Results
(e.g., duplicate sample sets)

FIGURE IV.15. Precision control chart.

TABLE IV.6

Accuracy Data

Date	Calibration range	Nominal (N)	Values	$\overline{X}$	N–$\overline{X}$
9/69	10–400 ppm	100 ppm	22.9, 21.5/	22.2	–0.7
	1.7–69.7 scale	22.9	22.7, 22.3	22.5	–0.4
10.69	10–400	100	21.6, 21.3/	21.5	0.0
	1.5–67.6	21.5			
2/70	10–400	100	23.6, 24.1/	23.9	–0.6
	1.4–62.5	24.5			
4/70	10–400	100	25.8, 26.5/	26.2	+0.2
	1.6–59.4	26.0	26.0, 26.7	26.4	+0.4
5/70	10–150	100	72.2, 70.2/	71.2	+1.2
	6.3–83.0	70.0			
6/70	10–150	100	71.0, 70.8/	71.1	+0.1
	6.6–85.0	71.0	71.0, 71.3	71.2	+0.2
7/70	10–150	60	14.9, 14.7/	14.8	–0.2
	1.8–33.5	15.0	15.1, 14.4	14.8	–0.2

3. Calculate the upper control limit (UCL) and lower control limit (LCL) by the equation:

$CL = \pm A_2 \overline{R}$ (A_2 from Table IV.4)

4. Calculate the warning limit (WL) by the equation:

$WL = \pm 2/3 A_2 \overline{R}$

5. Chart CL's and WL's on each side of the standard, which is set at zero, as shown in Figure IV.16 ("order" related to consecutive, or chronological order of the analyses) and Table IV.6.

6. Plot the difference between the nominal value and $\overline{X}$ and take action on points which fall outside of the control limits.

c. Control Charts for Individuals

In many instances a rational basis for subgrouping may not be available, or the analysis may be so infrequent as to require action on the basis of individual results. In such cases X charts are employed. However, the CL's must come from some subgrouping to obtain a measure of "within group" variability. This alternative has the advantage of displaying each result with respect to tolerance or specification limits (Figures IV.4, IV.5, IV.9, and IV.11). The disadvantages must be recognized when considering this approach:

1. The chart does not respond to changes in the average.

2. Changes in dispersion are not detected unless an R chart is included.

3. The distribution of results must approximate normal if the control limits remain valid.

Additional refinements, variations, and control charts for other variables will be found in standard statistics texts.[2,15,29-31]

d. Moving Averages and Ranges

The $\overline{X}$ control chart is more efficient for disclosing moderate changes in the average as the subgroup size increases. A logical compromise between the X and $\overline{X}$ approach would be application of the moving average. For a given series of analyses, the moving average is plotted. Such a set of data is shown in Table IV.7. The moving range serves well as a measure of acceptable variation when no rational basis for subgrouping is available or when results are infrequent or expensive to gather.

E. SAMPLING CRITERIA: SYSTEMS – WHEN, WHERE, HOW LONG, AND HOW OFTEN

In any analysis system the result will be reliable only to the extent the sample represents the true composition of the entire entity for which analytical information is required.

The analysis of the atmosphere, either indoors or out, is influenced by factors over which, with the exception of sampling, the investigator has

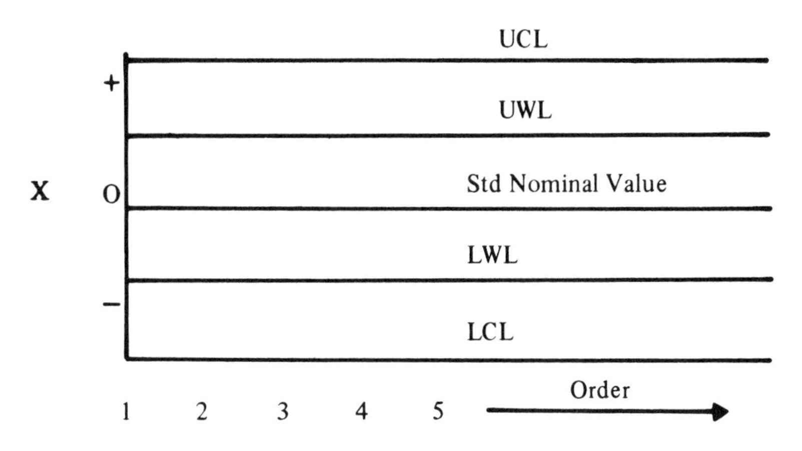

FIGURE IV.16. Accuracy control chart.

TABLE IV.7

Moving Average and Range Table (n ± 2)

Sample no.	Assay value	Sample nos. included	Moving average	Moving range
1	17.09	–	–	–
2	17.35	1–2	17.24	+0.26
3	17.40	2–3	17.38	+0.05
4	17.23	3–4	17.32	−0.17
5	17.09	4–5	17.16	−0.14
6	16.94	5–6	17.02	−0.15
7	16.68	6–7	16.81	−0.26
8	17.11	7–8	16.90	+0.43
9	18.47	8–9	17.79	+1.36
10	17.08	9–10	17.78	−1.39
11	17.08	10–11	17.08	0.00
12	16.92	11–12	17.00	−0.16
13	18.03	12–13	17.45	+1.11
14	16.81	13–14	17.42	−1.22
15	17.15	14–15	16.98	+0.34
16	17.34	15–16	17.25	+0.19
17	16.71	16–17	17.03	−0.73
18	17.28	17–18	17.00	+0.57
19	16.54	18–19	16.91	−0.74
20	17.30	19–20	16.92	+0.76

practically no control. Therefore, statistical principles must govern the adequacy of the sampling program, probability replaces certainty in the results, and statistical significance determines the soundness of the conclusions. Since the quality of the sampling program may well be the critical factor in an evaluation or control project, the objective involved must be kept uppermost in the investigator's mind. From a consideration of economy, the number of samples, duration of sampling, and sophistication of the equipment should be held to the minimum consistent with a reliable assessment of the problem. Exposure control criteria (threshold limit value or TLV) fall into four classes:

1. Instantaneous or short-term ceiling (designated C–, Reference 32) levels not to be exceeded: usually applied to irritants (e.g., TLV for formaldehyde – 5 ppm[32]).
2. Time: weighted average limits for the 8-hr day and 40-hr week with permissible excursions (e.g., mercury TLV 0.1 mg/m^3, Reference 32) which may be expressed as CXT values; C = concentration, T = time.
3. Nuisance: discomfort without significant hazard (e.g., ammonium chloride TLV 10 mg/m^3, Reference 32).
4. No contact: carcinogenic chemicals (e.g., β-naphthylamine).

Sampling for Class 1 compounds should provide practically instantaneous response for adequate warning. The longer the sampling duration, the greater is the damping effect on the range of excursions above and below the mean concentration level, until the effect of short-term excursions disappears entirely. This effect for SO_2 in the ambient atmosphere has been displayed graphically to confirm the effect of averaging time on the frequency distribution of concentration variations.[33]

A study of carbon monoxide (Class 2) content of the atmosphere in a warehouse clearly illustrates the influence of sampling technique on validity of conclusions. By random "grab sample" (instantaneous) sampling, frequent violation of the

TLV was indicated; however, by personnel monitoring (4-hr continuous sample-integrated CT value), compliance at all times was found.[3] Recommended averaging times for a large number of substances have been published.[34]

The objective of the study will determine the extent as well as type of sampling required and the accuracy and precision limits of the analysis. The objectives usually fall within four broad categories:

1. Preliminary investigation to determine whether a real or potential problem exists, usually a "walk-around" or "walk-through" survey during which "grab samples" are taken. Semiquantitative analytical procedures are sufficient to establish order of magnitude estimations. Direct reading colorimetric gas detector tubes are quite useful.[35] However, even in this case, the standard deviation must be recognized.

2. Detailed study of the work environment to establish conformance with existing TLV's which requires careful planning of the sampling program, the best available analytical procedures, and statistical analysis of the results. Large numbers of samples are frequently necessary.[4]

3. Detailed study as in 2 above, with additional supporting medical data to establish a TLV, and with biological monitoring (urine, blood, or breath analysis).

4. Permanent, long-term monitoring for exposure control and health conservation. Biological or personnel monitoring where possible is strongly recommended and a quality control program should be established.[4] Sampling frequency, location, and duration, as well as precision and accuracy requirements consistent with the hazard involved, should be based on the results from the detailed studies described in 2 and 3 above. The sample size will be determined by the limitations of the analytical method (sensitivity) and rate of sampling by collection limitations (efficiency). Additional information will be found in ASTM standards.[36]

In biological monitoring, the product of the biological half-life of the pollutant and the fraction absorbed as related to total resistance to its passage through the organism is a significant parameter. When sampling time is four times this parameter, attenuation of significant fluctuations is about the same in the samples and in the body. For times that are less than a factor of four, the sampling discloses fluctuations better than the biological response. Short sampling periods appear to provide unnecessary, fine detail for detection of significant biological response.

The variance (precision) of the overall results will be the sum of the variance of the analytical procedure (Va) and the sampling variance (Vs):

$$V_t = \frac{Vs}{k} + \frac{Va}{kn}$$

where k = number of samples, n = number of analyses per sample, and kn = total number of analyses. V_t should be minimized to produce the best possible precision and kn minimized to reduce the analytical cost. Increasing n reduces V_t to a greater extent than does increasing k.[15]

Other more sophisticated statistical techniques which can be helpful in setting up sampling programs include factorial design (ANOVA), Latin squares, simplex design, central composite design, block design, and chi-square test,[22] all of which may be found in elementary statistical text books.[2,15,16,19]

Jones and Brief developed a procedure by which the three American National Standards Institute (ANSI) exposure limits – peak, ceiling, and time-weighted average concentrations – for benzene could be compared with exposure data to establish conformance with the standards that were adopted by OSHA as consensus standards.[38,39] "For most industrial operations, the vapor concentrations closely follow a log-normal distribution which can be plotted on log-normal probability paper using Larsen's convention (% samples exceeding vs. log concentration in ppm). A line of limiting exposure is constructed on the same plot, and a visual comparison is made for compliance. Plotting values for the line of acceptable exposures are given for sampling periods of 10 to 240 min."[39] Equations for calculating these lines also are given. Two of the ANSI exposure criteria (1 and 3) were used to establish limiting lines of acceptable exposure:

1. Acceptable maximum for peak concentration is 50 ppm for no more than 10 min, and not more than once per work shift.

2. Acceptable ceiling concentration is 25 ppm.

3. Acceptable maximum time-weighted average concentration is 10 ppm.

The procedure for determining compliance with the time-weighted average standard (TLV) based on a small number of instantaneous ("grab") samples collected at random intervals during the work day has been published in the NIOSH criteria for carbon monoxide.[40] The following steps are recommended:

Given: the results of n samples with a mean m and a range (difference between least and greatest) r. If, for from 3 to 10 samples, m is greater than the total of:

A. The standard

B. The percentage of systematic instrument error multiplied times the standard

C. $\frac{t \times r}{n}$

then the true average concentration exceeds the standard ($p < 0.05$). The value of t is taken from the following table:

n (number of samples)	t (student's "t" test value)
3	2.35
4	2.13
5	2.01
6	1.94
7	1.89
8	1.86
9	1.83
10	1.81

For a large number of samples the procedure given in Section 3-2.2.1 of National Bureau of Standards (NBS) Handbook 91 shall be followed.*

F. ACKNOWLEDGMENTS FOR CHAPTER IV

The guidance derived from Section III, "Accuracy-Precision-Error," prepared by Audrey E. Donahue of the National Training Center, Water Quality Office, EPA (Cincinnati, Ohio) for *Quality Control in the Industrial Hygiene Laboratory,* published (May, 1971) by the National Institute for Occupational Safety and Health, Public Health Service, U.S. Department of Health, Education and Welfare, is gratefully acknowledged. Assistance from this source appears in Section B.1, "Detection," subsections b through e and g through i. Also, assistance derived from Section IV, "Quality Control," of *Quality Control in the Industrial Hygiene Laboratory,* prepared by W. D. Kelley and used in the preparation of Section D, "Quality Control Programs," is gratefully acknowledged.

Last, but not least, much credit and appreciation is due Mrs. Jean I. Cantino for her inexhaustible patience in typing the original manuscript of this chapter.

*From NIOSH Criteria for a Recommended Standard – Occupational Exposure to Carbon Monoxide, Appalachian Center for Occupational Safety and Health, 944 Chestnut Ridge Road, Morgantown, W. Va., 1972.

G. REFERENCES FOR CHAPTER IV

1. **Linch, A. L. and Pfaff, H. V.,** Carbon monoxide – evaluation of exposure by personnel monitor surveys, *Am. Ind. Hyg. Assoc. J.,* 32, 745, 1971.

1a. American Chemical Society, Guide for measures of precision and accuracy, *Anal. Chem.,* 35, 2262, 1963.

1b. **Kitagawa, T.,** Carbon monoxide detector tube no. 100, National Environmental Instruments, Inc., P.O. Box 590, Fall River, Mass., 1971.

2. **Youden, W. J.,** *Statistical Methods for Chemists,* John Wiley and Sons, New York, 1951.

3. American Society for Testing Materials, *ASTM Manual on Quality Control of Materials,* Special Technical Publication 15-C, ASTM, Philadelphia, 1951.

4. **Linch, A. L., Wiest, E. G., and Carter, M. D.,** Evaluation of tetraalkyl lead exposure by personnel monitor surveys, *Am. Ind. Hyg. Assoc. J.,* 31, 170, 1970.

5. **Saltzman, B. E.,** Preparation and analysis of calibrated low concentrations of sixteen toxic gases, *Anal. Chem.,* 33, 1100, 1961.

6. **Scarangelli, F. P., O'Keefe, A. E., Rosenberg, E., and Bell, J. P.,** Preparation of known concentrations of gases and vapors with permeating devices calibrated gravimetrically, *Anal. Chem.,* 42, 871, 1970.

7. **Katz, M., Ed.,** *Methods of Air Sampling and Analysis,* Part I, The Intersociety Committee on Methods for Air Sampling and Analysis, Am. Pub. Health Assoc. Publication Service, Washington, D.C., 1972.

8. **Grim, K. and Linch, A. L.,** Recent isocyanate-in-air analysis studies, *Am. Ind. Hyg. Assoc. J.,* 25, 285, 1964.

9. **Linch, A. L. and Corn, M.,** The standard midget impinger – design improvement and miniaturization, *Am. Ind. Hyg. Assoc. J.,* 26, 601, 1965.
10. **Davis, J. R. and Andelman, S. L.,** Urinary delta-amino-levulinic acid (ALA) levels in lead poisoning. 1. A modified method for the rapid determination of urinary delta-amino-levulinic acid using disposable ion-exchange chromatography columns, *Arch. Environ. Health,* 15, 53, 1967.
11. **Keenan, R. G., Byers, H. D., Saltzman, B. E., and Hyslop, F. L.,** Determination of lead in air and biological materials, the "USPHS" double extraction, mixed color dithizone method, *Am. Ind. Hyg. Assoc. J.,* 24, 481, 1963.
12. **Yeager, D. W., Cholak, J., and Henderson, E. W.,** Determination of lead in biological and related material by atomic absorption spectrophotometry, *Environ. Sci. Technol.,* 5, 1020, 1971.
13. **Woessner, W. W. and Cholak, J.,** Improvements in the rapid screening method for lead in urine, *A.M.A. Arch. Ind. Hyg. Occup. Med.,* 7, 249, 1953.
14. **Moss, R. and Browett, E. V.,** Determination of tetra-alkyl lead vapour and inorganic lead dust in air, *Analyst,* 91, 428, 1966.
15. **Bauer, E. L.,** *A Statistical Manual for Chemists,* 2nd ed., Academic Press, New York, 1971.
16. **Taras, M. J., Greenberg, A. E., Hoak, R. D., and Rand, M. C.,** *Standard Methods for the Examination of Water and Wastewater,* 13th ed., American Public Health Association, Washington, D.C., 1971.
17. American Society for Testing and Materials, Proposed procedure for determination of precision of committee D-19 methods, in *Manual on Industrial Water and Industrial Waste Water,* 2nd ed., ASTM, Philadelphia, 1966.
18. **Keppler, J. F., Maxfield, M. E., Moss, W. D., Tietjen, G., and Linch, A. L.,** Interlaboratory evaluation of the reliability of blood lead analysis, *Am. Ind. Hyg. Assoc. J.,* 31, 412, 1970.
19. **Hinchen, J. D.,** *Practical Statistics for Chemical Research,* Methuen and Co., London, 1969.
20. **Wetherhold, J. M., Linch, A. L., and Charsha, R. C.,** Hemoglobin analysis for aromatic nitro and amino compound exposure control, *Am. Ind. Hyg. Assoc. J.,* 20, 396, 1959.
21. **Steere, N. V., Ed.,** *Handbook of Laboratory Safety,* 2nd ed., Chemical Rubber, Cleveland, 1971.
22. **Wilcoxon, F.,** Some Rapid Approximate Statistical Procedures, Insecticide and Fungicide Section, American Cyanamid Co., Agricultural Chemicals Division, New York, 1949.
23. American Chemical Society, *Reagent Chemicals,* 4th ed., American Chemical Society Publications, Washington, D.C., 1968.
24. American Society for Testing and Materials, ASTM manual for conducting an interlaboratory study of a test method, Technical Publication No. 335, ASTM, Philadelphia, 1963. Available from University Microfilms, Ann Arbor, Mich.
25. **Weil, C. S.,** Critique of laboratory evaluation of the reliability of blood-lead analyses, *Am. Ind. Hyg. Assoc. J.,* 32, 304, 1971.
26. **Snee, R. D. and Smith, P. E.,** Statistical Analysis of Interlaboratory Studies, paper prepared for presentation at the American Industrial Hygiene Conference, San Francisco, May, 1971.
28. **Cralley, L. J., Berry, C. M., Palmes, E. D., Reinhardt, C. F., and Shipman, T. L.,** Guidelines for accreditation of industrial hygiene analytical laboratories, *Am. Ind. Hyg. Assoc. J.,* 31, 335, 1970.
29. **Duncan, A. J.,** *Quality Control and Industrial Statistics,* 3rd ed., R. D. Irwin, Homewood, Ill., 1965.
30. **Cowden, D. J.,** *Statistical Methods In Quality Control,* Prentice-Hall, Englewood Cliffs, N.J., 1957.
31. **Grant, E. L.,** *Statistical Quality Control,* Part 1, 3rd ed., McGraw-Hill, New York, 1964.
32. American Conference of Governmental Industrial Hygienists, *Threshold Limit Values of Airborne Contaminants and Physical Agents with Intended Changes Adopted by ACGIH for 1971,* ACGIH, Cincinnati, O.
33. **Stern, A. C.,** *Air Pollution, Volume 2, Analysis, Monitoring and Surveying,* 2nd ed., Academic Press, New York, 1968.
34. **Stern, A. C.,** *Air Pollution, Volume 3, Sources of Air Pollution and Their Control,* 2nd ed., Academic Press, New York, 1968.
35. **Saltzman, B. E.,** Section B-8, in *Air Sampling Instruments for Evaluation of Atmospheric Contaminants,* 4th ed., ACGIH, Cincinnati, O., 1972.
36. American Society for Testing and Materials, *Industrial Water; Atmospheric Analysis,* ASTM, Philadelphia, 1967, D1357-57, part 23, p. 770.
37. **Saltzman, B. E.,** Significance of sampling time in air monitoring, *J. Air Pollut. Control Assoc.,* 20, 660, 1970.
38. Williams-Steiger Occupational Safety and Health Act of 1970, *Fed. Register,* 37, 22142, October 18, 1972.
39. **Jones, A. R. and Brief, R. S.,** Evaluating benzene exposures, *Am. Ind. Hyg. Assoc. J.,* 32, 610, 1971.
40. National Institute for Occupational Safety and Health, *Criteria for a Recommended Standard – Occupational Exposure to Carbon Monoxide,* Appalachian Center for Occupational Safety and Health, 944 Chestnut Ridge Road, Morgantown, W. Va., 1972.

H. PREFERRED READING FOR CHAPTER IV

In addition to References 5 and 8 through 12, the following periodicals are recommended:

1. *American Industrial Hygiene Association Journal.*
2. *Journal of the Air Pollution Control Association.*
3. *Analytical Chemistry* (ACS).
4. *Environmental Science and Technology* (ACS).
5. *Air Pollution Manual,* Parts I and II, AIHA, Collingswood, N. J., 1972.
6. **Nelson, G. O.,** *Controlled Test Atmospheres: Principles and Techniques,* Ann Arbor Science, Ann Arbor, Mich., 1971. (Recommended for calibration references.)
7. **Leithe, W.,** *The Analysis of Air Pollutants,* Ann Arbor Science, Ann Arbor, Mich., 1970.

AUTHOR INDEX

N

P

Q

R

S

T

U

W

SUBJECT INDEX

D

L

M

N

O

P

Q

R

S

T

U

V

W

X

Z